Advances in Bacterial Cellulose

Fernando Dourado • Miguel Gama
Feng Hong • Rintu Banerjee • Mohan Das
Editors

Advances in Bacterial Cellulose

Towards Sustainable Production and Emerging Applications

Springer

Editors
Fernando Dourado
Centre of Biological Engineering
University of Minho - Campus of Gualtar
Braga, Portugal

Miguel Gama
Department of Biological Engineering
University of Minho - Campus de Gualtar
Braga, Portugal

Feng Hong
Department of Biological Science and
Biotechnology, College of Biological
Science and Medical Engineering
Donghua University
Shanghai, China

Rintu Banerjee
Department of Agricultural and Food
Engineering
Indian Institute of Technology Kharagpur
Kharagpur, West Bengal, India

Mohan Das
Department of Agricultural and Food
Engineering
Indian Institute of Technology Kharagpur
Kharagpur, West Bengal, India

ISBN 978-3-032-07041-8 ISBN 978-3-032-07042-5 (eBook)
https://doi.org/10.1007/978-3-032-07042-5

This Springer imprint is published by the registered company Springer Nature Switzerland AG
The registered company address is: Gewerbestrasse 11, 6330 Cham, Switzerland

Preface

The increasing global need for sustainability makes a shift urgent from petrochemical-derived materials. Bacterial cellulose (BC) stands out as a highly promising biopolymer, characterized by its exceptional purity, superior physicochemical properties, and distinctive nanofibrillar architecture. Since its early research, BC has gained growing industrial adoption in biomedical, food, and electronics sectors due to its sustainable and versatile nature. This book provides a comprehensive and updated exploration of bacterial cellulose, by exploring recent advancements in its production, characterization, and diverse applications, while addressing the challenges related to industrial scalability, economic feasibility, environmental sustainability, and safety considerations.

The book starts by elucidating the foundational biological strategies for optimizing BC production. It demonstrates how synthetic biology can serve as a transformative approach in the design of BC-based materials, emphasizing the modularization and standardization of genetic constructs to enable precise and efficient production processes. Further, the application of functional genomics and Genome-Scale Metabolic Models (GEMs) is presented as a robust methodology for elucidating gene functions, reconstructing metabolic pathways, and developing targeted metabolic engineering strategies to enhance BC yields.

A key focus is placed on innovative and cost-effective production methodologies. Particularly, the text explores the potential of nonconventional bacterial strains, such as *Enterobacter, Proteus, Rhodococcus, Bacillus, and Leifsonia*, as viable alternatives to the conventionally utilized *Komagataeibacter, Acetobacter*, and *Gluconacetobacter* species. These strains may offer solutions to overcome limitations such as oxygen dependency, suboptimal yields, and restricted substrate versatility. Also, the book examines the valorization of agroindustrial byproducts, encompassing sugar-rich, starch-based, cellulosic, and lignocellulosic residues, as well as crude glycerol, as sustainable, low-cost feedstocks for BC production. It highlights essential pretreatment techniques required to transform these waste streams into viable fermentation substrates.

The versatility of bacterial cellulose is here further demonstrated through its application across several fields, such as:

- Biomedical Applications: BC is used in 3D bioprinting bioinks allowing the creation of complex constructs for tissue engineering and regenerative medicine, particularly for cartilage and bone regeneration.
- Multiphase Systems: BC serves as a sustainable stabilizer in systems like Pickering emulsions and foams, with applications in the cosmetic and food industries, offering an eco-friendly alternative to traditional surfactants.
- Advanced Technological Applications: BC is utilized in flexible electronic devices, transparent conductive films, optical sensors, and energy systems such as batteries, supercapacitors, and nanogenerators, due to its strength and adaptable surface chemistry.

This book also addresses a detailed techno-economic analysis, evaluating the economic feasibility of the large-scale BC production, identifying major cost drivers, such as culture medium and purification processes, alongside capital and operational expenses. Alongside, a Life Cycle Assessment provides a comparative environmental analysis of BC relative to plant-derived micro- and nanocellulose, underscoring its ecological footprint and opportunities for optimization. The book finalizes with an assessment of the toxicological profile and regulatory framework for BC in food applications, where the safety of BC for human consumption is explored, and the complexities of applying conventional nanomaterial evaluation methods to this biopolymer are addressed.

Thus, by reviewing emergent topics overlooked in the literature, such as the non-conventional BC producing strains, applications in electronics and the regulatory framework for BC applications in the food industry, this book provides the scientific community with an updated state-of-the-art and a new perspective on the fascinating material, bacterial cellulose.

Braga, Portugal — Fernando Dourado
Braga, Portugal — Miguel Gama
Shanghai, China — Feng Hong
Kharagpur, West Bengal, India — Rintu Banerjee
Kharagpur, West Bengal, India — Mohan Das

Contents

List of Figures

List of Tables

Contributors

Rintu Banerjee Agricultural and Food Engineering Department, Indian Institute of Technology, Kharagpur, India

Jiantang Chen State Key Laboratory of Advanced Fiber Materials, College of Biological Science and Medical Engineering, Donghua University, Shanghai, China

Scientific Research Base for Bacterial Nanofiber Manufacturing and Composite Technology, China Textile Engineering Society, Shanghai, China

Shiyan Chen State Key Laboratory of Advanced Fiber Materials, College of Materials Science and Engineering, Donghua University, Shanghai, People's Republic of China

L. Costa CEB—Centre of Biological Engineering, University of Minho, Campus de Gualtar, Braga, Portugal

LABBELS–Associate Laboratory in Biotechnology and Bioengineering and Microelectromechanical Systems, Braga, Guimarães, Portugal

Mohan Das Agricultural and Food Engineering Department, Indian Institute of Technology, Kharagpur, India

Óscar Dias CEB—Centre of Biological Engineering, University of Minho, Campus de Gualtar, Braga, Portugal

LABBELS—Associate Laboratory in Biotechnology and Bioengineering and Microelectromechanical Systems, Braga, Guimarães, Portugal

Lucília Domingues CEB—Centre of Biological Engineering, University of Minho, Campus de Gualtar, Braga, Portugal

LABBELS—Associate Laboratory in Biotechnology and Bioengineering and Microelectromechanical Systems, Braga, Guimarães, Portugal

Fernando Dourado CEB—Centre of Biological Engineering, University of Minho, Campus de Gualtar, Braga, Portugal

LABBELS—Associate Laboratory in Biotechnology and Bioengineering and Microelectromechanical Systems, Braga, Guimarães, Portugal

Miguel Gama CEB—Centre of Biological Engineering, University of Minho, Campus de Gualtar, Braga, Portugal

LABBELS—Associate Laboratory in Biotechnology and Bioengineering and Microelectromechanical Systems, Braga, Guimarães, Portugal

Francisco Garrett CEB—Centre of Biological Engineering, University of Minho, Campus de Gualtar, Braga, Portugal

LABBELS—Associate Laboratory in Biotechnology and Bioengineering and Microelectromechanical Systems, Braga, Guimarães, Portugal

Feng F. Hong State Key Laboratory of Advanced Fiber Materials, College of Biological Science and Medical Engineering, Donghua University, Shanghai, China

Scientific Research Base for Bacterial Nanofiber Manufacturing and Composite Technology, China Textile Engineering Society, Shanghai, China

Jing Li Fujian Key Laboratory of Novel Functional Textile Fibers and Materials, Minjiang University, Fuzhou, People's Republic of China

Daniela Martins CEB—Centre of Biological Engineering, University of Minho, Campus de Gualtar, Braga, Portugal

LABBELS—Associate Laboratory in Biotechnology and Bioengineering and Microelectromechanical Systems, Braga, Guimarães, Portugal

Pedro Montenegro-Silva CEB—Centre of Biological Engineering, University of Minho, Campus de Gualtar, Braga, Portugal

LABBELS—Associate Laboratory in Biotechnology and Bioengineering and Microelectromechanical Systems, Braga, Guimarães, Portugal

Baneeprajnya Nayak Agricultural and Food Engineering Department, Indian Institute of Technology, Kharagpur, India

Miguel Pacheco CEB—Centre of Biological Engineering, University of Minho, Campus de Gualtar, Braga, Portugal

LABBELS—Associate Laboratory in Biotechnology and Bioengineering and Microelectromechanical Systems, Braga, Guimarães, Portugal

Ana Cristina Rodrigues CEB—Centre of Biological Engineering, University of Minho, Campus de Gualtar, Braga, Portugal

LABBELS—Associate Laboratory in Biotechnology and Bioengineering and Microelectromechanical Systems, Braga, Guimarães, Portugal

Ricardo Silva-Carvalho CEB—Centre of Biological Engineering, University of Minho, Campus de Gualtar, Braga, Portugal

LABBELS—Associate Laboratory in Biotechnology and Bioengineering and Microelectromechanical Systems, Braga, Guimarães, Portugal

Zhuotong Wu Fujian Agriculture and Forestry University, Fuzhou, People's Republic of China

Wenqiang Xu State Key Laboratory of Advanced Fiber Materials, College of Biological Science and Medical Engineering, Donghua University, Shanghai, China

Scientific Research Base for Bacterial Nanofiber Manufacturing and Composite Technology, China Textile Engineering Society, Shanghai, China

Fan Yang State Key Laboratory of Advanced Fiber Materials, College of Biological Science and Medical Engineering, Donghua University, Shanghai, China

Scientific Research Base for Bacterial Nanofiber Manufacturing and Composite Technology, China Textile Engineering Society, Shanghai, China

Chapter 1
Synthetic Biology as a Paradigm Shift in Bacterial Cellulose-Based Material Design and Production

Pedro Montenegro-Silva, Miguel Pacheco, Óscar Dias, Fernando Dourado, Miguel Gama, and Lucília Domingues

Abstract Bacterial cellulose (BC) is a high-performance biomaterial characterized by its exceptional properties; however, its production is subject to certain limitations. The emergence of synthetic biology has progressively facilitated the precise control over BC producing microorganisms, which is beneficial not only for enhancing yields but also for broadening their properties and applications. Innovations such as Clustered Regularly Interspaced Short Palindromic Repeats (CRISPRi) interference, quorum sensing circuits, and modular expression systems are augmenting BC biosynthesis and functionalization. Co-culture strategies, particularly those involving organisms more amenable to genetic modification, can enhance the incorporation of biosensing, enzymatic activity, and self-patterning, thereby paving the way for next-generation materials. The future integration of multi-omics, computational modeling, and automation is expected to refine BC bioengineering. As synthetic biology progresses, BC is evolving from a natural polymer to a programmable material with extensive industrial, environmental, and biomedical applications.

Keywords Bacterial cellulose (BC) · Synthetic biology · Engineered living materials (ELMs) · Clustered regularly interspaced short palindromic repeats (CRISPRi) · Genetic engineering

---

P. Montenegro-Silva · M. Pacheco · Ó. Dias · F. Dourado · M. Gama · L. Domingues (✉)
CEB—Centre of Biological Engineering, University of Minho, Campus de Gualtar, Braga, Portugal

LABBELS—Associate Laboratory in Biotechnology and Bioengineering and Microelectromechanical Systems, Braga, Guimarães, Portugal
e-mail: odias@ceb.uminho.pt; fdourado@ceb.uminho.pt; fmgama@deb.uminho.pt; luciliad@deb.uminho.pt

F. Dourado et al. (eds.), *Advances in Bacterial Cellulose*,
https://doi.org/10.1007/978-3-032-07042-5_1

1 Introduction

The most notable bacterial cellulose (BC) producers are bacteria within the *Komagataeibacter* genus (Wang et al. 2025; Zheng et al. 2024). BC is a natural biopolymer that, unlike plant-derived cellulose, is produced in a highly pure form due to the absence of hemicellulose and lignin, eliminating the need for energy-intensive downstream processing. The physicochemical properties of BC make it an exceptional biodegradable material with high water absorption capacity, crystallinity, and mechanical strength (Andriani et al. 2020; Mishra et al. 2022; Zhong 2020). The properties of BC make it an exceptional material for use in various advanced applications across several industries (Choi and Shin 2020; Gautam et al. 2010; Keshk 2014; Skočaj 2019).

BC is composed of glucose units linked by β-1,4-glycosidic bonds, which can be arranged into cellulose I (native; crystalline) or cellulose II (reconstituted; crystalline) (Potočnik et al. 2023). BC can form composites with other materials and be impregnated with functional agents, further expanding its utility in fields such as bioelectronics, environmental technology, and acoustics (Buldum and Mantalaris 2021; Shah et al. 2013; Vadanan et al. 2022). The BC production process and modulation of its properties have been mostly achieved through process optimization and in situ or ex situ modifications (Buldum and Mantalaris 2021). However, the continuous progress made in synthetic biology and systems biology has enabled the development of novel microbial strains and co-culture systems with superior yields and novel material functionalities (e.g., antibacterial and electrical conductivity) (Jin et al. 2022; Malcı et al. 2024; Stumpf et al. 2018), thus allowing BC to stand out as a versatile platform to generate novel advanced materials with application across diverse industries (Gautam et al. 2010; Jedrzejczak-Krzepkowska et al. 2016; Jin et al. 2022; Malcı et al. 2024; Rodrigues et al. 2019; Stumpf et al. 2018). The biotechnological field of BC production and functionalization has been largely transformed by synthetic biology. The emerging interdisciplinary field applies principles such as modularity, standardization, and programmable DNA design, facilitating the development of engineered BC producers with improved traits (Andrianantoandro et al. 2006; Khalil and Collins 2010; Malcı et al. 2024; Singh et al. 2020). The adoption of BC has been hindered by high costs associated with synthetic culture media, which has been partially mitigated through the utilization of industrial by-products and fermentation optimization techniques, achieving yields as high as 10 g L^{-1} and productivity rates up to 0.4 g L^{-1} h^{-1} (Cacicedo et al. 2016; El-Gendi et al. 2022; Jozala et al. 2016; Rodrigues et al. 2019). Despite the embryonic stage of the synthetic biology tools for *Komagataeibacter*, when compared with other organisms, the availability of genetic tools has increased considerably (e.g., plasmid backbones, promoters, terminators, CRISPRi interference (Clustered Regularly Interspaced Repeats; CRISPRi)), enabling targeted modifications for yield improvements, by-product reduction, and modulation of the material's properties (Ryngajłło et al. 2019; Teh et al. 2019). The development of BC-based composites through co-culture systems and material incorporation was a critical innovation in BC functionality (Gilbert et al. 2021; Jin et al. 2022; Stumpf et al. 2018). The relevance lies in the utilization of more tractable organisms for the production of functional components such as alginate, collagen, or gelatin, whose incorporation into BC matrices can enhance properties like mechanical strength and antimicrobial activity (Báez et al. 2005; Campebell et al. 2025; Gilbert et al. 2021; Hay et al. 2013; Jin et al. 2022; Noh et al. 2019). Engineered Living Materials (ELMs) are responsive, multifunctional composites for diverse applications, including medicine, architecture, and electronics (Bose et al. 2020; Cao et al. 2017; Heveran et al. 2019). ELMs can even be programmed with easy-to-assemble modular genetic circuits for enhanced functionality, specificity, and spatiotemporal control (Andrianantoandro et al. 2006; Hernández-Arriaga et al. 2022; Keasling 2012; Khalil and Collins 2010; Malcı et al. 2024; Nielsen and Keasling 2011; Ryngajłło et al. 2020; Singh et al. 2020; Stephanopoulos 2012). BC research is currently aiming at enabling the cost-effective production of tailor-made BC products, which requires optimization of microbial strains and bioprocesses, and can be revolutionized by multidisciplinary advances in bioreactor design, co-culture strategies, and synthetic biology (Andriani et al. 2020; Mishra et al. 2022; Ryngajłło et al. 2020) (Goosens et al. 2021; Jin et al. 2022; Liu et al. 2020; Malcı et al. 2024; Potočnik et al. 2023; Singh et al. 2020; Teh et al. 2019; Vadanan et al. 2022). Such advances are gradually driving the research field to meet the demand for improved materials in industries ranging from biomedical engineering to green electronics (Andriani et al. 2020; Cacicedo et al. 2016; El-Gendi et al. 2022; Mishra et al. 2022; Stumpf et al. 2018; Wang et al. 2025; Zheng et al. 2024; Zhong 2020).

This chapter explores the impact of synthetic biology in the development of high yield BC producing strains, and BC-based materials with increased functionality. It starts by briefly introducing the most relevant biosynthetic pathways, and the advances in genetic and metabolic engineering as innovation driving forces. Key questions addressed include: How can synthetic biology tools and approaches be leveraged to improve BC yield and properties? What strategies are most effective for tailoring BC to meet the requirements of advanced applications, such as biomedical devices and smart materials? By addressing these questions, this chapter highlights the paradigm shift in material design and production enabled by synthetic biology, paving the way for next-generation BC-based materials.

2 Bacterial Cellulose Biosynthesis

Komagataeibacter species synthesize BC in a highly efficient process that polymerizes glucose into nanostructured cellulose materials with unique mechanical and functional properties. The first committal step in BC biosynthesis is the conversion of glucose-6-phosphate to glucose-1-phosphate by phosphoglucomutase. The phosphorylated sugar is further converted to uridine diphosphate (UDP)-glucose, by UDP-glucose pyrophosphorylase, which is the substrate of the bacterial cellulose synthase (BCS) complex—the transmembrane heteromeric complex responsible for the simultaneous polymerization of UDP-glucose monomers into β-1,4-glucan chains, and respective extrusion to the extracellular environment (Buldum and Mantalaris 2021; Jacek et al. 2019a; Li et al. 2022; Römling and Galperin 2015; Ryngajłło et al. 2020; Valla et al. 1989).

The genetic composition of the *bcs* operon varies across bacterial taxa and, while maintaining the core biosynthetic mechanism, such variation translates into substantial differences in terms of strain productivity and physicochemical properties. Within the *Komagataeibacter* genus, the *bcsABCD* operon encodes the key subunits responsible for cellulose biosynthesis (Römling and Galperin 2015). BcsA forms a channel that spans across the inner membrane, contains the catalytic glycosyltransferase domain and a PilZ-like domain, which binds cyclic diguanylic acid (c-di-GMP), an essential allosteric activator (Fujiwara et al. 2013; Römling 2012; Römling et al. 2013; Römling and Galperin 2015). BcsB acts as a chaperone, stabilizing BcsA activity and aiding glucan chain translocation through the inner membrane. BcsC forms an outer membrane pore to guide the nascent glucan chain outside the cell, while BcsD organizes glucan chains into crystalline fibrils, enhancing structural alignment and overall cellulose production efficiency (Buldum and Mantalaris 2021; Gullo et al. 2018; Hu et al. 2010; Jacek et al. 2019a; Römling and Galperin 2015; Ryngajłło et al. 2020) (Fig. 1.1a).

Regulation of cellulose biosynthesis involves both genetic and environmental factors. The levels of c-di-GMP are modulated by diguanylate cyclases (DGCs) and phosphodiesterases (PDEs), encoded by several operons such as *cdg1*, *cdg2*, and *cdg3*. This layer of regulation allows the bacteria to precisely adjust the activity of the BCS complex in response to environmental cues (e.g., oxygen availability) (Chang et al. 2001; Römling 2002, 2012; Römling et al. 2013; Römling and Galperin 2015). The *bcsABCD* operon contains accessory genes that, while not being essential, are important for the cellulose's structural integrity and/or the strains' maximum productivity (Buldum and Mantalaris 2021; Jacek et al. 2019a; Römling 2002, 2012; Römling et al. 2013; Römling and Galperin 2015; Ryngajłło et al. 2020).

The fine coordination between the simultaneous polymerization and secretion processes is a major factor affecting the maximum BC productivity and the material's properties. During this process, the nascent glucan chains are gradually assembled into protofibrils and further into larger microfibrils, and further into hydrogen-bonded cellulose ribbons. The delicately assembled nanostructure is responsible for the distinctive mechanical strength, water retention capacity, and biocompatibility of BC (Andriani et al. 2020; Cacicedo et al. 2016; Iguchi et al. 2000; Jin et al. 2022; Jozala et al. 2016; Portela et al. 2019; Zhong 2020).

Despite the remarkable efficiency of natural BC synthesis in *Komagataeibacter* species, several challenges hinder its optimization for industrial applications. By-product formation is one of the major bottlenecks (e.g., gluconic acid, exopolysaccharide (EPS)) due to reduction of the process cost-efficiency (Griffin et al. 1996; Ishida et al. 2002; Kuo et al. 2015; Liu et al. 2020; Montenegro-Silva

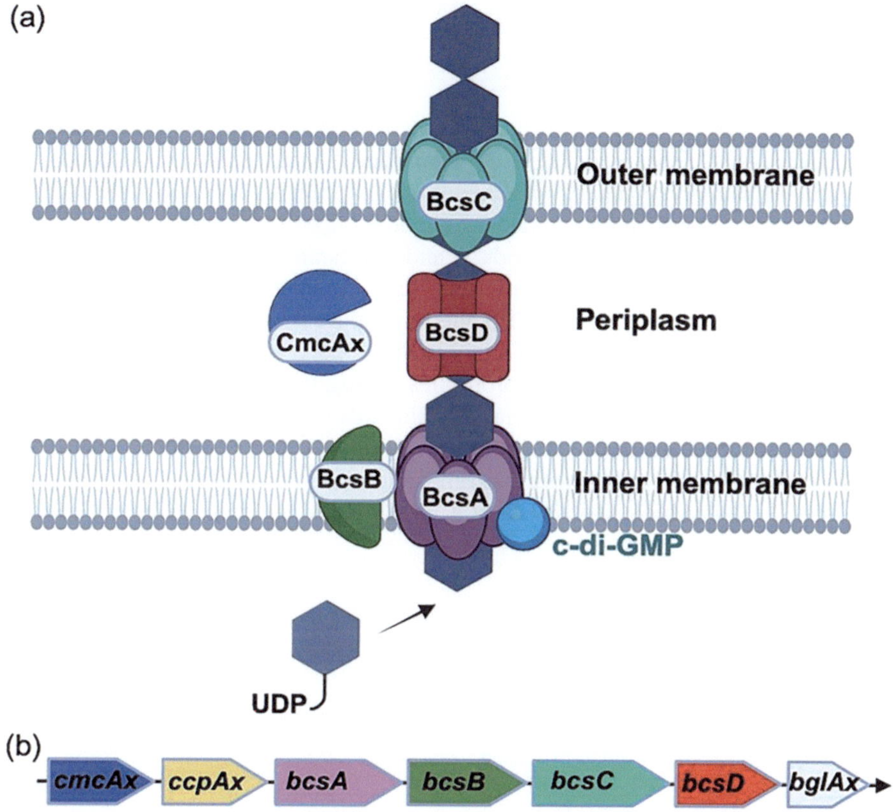

Fig. 1.1 (**a**) Proposed organization of the Type I BCS complex (Gullo et al. 2018; Jacek et al. 2019a; Römling and Galperin 2015). (**b**) Type I bacterial cellulose operon (Römling and Galperin 2015) (Figure and legend from Römling and Galperin 2015)

et al. 2024; Peng et al. 2024; Shigematsu et al. 2005; Griffin et al. 1996; Ishida et al. 2002). Additionally, *Komagataeibacter* strains are highly sensitive to environmental factors (e.g., oxygen availability, pH, and temperature) which potentially affects the intracellular concentration of regulatory molecules such as c-di-GMP (Andriani et al. 2020; Buldum and Mantalaris 2021; Jacek et al. 2019a; Römling et al. 2013; Ryngajłło et al. 2020). Another challenge lies in the high energy and precursor demands of BC production, which can strain cellular resources, particularly under suboptimal growth conditions (Buldum and Mantalaris 2021; Ryngajłło et al. 2020; Zhong 2020). These challenges pose an opportunity to leverage synthetic biology and genetic engineering for targeted strategies to suppress by-product formation, optimize metabolic fluxes, and improve the robustness of the production process (Jin et al. 2022; Malcı et al. 2024; Ryngajłło et al. 2020; Singh et al. 2020).

3 Synthetic Biology and Genetic Engineering of BC Producing Strains

Advances in genetic engineering and synthetic biology have revolutionized the engineering of BC producers (Fig. 1.2). Traditional genetic engineering focuses on manipulating individual genes, while metabolic engineering focuses on the optimization metabolic flows and networks. Synthetic biology, on the other hand, applies engineering principles, through a modular and programmable approach that enables precise control over cellular behavior, thus facilitating the generation of organisms with tailor-made traits (Keasling 2012; Nielsen and Keasling 2011; Stephanopoulos 2012). A cornerstone of synthetic biology in this context is the use of ELMs— biomaterials that integrate functional properties during the growth process of living organisms like BC-producing bacteria (Keasling 2012). ELMs offer a unique advantage by combining biological self-assembly with engineered functionalities, allowing researchers to design BC with customizable mechanical, chemical, or sensory properties during its biosynthesis (Jin et al. 2022; Keasling 2012; Malcı et al. 2024; Singh et al. 2020).

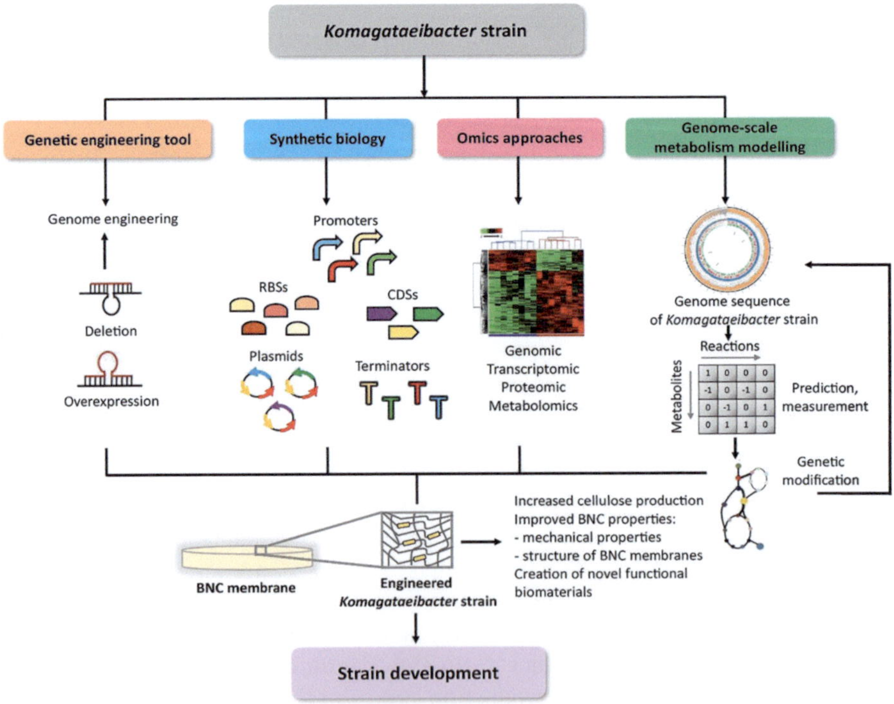

Fig. 1.2 Diagram shows the molecular strategies and techniques for *Komagataeibacter* strains development (Figure from Malcı et al. 2024, ACS Synthetic Biology. Used under CC-BY 4.0 license) (Malcı et al. 2024)

Moreover, ELMs can also be generated through co-culture strategies that pair BC-producing bacteria with engineered model organisms such as the bacteria *Escherichia coli* or the yeast *Saccharomyces cerevisiae* (Birnbaum et al. 2021; Gilbert et al. 2021). The tractability of model organisms can be leveraged to add functionality (e.g., production of cross-linking agents and dyes) and further expand the versatility of BC-based products (Gilbert et al. 2021; Gilbert and Ellis 2019; Jin et al. 2022; Malcı et al. 2024; Singh et al. 2020).

This section will focus on the synthetic biology and genetic/metabolic engineering tools that are driving innovations in BC production and also explore key engineering approaches to enhance BC yield, customize its properties, and co-culture strategies to create multifunctional BC-based ELMs for advanced applications.

3.1 *Overview of Synthetic Biology Tools (e.g., Plasmids, CRISPRi, Gene Circuits, Promoters)*

Synthetic biology has had a large impact in the field of BC production due to the introduction of flexible, modular, and programmable tools to engineer *Komagataeibacter* species (Goosens et al. 2021; Huang et al. 2020; Teh et al. 2019). The novel tools accelerated the assembly of standardized parts into plasmid backbones (e.g., transcription units, gene circuits, and CRISPRi), which can be used to generate productive strains that are able to synthesize BC with tuneable properties (Florea et al. 2016; Goosens et al. 2021; Huang et al. 2020; Teh et al. 2019) (Fig. 1.3).

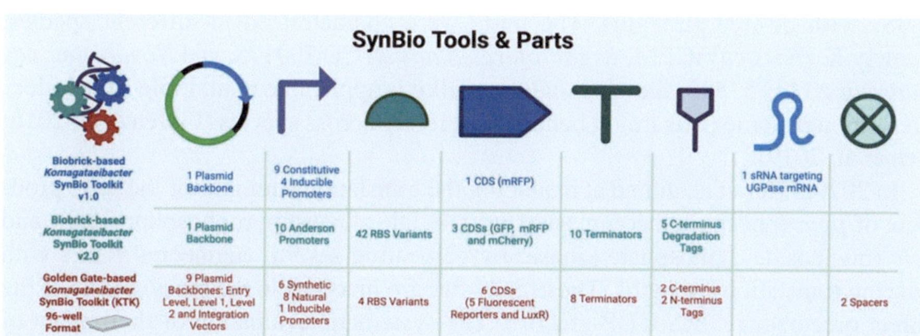

Fig. 1.3 SynBio and computational tools for engineering BC producers. The genetic toolkits developed for *Komagataeibacter*. The first toolkit is based on Biobrick assembly with constitutive and inducible promoters and a small RNA (sRNA) targeting UGPase mRNA. The second version of the BioBrick-based toolkit, offering a wide array of promoters, RBSs, coding sequences (CDSs), terminators, and protein tags, tested and characterized in *K. xylinus*, *K. hansenii*, and *K. rhaeticus*. The Golden Gate-based *Komagataeibacter* SynBio toolkit (KTK) enables the hierarchical assembly of transcription units and multicassette plasmids, available as a cost-effective 96-well plate on Addgene (Figure adapted from Malcı et al. 2024, ACS Synthetic Biology. Used under CC-BY 4.0 license) (Malcı et al. 2024)

For more than a decade, the targeted genetic modification of *Komagataeibacter* species was relatively rare and primarily relied on two expression vector backbones: pSA and pBBR122 (Nakai et al. 1999; Setyawati et al. 2007; Tonouchi et al. 1994; Yadav et al. 2010). More recently, the plasmid repertoire has been expanded with the plasmids pTI99, pTSa, and pSEVA331Bb. These vectors have been employed for both endogenous and heterologous gene expression (Fang et al. 2015; Florea et al. 2016; Gwon et al. 2019; Jacek et al. 2019b). Gene disruption was typically achieved through homologous recombination using non-replicating plasmid-based disruption cassette carriers such as pUC18/pUC19, pT7Blue, pET14, or pHSG399 (Bae et al. 2004; Ishida et al. 2002; Jacek et al. 2019b; Kuo et al. 2015; Shigematsu et al. 2005).

In 2016, Florea et al. have made the first effort to bring modularization and standardization of genetic constructs to the field. The researchers started by engineering the plasmids pSEVA321 and pSEVA331 to make them compatible with BioBrick standard and proceeded with the characterization of 9 constitutive promoters, 2 inducible promoters (P_{Lux} and P_{Tet}), and 3 fluorescent proteins. Additionally, a small RNA tool for genetic knockdowns was also developed and used for translational modulation of UDP-glucose pyrophosphorylase transcripts, allowing control over BC production levels (Florea et al. 2016) (Fig. 1.3).

Three years later, the tool repertoire for *Komagataibacter* was further expanded by Teh et al. with libraries consisting of several constitutive and inducible synthetic promoters, 10 terminators, 42 ribosome-binding sites (RBSs), and 5 degradation tags to fine-tune transcription and translation levels. The inducible promoters P_{BAD} and P_{Lux} were comparatively characterized with the first showing lower induction levels, but also less leakiness. Additionally, RBS characterization allowed the identification of important sequence features that could facilitate the design of novel RBSs with desired strengths. The parts were characterized in different species, namely *K. rhaeticus* iGEM, *K. sucrofermentans* ATCC 700178, and *Novacetimonas hansenii* ATCC 53582, showing that the toolkit is applicable in multiple BC producers, although some parts might behave differently across species (Florea et al. 2016; Teh et al. 2019).

In 2020a, Hur et al. aimed at balancing the translation strengths of the gene products of *pgm* (phosphoglucomutase), *galU* (UDP-glucose pyrophosphorylase), and *ndp* (nucleoside diphosphate kinase) by generating several engineered RBSs with varying translation strengths. These enzymes are involved in the biosynthesis of the direct precursor of BC, UDP-glucose. The systematic evaluation of the impact of different RBS strengths allowed to increase BC production by balancing protein expression levels in *K. xylinus* DSM 2325 (Hur et al. 2020a).

In 2021, Goosens et al. developed a Golden Gate-based system, named Komagataeibacter Tool Kit (KTK), designed for standardized, modular, and hierarchical assembly of diverse genetic parts (Goosens et al. 2021). The KTK features a user-friendly and expandable plasmid library of genetic parts, which includes promoters, RBSs, terminators, fusion tags, and reporter proteins, along with replicating plasmids. Using one-pot Golden Gate reactions, researchers can rapidly assemble these parts into transcriptional units and further combine them into complex multigene constructs (Goosens et al. 2021) (Fig. 1.3). Due to its standardized and modular nature, KTK can be a driving force in BC research by facilitating collaboration between research groups and improving data reproducibility.

The development of CRISPRi has simultaneously added a new layer of precision to gene regulation and enabled the easy and reprogrammable knock down of essential genes, which is useful due to the lethality associated with eliminating such targets (Byun et al. 2023; Gilbert et al. 2013; Peters et al. 2016). CRISPRi uses a catalytically inactive Cas9 (dCas9) protein to repress transcription without introducing double-strand breaks (Byun et al. 2023; Gilbert et al. 2013; Peters et al. 2016). This technique has been employed to downregulate key genes like *acsAB* and *galU* in *K. xylinus* CGMCC 2955 and *N. hansenii* ATCC 53582, respectively, resulting in tailored alterations in BC production and material properties (Huang et al. 2020; Teh et al. 2019). As an example, transcriptomic data hints that galU is critical for controlling the carbon flux between BC biosynthesis and the pentose phosphate pathway, thus influencing the rate of BC biosynthesis. By using CRISPRi to control the galU expression levels, Huang et al. (2020) were able to understand its relationship with porosity and crystallinity (Huang et al. 2020).

Gene circuits and quorum sensing systems have also been explored to expand the toolkit for BC engineering. In 2016, Florea et al. (2016) adapted an acyl-homoserine lactone (AHL)-based system for inter-strain communication and regulation of BC biosynthesis (Florea et al. 2016). Building on this, another tool was developed specifically for co-culture systems, where "sender" cells produce AHL to trigger a response in "receiver" cells, enabling spatial patterning and underlining the versatility of synthetic biology to develop intelligent biomaterials (Walker et al. 2019).

Overall, the integration of synthetic biology tools such as plasmids, CRISPRi systems, and gene circuits has revolutionized the genetic engineering of BC-producing bacteria. These tools not only enhance BC yields but also allow for the customization of its structural and functional properties, paving the way for novel applications in material science and biomanufacturing. As these technologies continue to evolve, their application across various *Komagataeibacter* species is expected to unlock even greater potential for sustainable and high-performance BC production.

3.2 *Key Genetic Approaches to Enhance BC Yield*

Broadening the range of substrates for BC production enhances both cost-effectiveness and versatility by allowing the use of diverse carbon sources and feedstocks. For example, overexpressing sucrose synthase enables the efficient conversion of sucrose into UDP-glucose. This process, coupled with UDP recycling, sustains cellulose biosynthesis and prevents the accumulation of metabolic intermediates (Nakai et al. 1999). Similarly, modifying the mannose catabolic pathway of *K. xylinus* ATCC 23770, by expressing mannose kinase and phosphomannose isomerase, has improved the utilization of mannose-rich biomass. These modifications resulted in an 84% increase in BC yield and enhanced mechanical properties, including greater tensile strength and elongation (Yang et al. 2023). Furthermore, introducing the *lacZ* gene in *K. xylinus* ITz3 enables lactose metabolism, allowing the use of lactose-rich substrates such as whey, leading to a 28-fold increase in BC production (Battad-Bernardo et al. 2004). There have also been efforts to improve glucose utilization through heterologous expression of the glucose facilitator protein from *Zymomonas mobilis* and the phosphoenolpyruvate-dependent glucose phosphotransferase system (PTS_{Glc}), with the latter being more efficient (Liu et al. 2020; Peng et al. 2024). Restoring the PTS_{Glc} system in *K. xylinus* CGMCC 2955 required the balanced expression of one operon and one additional gene, while restoring glycolysis required pfkA overexpression (Peng et al. 2024). These genetic improvements demonstrate the potential for leveraging diverse and economically viable carbon sources to enhance BC yields and industrial scalability (Battad-Bernardo et al. 2004; Nakai et al. 1999; Peng et al. 2024; Yang et al. 2023).

In addition to optimizing substrate utilization, environmental and operational conditions play a crucial role in enhancing BC production (e.g., oxygen availability, static/agitated culture). The expression of the *Vitreoscilla* hemoglobin in *K. xylinus* BCRC 12334 has been shown to increase BC synthesis by up to 70%. This effect is particularly pronounced under oxygen-limited conditions, likely due to improved adenosine triphosphate (ATP) generation, which redirects carbon flux towards BC biosynthesis. Furthermore, lowering oxygen tension in sealed cultures reduces the oxygen-intensive incomplete oxidation of glucose. While this process contributes to ATP generation via the electron transport chain, it also leads to the extracellular accumulation of gluconic acids, which in turn lowers the culture's pH (Setyawati et al. 2007). This incomplete oxidation process is inherently linked to oxygen availability, as oxygen serves as the final electron acceptor in the periplasmic oxidation of sugars and sugar alcohols.

However, this oxygen-dependent periplasmic oxidation is a double-edged sword. While it provides an ecological advantage for the genus, it also reduces carbon availability for BC synthesis and contributes to acidification, which forces BCS to function under suboptimal conditions. That is the case for the supra-mentioned partial oxidation of glucose, which competes with BC production. To address this inefficiency, several studies report significant improvements in BC production following the knockout of the *gcd* gene, in various *Komagataeibacter* strains, which encodes a pyrroloquinoline quinone (PQQ)-dependent glucose dehydrogenase, thus preventing the oxidation of glucose into gluconic acid (Kuo et al. 2015; Liu et al. 2020; Montenegro-Silva et al. 2024; Shigematsu et al. 2005).

In sharp contrast, ethanol supplementation is a recurring strategy to boost BC production across the literature (Kornmann et al. 2003; Montenegro-Silva et al. 2024; Ryngajłło et al. 2019). Unlike glucose oxidation, ethanol oxidation produces acetic acid, which is further metabolized in the tricarboxylic acid cycle without a pH drop as drastic as in the former example (Montenegro-Silva et al. 2024; Ryngajłło et al. 2019). In *K. sucrofermentans* ATCC 700178, knocking out the gene encoding the PQQ-dependent alcohol dehydrogenase produces a strain that retains the ability to metabolize ethanol. However, BC production no longer benefits from ethanol supplementation. The mutant also differs from the parental strain in its non-diauxic growth in the presence of ethanol and extracellular accumulation of only insignificant amounts of acetic acid (Montenegro-Silva et al. 2024).

Addressing the occurrence of non-producing (*Cel−*) mutants, often caused by insertion sequences disrupting the *bcsA* gene, is also key for maintaining high BC productivity. Genetic modifications to stabilize the *bcs* operon have reduced IS element insertion events, increasing BC yield by 1.7-fold (Hur et al. 2020b). Additionally, the spontaneous conversion of *Cel+* to *Cel−* phenotypes, frequently observed in cellulose-producing species, has been linked to stress factors like increased oxygen concentration, agitation speed, and temperature (Jung et al. 2007; Matsutani et al. 2015). Furthermore, transcriptomic analysis has demonstrated that ethanol supplementation downregulates genes encoding transposases of the IS110 family, contributing to greater genomic stability in *K. xylinus* E25 (Ryngajłło et al. 2019). Both genetic and supplementation approaches result in higher BC production and are useful to increase the process' reliability due to increased genetic stability (Hur et al. 2020b).

Although another chapter explores systems biology and multi-omics approaches in greater detail, their role in guiding metabolic engineering strategies is worth highlighting briefly here. For example, the genome-scale metabolic model (GEM) KxyMBEL1810, for *K. xylinus* DSM 2325 has identified key overexpression targets, including *pgi* (glucose-6-phosphate isomerase) and *gnd* (phosphogluconate dehydrogenase) (Jang et al. 2019). These enzymes, selected based on random sampling and correlation analysis, play crucial roles in glycolysis and the pentose phosphate pathway, respectively. Overexpression of either *pgi* or *gnd* genes resulted in improved BC production, with the former leading to a 115.8% increase (Jang et al. 2019).

In another study, another GEM of *K. xylinus* DSM 2325 was reconstructed through genome sequencing and analysis, identifying the introduction of the *pfkA* gene, under control of the tac promoter, as a critical factor for improving BC yield since it restores the Embden-Meyerhof-Parnas pathway, which is commonly absent in bacteria within the genus (Gwon et al. 2019). Simulated flux balance analysis revealed that expressing *pfkA* not only increased ATP flux but also enhanced other central metabolic fluxes. Further improvement was achieved by incorporating the catabolite repressor protein (CRP) regulator, which was expected to modulate the expression of carbon metabolism-related genes and coincided with a reduction in gluconic acid accumulation (Gwon et al. 2019). To further optimize production, computational fluid dynamics (CFD) was employed to refine fermentation parameters, including impeller configuration, pH, and oxygen levels, adjustments that also correlated with an increase in BC yield. Consequently, BC yield improved from 9.5 to 28.4% in the resulting S.Koma-pfkA/crp strain, reaching up to 31.5% under optimized fermentation conditions (g BC per g glucose). These findings underscore the potential of rational metabolic engineering, coupled with fermentation optimization, to significantly enhance BC production for industrial applications (Gwon et al. 2019).

Genetic engineering has significantly improved BC yield by optimizing metabolic pathways, substrate utilization, and fermentation conditions. Future advances in synthetic biology and multi-omics may help further refine these strategies, or discovering novel ones, potentially enabling scalable and sustainable BC production.

3.3 Key Genetic Approaches to Modulate and Functionalize BC

Traditional ex situ methods for functionalizing BC involve dissolving or disintegrating its matrix to blend it with functional molecules, or additives, such as polymers, nanoparticles, or enzymes. While this approach allows the control over the material's composition, it disrupts BC's nanostructure, reducing its mechanical integrity and biocompatibility. A less invasive modification route consists in immersing the BC pellicles in solutions containing additives, but this route typically results in less uniform materials. These limitations were major drivers towards in situ and *in vivo* strategies that homogeneously incorporate additives during BC biosynthesis to preserve its structural properties (Fonseca et al. 2021; Jin et al. 2022). In situ approaches involve adding functional molecules directly to the culture medium during BC production, thus allowing bacteria to incorporate the molecules into the cellulose matrix as it forms, minimizing disruptions to the nanostructure. However, this method is often limited by the high cost of purified additives, which may be lost, to some extent, during the BC washing/purification. A more recent advancement, *in vivo* biomanufacturing, leverages genetically engineered co-culture systems. For example, certain *Komagataeibacter* species have been co-cultured with engineered *E. coli* to produce composites with enhanced molecular sequestration and probiotic functionality (Birnbaum et al. 2021; Qiao et al. 2023). These systems can potentially reduce costs and allow greater control over composite properties, by adjusting parameters such as inoculation ratios and carbon source feed rates. The production of composites can be achieved also by genetically modifying the BC producing strains.

3.3.1 Modulation and Functionalization of BC by Engineering BC Producers

An interesting case of mechanical properties' modulation through genetic engineering consisted in the overexpression of *motA* and *motB* in *N. hansenii* ATCC 23769, which resulted in elongated cells with increased motility and productivity (Jacek et al. 2019b, 2019c). The resulting structure, with thicker fibers and larger pores, suggests that ribbon formation is affected by cell elongation, possibly due to life cycle-related fiber rearrangement. Additionally, the upregulation of genes related to stress response, translation, and biofilm formation was also observed, which further supports the role of *motA* and *motB* in regulating both cellular functions and BC architecture (Jacek et al. 2019b, 2019c).

To improve biodegradation, an engineered strain of *K. xylinus* ATCC 10245 was modified to incorporate N-acetylglucosamine (GlcNAc) residues into its BC matrix, creating a BC-chitin composite. This was achieved by introducing and expressing a synthetic operon from *C. albicans* containing the *AGM1*, *UAP1*, and *NAG5* genes, which drive the biosynthesis of UDP-GlcNAc. The BCS complex was able to efficiently utilize this activated sugar precursor during cellulose synthesis, leading to the formation of a hybrid polymer with a disrupted crystalline structure. The modified BC exhibited improved susceptibility to enzymatic degradation by lysozyme and cellulase, facilitating potential applications in the biomedical fields (Yadav et al. 2010). Another example is a BC/Curdlan nanocomposite produced by a genetically modified *N. hansenii* ATCC 23769 strain able to co-synthesize these two biopolymers *in vivo*, due to the heterologous expression of the crdS gene from *Agrobacterium* sp. ATCC 31749, responsible for curdlan biosynthesis. The engineered strain simultaneously produced cellulose nanofibers and β-1,3-glucan without additional secretion machinery, producing a less water-permeable composite with altered surface morphology that preserves the integrity of the BC matrix (Fang et al. 2015). Another good example is the fabrication of nanocomposite materials through the simultaneous production of BC and hyaluronan (HA) in a genetically engineered *N. hansenii* ATCC 23769 strain. The approach required the introduction of key genes, including *PmHAS* from *Pasteurella multocida* and UDP-glucose dehydrogenase from *Sinorhizobium meliloti*, which are essential for HA biosynthesis. Fluorescence microscopy analysis suggested that HA is co-secreted with BC through the bacterium's native cellulose secretion machinery, aligning with the previous findings where *N. hansenii* ATCC 23769 was engineered to produce curdlan-based nanocomposites, reinforcing the hypothesis that its transmembrane channels facilitate the secretion of heterologous polysaccharides. Notably, the HA/BC nanocomposites exhibited enhanced structural and biofunctional properties compared to conventional in situ nanocomposites, demonstrating improved surface morphology, stability, and adhesion between HA and BC. These advancements highlight the potential of *N. hansenii* ATCC 23769 as a versatile microbial platform for the biosynthesis of innovative biomaterials with applications in biomedicine (Takahama et al. 2021).

The production of dyed BC by synthetic biology was elegantly presented by Walker et al. This approach potentially represents an alternative to conventional dyeing processes, which often rely on chemical-intensive methods, by engineering a BC producer to produce self-pigmented materials (Walker et al. 2025). The authors modified the *K. rhaeticus* iGEM strain to synthesize eumelanin through the expression of a heterologous tyrosinase. Due to the stability and low solubility in water of the black pigment, the resulting BC membranes retained their coloration after autoclaving, posing as a demonstration of the pigmentation robustness. Additionally, the researchers were able to scale-up the production process in order to prototype fashion products such as wallets and shoe uppers. Given the stability of eumelanin and its additional properties, including UV protection and heat resistance, melanated

BC holds potential as a sustainable textile material (Walker et al. 2025). The current method involves a two-step process requiring a development buffer, awaiting future optimizations that could streamline pigment synthesis under the naturally acidic conditions of *K. rhaeticus* iGEM, reducing water consumption and enhancing sustainability. Further exploration of acid-tolerant tyrosinases and alternative melanin biosynthesis pathways could improve efficiency and facilitate industrial-scale implementation (Walker et al. 2025). The researchers further explored spatial control over pigment deposition by implementing a blue-light sensitive T7 RNA polymerase system, which enabled patterned gene expression on the cellulose pellicle (Baumschlager et al. 2017; Walker et al. 2025). Although the patterning approach faced challenges such as elevated background pigmentation, this work lays the foundation for creating advanced BC-based biofabricated textiles with customizable designs.

Even though this chapter enumerates several toolkits and strategies to engineer BC-producing bacteria, other microorganisms can be easier and more flexible to genetically engineer or be better chassis for the desired phenotype. For that reason, sometimes it is easier to co-culture BC producers with microorganisms that have more tools available and/or are more amenable to genetic modification. For instance, *Komagataeibacter* spp. are not as proficient as yeast at secreting recombinant proteins. A co-culture of *K. rhaeticus* iGEM with *S. cerevisiae* yCelMix, an engineered yeast strain that secretes a cellulose-degrading enzyme cocktail, has been used as a strategy for modulation of the mechanical and viscoelastic properties of BC (C. Gilbert et al. 2021). Additionally, a co-culture strategy has also been used for the in situ production of HA/BC nanocomposites, like in the previously mentioned example. However, instead of using a recombinant *N. hansenii* strain, the wild strain was co-cultured with a *Lactococcus lactis* strain that was engineered for HA production. The authors also demonstrated that the HA yield was dependent on the pH of the culture, enabling the use of pH as a modulating parameter for the final physical properties of the material (Liu and Catchmark 2020).

3.3.2 Engineered Living Materials

Engineered living materials (ELM) are another interesting and potentially disruptive strategy for the creation of BC-based smart products. Such ELMs can consist of organisms from one or multiple species, with the first major proof of concept dating to 2014 (Qin et al. 2014). In 2019, Walker et al. developed a self-patterning ELM by adapting a Lux quorum sensing system from the Registry of Standard Biological parts for use in *K. rhaeticus* iGEM. The ELM was composed of two *K. rhaeticus* iGEM-derived strains—"sender" and "receiver"—were the "sender" strain produced AHL through the constitutive expression of an AHL synthase, and the "receiver" strain produced red fluorescence in response (Walker et al. 2019). A follow-up paper reported the production of "sender" and "receiver" BC spheroids, which serve as modular building blocks for constructing three-dimensional structures and patterned ELMs. These spheroids can be genetically functionalized to serve as dynamic biosensors, producing visible or hidden signals in response to environmental stimuli. Potential applications include smart wound dressings that monitor tissue repair or sweat-analyzing patches, capable of detecting biomarkers related to various physiological conditions. Furthermore, BC spheroids exhibited potential for material regeneration by fusing damaged BC components or assembling separate cellulose and synthetic sections into cohesive structures (Caro-Astorga et al. 2021).

As previously mentioned, non-model organisms, such as BC producers, often lack the essential tools needed to expand the potential applications of their capabilities. In such cases, it might be useful to use BC producers, exclusively or not, as responsible for matrix production, while an engineered strain of a model organism brings functionality to the living material (Birnbaum et al. 2021; Gilbert et al. 2021; Rodrigo-Navarro et al. 2021). As a proof of concept, an IPTG-responsive BC membrane was developed by co-culturing *K. xylinus* ATCC 10245 with engineered *E. coli* strains. In this case, two *E. coli* strains were individually expressing the green and red fluorescent proteins (GFP and RFP, respectively) upon induction (Qin et al. 2014). Furthermore, a modular positive-feedback genetic amplifier was designed to enhance reporter gene signals, improving the detection sensitivity of transcriptional responses. This system, based on the LuxR system from *Vibrio fischeri*, enables amplified GFP expression upon IPTG induction in *E. coli* BL21 star (DE3). When *E. coli* is incorporated in BC membranes, the genetic amplifier significantly increased fluorescence output, even at low bacterial loads, demonstrating its potential for enhancing biosensing applications (Qin et al. 2014).

Although the previous example was successful, BC's permeability allows microorganisms to escape. Moreover, the simplest biosensor implementations do not allow the distinction between the absence of the target analyte and cell death (Drachuk et al. 2013, 2017). To tackle this, *K. xylinus* and *E. coli* cells were "trapped" in the BC scaffold by addition of reconstituted silk fibroin (SF) to the membrane. Additionally, a biosensor module including the FimS promoter, flanked by inverted repeats, two fluorescent proteins (GFPa1 and mKate2), and a riboswitch-controlled FimE recombinase was developed (Harbaugh et al. 2017). *E. coli* constitutively expresses GFPa1 unless theophylline is present. When theophylline binds, the riboswitch activates FimE expression, resulting in the inversion of the FimS promoter, switching off GFPa1 expression, and activating mKate2. These results demonstrate the potential of BC-SF composite scaffolds for stabilizing recombinant *E. coli* in biotechnological applications. The improved metabolic activity and stress resistance enhance biosensor performance. Further optimization could tailor scaffold properties for specific uses in biosensing and biomedicine (Drachuk et al. 2017).

Still using *E. coli* to provide functionality to the BC-based ELMs, there is another example of an innovation but, this time, at the materials' geometry. Instead of conventional pellicles, hollow spherical BC capsules were designed to create topologically distinct regions, where *N. hansenii* ATCC 53582 forms a dense cellulose outer shell, while engineered *E. coli* occupies the inner core, enabling programmable genetic functionalities (Birnbaum et al. 2021). These capsules were produced by incubating droplets of fresh media containing *N. hansenii* on top of superhydrophobic PTFE powder, allowing for spatial separation of bacterial populations. Capsules expressing curli-tethered nanobodies efficiently captured green fluorescent protein (GFP) from solution, demonstrating high specificity (Birnbaum et al. 2021). Additionally, capsules engineered to display α-amylase exhibited significant enzymatic activity, breaking down starch-based substrates (Birnbaum et al. 2021). Another ELM was engineered for biomineralization, where *E. coli* HB101, transformed with pBR322-Ure, catalyzed urea hydrolysis, leading to carbonate ion formation and $CaCO_3$ precipitation within the cellulose matrix. When incubated in urea and $CaCl_2$-supplemented media, these capsules showed rapid mineralization, as indicated by a decrease in soluble calcium levels, increased mechanical stiffness, and a visible transformation into yellowish, pebble-like structures. X-ray powder diffraction (XRPD) analysis confirmed the presence of calcite and vaterite, suggesting that the BC network influences the mineralization process, making the material structurally adaptive and self-modifying (Birnbaum et al. 2021).

The Syn-SCOBY approach demonstrates the possibility of a modular co-culture system that integrates engineered *S. cerevisiae* into BC-based ELMs, enabling them to sense and respond to environmental stimuli (Gilbert et al. 2021). Yeast biosensors were developed to detect β-estradiol (BED) and express GFP as a reporter, maintaining functionality even after drying and long-term storage. Additionally, a yeast strain secreting laccase under a BED-inducible promoter enabled the biodegradation of BED, showcasing the potential for dynamic, responsive biomaterials (Gilbert et al. 2021). Beyond chemical sensing, optogenetics was used to create light-responsive BC materials by co-culturing *K. rhaeticus* iGEM with *S. cerevisiae* engineered to express NanoLuc luciferase under a blue-light-inducible promoter. Light patterning techniques, such as masking and projection, allowed for precise spatial control over bioluminescence, with sharper patterns in yeast displaying surface-anchored NanoLuc compared to those secreting a cellulose-binding version (Gilbert et al. 2021). This spatial programming of protein functionalization in BC materials highlights the potential for multicolor optogenetic systems, enabling spatially controlled enzymatic cascades for advanced biosensing and biocatalysis applications (Gilbert et al. 2021).

The future of BC functionalization lies in advancing synthetic biology and co-culture techniques to expand the repertoire of functional molecules integrated into the matrix. As researchers develop more efficient protein secretion mechanisms in BC-producing bacteria and refine metabolic engineering strategies, the potential for creating tailored BC composites with multifunctional properties will continue to grow. This progress will open new avenues for applications in medicine, environmental remediation, and next-generation smart materials.

4 Conclusions and Future Perspectives

While BC has been gaining attractiveness due to its outstanding characteristics, its production methods are facing significant challenges with by-product formation, metabolic competition, and the high precursor demands for cellulose biosynthesis, to name a few (Cacicedo et al. 2016; El-Gendi et al. 2022; Mishra et al. 2022; Wang et al. 2025). Synthetic biology holds the potential to be a disruptive approach to overcoming these limitations, not only enabling the creation of more productive strains, but also the modulation of the material's properties and functionalization, expanding its application in several industries (Andrianantoandro et al. 2006; Jin et al. 2022; Khalil and Collins 2010; Malcı et al. 2024; Nielsen and Keasling 2011; Singh et al. 2020; Stephanopoulos 2012).

The development of synthetic biology tools has enabled a plethora of strategies that aim at fine tuning metabolic pathways, optimizing substrate utilization and regulating gene expression with high precision (Buldum and Mantalaris 2021; Hur et al. 2020a; Peng et al. 2024; Ryngajłło et al. 2020). For example, CRISPRi has been used to modulate the activity of key metabolic steps allowing the fine tuning of BC's porosity and crystallinity (Huang et al. 2020; Teh et al. 2019). Moreover, improving UDP-glucose biosynthesis through the overexpression of genes such as *pfkA*, *gnd*, and *pgi* resulted in increased cellulose yields (Gwon et al. 2019). Additionally, the implementation of quorum sensing mechanisms and synthetic gene circuits has further enabled autonomous control over BC biosynthesis, allowing for the spatial and temporal patterning of cellulose production (Caro-Astorga et al. 2021; Walker et al. 2019).

Genetic engineering has enabled the fine tuning of mechanical properties of BC and also the direct incorporation of functional molecules into the BC matrix (Birnbaum et al. 2021; Drachuk et al. 2017; Fang et al. 2015; Takahama et al. 2021). A notable example is the introduction of motility genes which resulted in modified cellulose fiber morphology and increased porosity (Jacek et al. 2019b, 2019c). Similarly, engineered BC composites obtained by in situ co-production of hyaluronan or curdlan into the cellulose matrix excel in its biodegradability, biofunctionality, and mechanical performance (Yadav et al. 2010). The potential of BC is expanding to an increasing number of industries (e.g., electronics, regenerative medicine, and sustainable biofabrication) due to the continuous innovations in BC functionalization research (Andriani et al. 2020; Cacicedo et al. 2016; Choi and Shin 2020; Zheng et al. 2024; Zhong 2020).

Co-culture systems are among the most significant advances in BC research since it opens the possibility to use more tractable organisms for in situ functionalization (Drachuk et al. 2017; Gilbert et al. 2021; Jin et al. 2022; Malcı et al. 2024; Singh et al. 2020). Due to the availability of well-established synthetic biology toolkits, *Escherichia coli* and *Saccharomyces cerevisiae* have been used in co-culture strategies for BC functionalization (Drachuk et al. 2017; Gilbert et al. 2021; Jin et al. 2022). The co-culture strategies enabled the creation of BC-based living materials with enzyme activity, biosensing capabilities, and self-patterning properties, thus demonstrating the feasibility of using BC as a platform for smart and interactive materials (Drachuk et al. 2017; Gilbert et al. 2021; Jin et al. 2022; Malcı et al. 2024; Singh et al. 2020).

The future of BC research lies at the intersection of synthetic biology, systems biology, and bioprocess engineering (Malcı et al. 2024; Potočnik et al. 2023; Ryngajłło et al. 2020; Singh et al. 2020; Vadanan et al. 2022). By integrating multiomics approaches (e.g., genomics, transcriptomics, and metabolomics), the community might obtain deeper insights into the regulatory mechanisms over BC biosynthesis, which can potentially translate into more precise metabolic engineering strategies (Ryngajłło et al. 2020). Computational modeling and genome-scale metabolic network reconstruction methodologies are being continuously developed aiming at facilitating the rational design of microbial BC factories through the predictive optimization of metabolic pathways. Moreover, advances in automation (e.g., high-throughput screening) can accelerate the development of superior strains (Buldum and Mantalaris 2021; Ryngajłło et al. 2020).

Looking ahead, we speculate that synthetic biology is set to play an even greater role in the advancement of BC-based materials. The potential applications of engineered BC should continue to expand in diverse fields (e.g., tissue engineering, sustainable packaging, and biophotonic devices) as research continuously innovates genetic and bioprocessing strategies. The ability to program self-assembled multifunctional microbial communities is closer to becoming a reality in the industry, thus strengthening the role of BC-based materials as a paradigm shift in biofabrication due to their potential to outperform their synthetic counterparts (Jin et al. 2022; Khalil and Collins 2010; Malcı et al. 2024; Nielsen and Keasling 2011; Potočnik et al. 2023; Singh et al. 2020; Stephanopoulos 2012; Vadanan et al. 2022).

In conclusion, BC is gradually becoming a bioengineering platform for future materials. The success depends on an interdisciplinary effort to integrate methodologies from synthetic biology, metabolic engineering, and materials science, with the potential to set a new era in biofabrication where living systems can be leveraged to create sustainable, responsive, high-performance materials tailored to meet the evolving demands of modern industries (Malcı et al. 2024; Potočnik et al. 2023; Singh et al. 2020).

References

Andrianantoandro E, Basu S, Karig DK, Weiss R (2006) Synthetic biology: new engineering rules for an emerging discipline. Mol Syst Biol 2:2006.0028. https://doi.org/10.1038/msb4100073

Andriani D, Apriyana AY, Karina M (2020) The optimization of bacterial cellulose production and its applications: a review. Cellulose 27(12):6747–6766). Springer. https://doi.org/10.1007/s10570-020-03273-9

Bae SO, Sugano Y, Ohi K, Shoda M (2004) Features of bacterial cellulose synthesis in a mutant generated by disruption of the diguanylate cyclase 1 gene of Acetobacter xylinum BPR 2001. Appl Microbiol Biotechnol 65(3):315–322. https://doi.org/10.1007/s00253-004-1593-7

Báez J, Olsen D, Polarek JW (2005) Recombinant microbial systems for the production of human collagen and gelatin. Appl Microbiol Biotechnol 69(3):245–252. https://doi.org/10.1007/s00253-005-0180-x

Battad-Bernardo E, McCrindle SL, Couperwhite I, Neilan BA (2004) Insertion of an E. coli lacZ gene in Acetobacter xylinus for the production of cellulose in whey. FEMS Microbiol Lett 231(2):253–260. https://doi.org/10.1016/S0378-1097(04)00007-2

Baumschlager A, Aoki SK, Khammash M (2017) Dynamic blue light-inducible T7 RNA polymerases (Opto-T7RNAPs) for precise spatiotemporal gene expression control. ACS Synth Biol 6(11):2157–2167. https://doi.org/10.1021/acssynbio.7b00169

Birnbaum DP, Manjula-Basavanna A, Kan A, Tardy BL, Joshi NS (2021) Hybrid living capsules autonomously produced by engineered bacteria. Adv Sci 8(11):2004699. https://doi.org/10.1002/advs.202004699

Bose S, Volpatti LR, Thiono D, Yesilyurt V, McGladrigan C, Tang Y, Facklam A, Wang A, Jhunjhunwala S, Veiseh O, Hollister-Lock J, Bhattacharya C, Weir GC, Greiner DL, Langer R, Anderson DG (2020) A retrievable implant for the long-term encapsulation and survival of therapeutic xenogeneic cells. Nat Biomed Eng 4(8):814–826. https://doi.org/10.1038/s41551-020-0538-5

Buldum G, Mantalaris A (2021) Systematic understanding of recent developments in bacterial cellulose biosynthesis at genetic, bioprocess and product levels. Int J Mol Sci 22(13):7192. MDPI. https://doi.org/10.3390/ijms22137192

Byun G, Yang J, Seo SW (2023) CRISPRi-mediated tunable control of gene expression level with engineered single-guide RNA in Escherichia coli. Nucleic Acids Res 51(9):4650–4659. https://doi.org/10.1093/nar/gkad234

Cacicedo ML, Castro MC, Servetas I, Bosnea L, Boura K, Tsafrakidou P, Dima A, Terpou A, Koutinas A, Castro GR (2016) Progress in bacterial cellulose matrices for biotechnological applications. Bioresour Technol 213:172–180. Elsevier Ltd. https://doi.org/10.1016/j.biortech.2016.02.071

Campebell RC, Oliveira AB, Fagundes JLA, Fortes BNA, Veado HC, Macedo IL, Dallago BSL, Barud HS, Adorno J, Salvador PAV, Santos PS, Castro MB (2025) Evaluation of bacterial cellulose/alginate-based hydrogel and frog skin dressings in equine skin wound healing. Gels 11(2):107. https://doi.org/10.3390/gels11020107

Cao Y, Feng Y, Ryser MD, Zhu K, Herschlag G, Cao C, Marusak K, Zauscher S, You L (2017) Programmable assembly of pressure sensors using pattern-forming bacteria. Nat Biotechnol 35(11):1087–1093. https://doi.org/10.1038/nbt.3978

Caro-Astorga J, Walker KT, Herrera N, Lee KY, Ellis T (2021) Bacterial cellulose spheroids as building blocks for 3D and patterned living materials and for regeneration. Nat Commun 12(1):5027. https://doi.org/10.1038/s41467-021-25350-8

Chang AL, Tuckerman JR, Gonzalez G, Mayer R, Weinhouse H, Volman G, Amikam D, Benziman M, Gilles-Gonzalez M-A (2001) Phosphodiesterase A1, a regulator of cellulose synthesis in Acetobacter xylinum, is a heme-based sensor. Biochemistry 40(12):3420–3426. https://doi.org/10.1021/bi0100236

Choi SM, Shin EJ (2020) The nanofication and functionalization of bacterial cellulose and its applications. Nanomaterials 10(3):406. MDPI AG. https://doi.org/10.3390/nano10030406

Drachuk I, Shchepelina O, Harbaugh S, Kelley-Loughnane N, Stone M, Tsukruk VV (2013) Cell surface engineering with edible protein nanoshells. Small 9(18):3128–3137. https://doi.org/10.1002/smll.201202992

Drachuk I, Harbaugh S, Geryak R, Kaplan DL, Tsukruk VV, Kelley-Loughnane N (2017) Immobilization of recombinant E. coli cells in a bacterial cellulose-silk composite matrix to preserve biological function. ACS Biomater Sci Eng 3(10):2278–2292. https://doi.org/10.1021/acsbiomaterials.7b00367

El-Gendi H, Taha TH, Ray JB, Saleh AK (2022) Recent advances in bacterial cellulose: a low-cost effective production media, optimization strategies and applications. Cellulose 29(14):7495–7533. Springer Science and Business Media B.V. https://doi.org/10.1007/s10570-022-04697-1

Fang J, Kawano S, Tajima K, Kondo T (2015) In vivo Curdlan/cellulose bionanocomposite synthesis by genetically modified Gluconacetobacter xylinus. Biomacromolecules 16(10):3154–3160. https://doi.org/10.1021/acs.biomac.5b01075

Florea M, Hagemann H, Santosa G, Abbott J, Micklem CN, Spencer-Milnes X, De Arroyo Garcia L, Paschou D, Lazenbatt C, Kong D, Chughtai H, Jensen K, Freemont PS, Kitney R, Reeve B, Ellis T (2016) Engineering control of bacterial cellulose production using a genetic toolkit and a new cellulose producing strain. Proc Natl Acad Sci U S A 113(24):E3431–E3440. https://doi.org/10.1073/pnas.1522985113

Fonseca DFS, Vilela C, Pinto RJB, Bastos V, Oliveira H, Catarino J, Faísca P, Rosado C, Silvestre AJD, Freire CSR (2021) Bacterial nanocellulose-hyaluronic acid microneedle patches for skin applications: in vitro and in vivo evaluation. Mater Sci Eng C 118:111350. https://doi.org/10.1016/j.msec.2020.111350

Fujiwara T, Komoda K, Sakurai N, Tajima K, Tanaka I, Yao M (2013) The c-di-GMP recognition mechanism of the PilZ domain of bacterial cellulose synthase subunit A. Biochem Biophys Res Commun 431(4):802–807. https://doi.org/10.1016/j.bbrc.2012.12.103

Gautam S, Bundela P, Pandey A, Awasthi M, Sarsaiya S (2010) A review on systematic study of cellulose. India J Appl Nat Sci 2(2):330–343. www.ansfoundation.org

Gilbert C, Ellis T (2019) Biological engineered living materials: growing functional materials with genetically programmable properties. ACS Synth Biol 8(1):1–15). American Chemical Society. https://doi.org/10.1021/acssynbio.8b00423

Gilbert LA, Larson MH, Morsut L, Liu Z, Brar GA, Torres SE, Stern-Ginossar N, Brandman O, Whitehead EH, Doudna JA, Lim WA, Weissman JS, Qi LS (2013) CRISPR-mediated modular RNA-guided regulation of transcription in eukaryotes. Cell 154(2):442–451. https://doi.org/10.1016/j.cell.2013.06.044

Gilbert C, Tang T-C, Ott W, Dorr BA, Shaw WM, Sun GL, Lu TK, Ellis T (2021) Living materials with programmable functionalities grown from engineered microbial co-cultures. Nat Mater 20(5):691–700. https://doi.org/10.1038/s41563-020-00857-5

Goosens VJ, Walker KT, Aragon SM, Singh A, Senthivel VR, Dekker L, Caro-Astorga J, Buat MLA, Song W, Lee K-Y, Ellis T (2021) Komagataeibacter tool kit (KTK): A modular cloning system for multigene constructs and programmed protein secretion from cellulose producing bacteria. ACS Synth Biol 10(12):3422–3434. https://doi.org/10.1021/acssynbio.1c00358

Griffin AM, Morris VJ, Gasson MJ (1996) Genetic analysis of the acetan biosynthetic pathway in Acetobacter xylinum: nucleotide sequence analysis of the aceB aceC aceD and aceE genes. DNA Seq 6(5):275–284. https://doi.org/10.3109/10425179609020874

Gullo M, La China S, Falcone PM, Giudici P (2018) Biotechnological production of cellulose by acetic acid bacteria: current state and perspectives. Appl Microbiol Biotechnol 102(16):6885–6898). Springer Verlag. https://doi.org/10.1007/s00253-018-9164-5

Gwon H, Park K, Chung SC, Kim RH, Kang JK, Ji SM, Kim NJ, Lee S, Ku JH, Do EC, Park S, Kim M, Shim WY, Rhee HS, Kim JY, Kim J, Kim TY, Yamaguchi Y, Iwamuro R et al (2019) A safe and sustainable bacterial cellulose nanofiber separator for lithium rechargeable batteries. Proc Natl Acad Sci U S A 116(39):19288–19293. https://doi.org/10.1073/pnas.1905527116

Harbaugh SV, Goodson MS, Dillon K, Zabarnick S, Kelley-Loughnane N (2017) Riboswitch-based reversible dual color sensor. ACS Synth Biol 6(5):766–781. https://doi.org/10.1021/acssynbio.6b00199

Hay ID, Rehman ZU, Moradali MF, Wang Y, Rehm BHA (2013) Microbial alginate production, modification and its applications. Microb Biotechnol 6(6):637–650. https://doi.org/10.1111/1751-7915.12076

Hernández-Arriaga AM, Campano C, Rivero-Buceta V, Prieto MA (2022) When microbial biotechnology meets material engineering. Microb Biotechnol 15(1):149–163). John Wiley and Sons Ltd. https://doi.org/10.1111/1751-7915.13975

Heveran CM, Liang L, Nagarajan A, Hubler MH, Gill R, Cameron JC, Cook SM, Srubar WV (2019) Engineered ureolytic microorganisms can tailor the morphology and nanomechanical properties of microbial-precipitated calcium carbonate. Sci Rep 9(1):14721. https://doi.org/10.1038/s41598-019-51133-9

Hu S-Q, Gao Y-G, Tajima K, Sunagawa N, Zhou Y, Kawano S, Fujiwara T, Yoda T, Shimura D, Satoh Y, Munekata M, Tanaka I, Yao M (2010) Structure of bacterial cellulose synthase subunit D octamer with four inner passageways. Proc Natl Acad Sci U S A 107(42):17957–17961. https://doi.org/10.1073/pnas.1000601107/-/DCSupplemental

Huang L-H, Liu Q-J, Sun X-W, Li X-J, Liu M, Jia S-R, Xie Y-Y, Zhong C (2020) Tailoring bacterial cellulose structure through CRISPR interference-mediated downregulation of galU in Komagataeibacter xylinus CGMCC 2955. Biotechnol Bioeng 117(7):2165–2176. https://doi.org/10.1002/bit.27351

Hur DH, Choi WS, Kim TY, Lee SY, Park JH, Jeong KJ (2020a) Enhanced production of bacterial cellulose in Komagataeibacter xylinus via tuning of biosynthesis genes with synthetic RBS. J Microbiol Biotechnol 30(9):1430–1435. https://doi.org/10.4014/jmb.2006.06026

Hur DH, Rhee HS, Lee JH, Shim WY, Kim TY, Lee SY, Park JH, Jeong KJ (2020b) Enhanced production of cellulose in Komagataeibacter xylinus by preventing insertion of IS element into cellulose synthesis gene. Biochem Eng J 156:107527. https://doi.org/10.1016/j.bej.2020.107527

Iguchi M, Yamanaka S, Budhiono A (2000) Bacterial cellulose a masterpiece of nature's arts. J Mater Sci 35:261–270

Ishida T, Sugano Y, Nakai T, Shoda M (2002) Effects of acetan on production of bacterial cellulose by Acetobacter xylinum. Biosci Biotechnol Biochem 66(8):1677–1681

Jacek P, Dourado F, Gama M, Bielecki S (2019a) Molecular aspects of bacterial nanocellulose biosynthesis. Microb Biotechnol 12(4):633–649. https://doi.org/10.1111/1751-7915.13386

Jacek P, Kubiak K, Ryngajłło M, Rytczak P, Paluch P, Bielecki S (2019b) Modification of bacterial nanocellulose properties through mutation of motility related genes in Komagataeibacter hansenii ATCC 53582. New Biotechnol 52:60–68. https://doi.org/10.1016/j.nbt.2019.05.004

Jacek P, Ryngajłło M, Bielecki S (2019c) Structural changes of bacterial nanocellulose pellicles induced by genetic modification of Komagataeibacter hansenii ATCC 23769. Appl Microbiol Biotechnol 103(13):5339–5353. https://doi.org/10.1007/s00253-019-09846-4

Jang WD, Kim TY, Kim HU, Shim WY, Ryu JY, Park JH, Lee SY (2019) Genomic and metabolic analysis of Komagataeibacter xylinus DSM 2325 producing bacterial cellulose nanofiber. Biotechnol Bioeng 116(12):3372–3381. https://doi.org/10.1002/bit.27150

Jedrzejczak-Krzepkowska M, Kubiak K, Ludwicka K, Bielecki S (2016) Chapter 2—Bacterial nanocellulose synthesis, recent findings. In: Gama M, Dourado F, Bielecki S (eds) Bacterial nanocellulose. Elsevier, pp 19–46. https://doi.org/10.1016/B978-0-444-63458-0.00002-0

Jin K, Jin C, Wu Y (2022) Synthetic biology-powered microbial co-culture strategy and application of bacterial cellulose-based composite materials. Carbohydr Polym 283:119171. Elsevier Ltd. https://doi.org/10.1016/j.carbpol.2022.119171

Jozala AF, de Lencastre-Novaes LC, Lopes AM, de Carvalho Santos-Ebinuma V, Mazzola PG, Pessoa-Jr A, Grotto D, Gerenutti M, Chaud MV (2016) Bacterial nanocellulose production and application: a 10-year overview. Appl Microbiol Biotechnol 100(5):2063–2072. Springer Verlag. https://doi.org/10.1007/s00253-015-7243-4

Jung JY, Khan T, Park JK, Chang HN (2007) Production of bacterial cellulose by Gluconacetobacter hansenii using a novel bioreactor equipped with a spin filter. Korean J Chem Eng 24(2):265–271. https://doi.org/10.1007/s11814-007-5058-4

Keasling JD (2012) Synthetic biology and the development of tools for metabolic engineering. Metab Eng 14(3):189–195. https://doi.org/10.1016/j.ymben.2012.01.004

Keshk SMAS (2014) Vitamin C enhances bacterial cellulose production in Gluconacetobacter xylinus. Carbohydr Polym 99:98–100. https://doi.org/10.1016/j.carbpol.2013.08.060

Khalil AS, Collins JJ (2010) Synthetic biology: applications come of age. Nat Rev Genet 11(5):367–379. https://doi.org/10.1038/nrg2775

Kornmann H, Duboc P, Niederberger P, Marison I, Von Stockar U (2003) Influence of residual ethanol concentration on the growth of Gluconacetobacter xylinus I 2281. Appl Microbiol Biotechnol 62(2–3):168–173. https://doi.org/10.1007/s00253-003-1299-2

Kuo CH, Teng HY, Lee CK (2015) Knock-out of glucose dehydrogenase gene in Gluconacetobacter xylinus for bacterial cellulose production enhancement. Biotechnol Bioprocess Eng 20(1):18–25. https://doi.org/10.1007/s12257-014-0316-x

Li G, Wang L, Deng Y, Wei Q (2022) Research progress of the biosynthetic strains and pathways of bacterial cellulose. J Ind Microbiol Biotechnol 49(1):kuab071. Oxford University Press. https://doi.org/10.1093/jimb/kuab071

Liu K, Catchmark JM (2020) Bacterial cellulose/hyaluronic acid nanocomposites production through co-culturing Gluconacetobacter hansenii and Lactococcus lactis under different initial pH values of fermentation media. Cellulose 27(5):2529–2540. https://doi.org/10.1007/s10570-019-02924-w

Liu L-P, Yang X, Zhao X-J, Zhang K-Y, Li W-C, Xie Y-Y, Jia S-R, Zhong C (2020) A lambda red and FLP/FRT-mediated site-specific recombination system in Komagataeibacter xylinus and its application to enhance the productivity of bacterial cellulose. ACS Synth Biol 9(11):3171–3180. https://doi.org/10.1021/acssynbio.0c00450

Malcı K, Li IS, Kisseroudis N, Ellis T (2024) Modulating microbial materials—engineering bacterial cellulose with synthetic biology. ACS Synth Biol 13:3857–3875. American Chemical Society. https://doi.org/10.1021/acssynbio.4c00615

Matsutani M, Ito K, Azuma Y, Ogino H, Shirai M, Yakushi T, Matsushita K (2015) Adaptive mutation related to cellulose producibility in Komagataeibacter medellinensis (Gluconacetobacter xylinus) NBRC 3288. Appl Microbiol Biotechnol 99(17):7229–7240. https://doi.org/10.1007/s00253-015-6598-x

Mishra S, Singh PK, Pattnaik R, Kumar S, Ojha SK, Srichandan H, Parhi PK, Jyothi RK, Sarangi PK (2022) Biochemistry, synthesis, and applications of bacterial cellulose: a review. Front Bioeng Biotechnol 10:780409. Frontiers Media S.A. https://doi.org/10.3389/fbioe.2022.780409

Montenegro-Silva P, Ellis T, Dourado F, Gama M, Domingues L (2024) Enhanced bacterial cellulose production in Komagataeibacter sucrofermentans: impact of different PQQ-dependent dehydrogenase knockouts and ethanol supplementation. Biotechnol Biofuels Bioprod 17(1):35. https://doi.org/10.1186/s13068-024-02482-9

Nakai T, Tonouchi N, Konishi T, Kojima Y, Tsuchida T, Yoshinaga F, Sakai F, Hayashi T (1999) Enhancement of cellulose production by expression of sucrose synthase in Acetobacter xylinum (cellulose synthesis sucrose metabolism). Proc Natl Acad Sci U S A 96:14–18. www.pnas.org

Nielsen J, Keasling JD (2011) Synergies between synthetic biology and metabolic engineering. Nat Biotechnol 29(8):693–695. https://doi.org/10.1038/nbt.1937

Noh YK, Dos Santos Da Costa A, Park YS, Du P, Kim I-H, Park K (2019) Fabrication of bacterial cellulose-collagen composite scaffolds and their osteogenic effect on human mesenchymal stem cells. Carbohydr Polym 219:210–218. https://doi.org/10.1016/j.carbpol.2019.05.039

Peng Z, Lv Z, Liu J, Wang Y, Zhang T, Xie Y, Jia S, Xin B, Zhong C (2024) Engineering PTS-based glucose metabolism for efficient biosynthesis of bacterial cellulose by Komagataeibacter xylinus. Carbohydr Polym 343:122459. https://doi.org/10.1016/j.carbpol.2024.122459

Peters JM, Colavin A, Shi H, Czarny TL, Larson MH, Wong S, Hawkins JS, Lu CHS, Koo B-M, Marta E, Shiver AL, Whitehead EH, Weissman JS, Brown ED, Qi LS, Huang KC, Gross CA (2016) A comprehensive, CRISPR-based functional analysis of essential genes in bacteria. Cell 165(6):1493–1506. https://doi.org/10.1016/j.cell.2016.05.003

Portela R, Leal CR, Almeida PL, Sobral RG (2019) Bacterial cellulose: a versatile biopolymer for wound dressing applications. Microb Biotechnol 12(4):586–610. John Wiley and Sons Ltd. https://doi.org/10.1111/1751-7915.13392

Potočnik V, Gorgieva S, Trček J (2023) From nature to lab: sustainable bacterial cellulose production and modification with synthetic biology. Polymers 15(16):3466. https://doi.org/10.3390/polym15163466

Qiao J-Y, Zeng R-Y, Chen Q-W, Zheng D, Hao P-L, Liu X-H, Zhao F, Zhang X-Z (2023) Autonomously assembled living capsules by microbial coculture for enhanced bacteriotherapy of inflammatory bowel disease. Nano Lett 23(10):4375–4383. https://doi.org/10.1021/acs.nanolett.3c00657

Qin G, Panilaitis BJ, Kaplan ZSDL (2014) A cellulosic responsive "living" membrane. Carbohydr Polym 100:40–45. https://doi.org/10.1016/j.carbpol.2013.06.019

Rodrigo-Navarro A, Sankaran S, Dalby MJ, del Campo A, Salmeron-Sanchez M (2021) Engineered living biomaterials. Nat Rev Mater 6(12):1175–1190. Nature Research. https://doi.org/10.1038/s41578-021-00350-8

Rodrigues AC, Fontão AI, Coelho A, Leal M, Soares da Silva FAG, Wan Y, Dourado F, Gama M (2019) Response surface statistical optimization of bacterial nanocellulose fermentation in static culture using a low-cost medium. New Biotechnol 49:19–27. https://doi.org/10.1016/j.nbt.2018.12.002

Römling U (2002) Molecular biology of cellulose production in bacteria. Res Microbiol 153:205–212. www.elsevier.com/locate/resmicMini-review

Römling U (2012) Cyclic di-GMP, an established secondary messenger still speeding up. Environ Microbiol 14(8):1817–1829. https://doi.org/10.1111/j.1462-2920.2011.02617.x

Römling U, Galperin MY (2015) Bacterial cellulose biosynthesis: diversity of operons, subunits, products, and functions. Trends Microbiol 23(9):545–557). Elsevier Ltd. https://doi.org/10.1016/j.tim.2015.05.005

Römling U, Galperin MY, Gomelsky M (2013) Cyclic di-GMP: the first 25 years of a universal bacterial second messenger. Microbiol Mol Biol Rev 77(1):1–52. https://doi.org/10.1128/mmbr.00043-12

Ryngajłło M, Jacek P, Cielecka I, Kalinowska H, Bielecki S (2019) Effect of ethanol supplementation on the transcriptional landscape of bionanocellulose producer Komagataeibacter xylinus E25. Appl Microbiol Biotechnol 103(16):6673–6688. https://doi.org/10.1007/s00253-019-09904-x

Ryngajłło M, Jędrzejczak-Krzepkowska M, Kubiak K, Ludwicka K, Bielecki S (2020) Towards control of cellulose biosynthesis by Komagataeibacter using systems-level and strain engineering strategies: current progress and perspectives. Appl Microbiol Biotechnol 104(15):6565–6585. https://doi.org/10.1007/s00253-020-10671-3

Setyawati MI, Chien LJ, Lee CK (2007) Expressing Vitreoscilla hemoglobin in statically cultured Acetobacter xylinum with reduced O2 tension maximizes bacterial cellulose pellicle production. J Biotechnol 132(1):38–43. https://doi.org/10.1016/j.jbiotec.2007.08.012

Shah N, Ul-Islam M, Khattak WA, Park JK (2013) Overview of bacterial cellulose composites: a multipurpose advanced material. Carbohydr Polym 98(2):1585–1598. https://doi.org/10.1016/j.carbpol.2013.08.018

Shigematsu T, Takamine K, Kitazato M, Morita T, Naritomi T, Morimura S, Kida K (2005) Cellulose production from glucose using a glucose dehydrogenase gene (gdh)-deficient mutant of Gluconacetobacter xylinus and its use for bioconversion of sweet potato pulp. J Biosci Bioeng 99(4):415–422. https://doi.org/10.1263/jbb.99.415

Singh A, Walker KT, Ledesma-Amaro R, Ellis T (2020) Engineering bacterial cellulose by synthetic biology. Int J Mol Sci 21(23):9185. https://doi.org/10.3390/ijms21239185

Skočaj M (2019) Bacterial nanocellulose in papermaking. Cellulose 26(11):6477–6488). Springer Netherlands. https://doi.org/10.1007/s10570-019-02566-y

Stephanopoulos G (2012) Synthetic biology and metabolic engineering. ACS Synth Biol 1(11):514–525. https://doi.org/10.1021/sb300094q

Stumpf TR, Yang X, Zhang J, Cao X (2018) In situ and ex situ modifications of bacterial cellulose for applications in tissue engineering. Mater Sci Eng C 82:372–383. https://doi.org/10.1016/j.msec.2016.11.121

Takahama R, Kato H, Tajima K, Tagawa S, Kondo T (2021) Biofabrication of a hyaluronan/bacterial cellulose composite nanofibril by secretion from engineered Gluconacetobacter. Biomacromolecules 22(11):4709–4719. https://doi.org/10.1021/acs.biomac.1c00987

Teh MY, Ooi KH, Danny Teo SX, Bin Mansoor ME, Shaun Lim WZ, Tan MH (2019) An expanded synthetic biology toolkit for gene expression control in Acetobacteraceae. ACS Synth Biol 8(4):708–723. https://doi.org/10.1021/acssynbio.8b00168

Tonouchi N, Tsuchida T, Yoshinaga F, Horinouchi S, Beppu T (1994) A host–vector system for a cellulose-producing Acetobacter strain. Biosci Biotechnol Biochem 58(10):1899–1901. https://doi.org/10.1271/bbb.58.1899

Vadanan SV, Basu A, Lim S (2022) Bacterial cellulose production, functionalization, and development of hybrid materials using synthetic biology. Polym J 54(4):481–492. https://doi.org/10.1038/s41428-021-00606-8

Valla S, Coucheron DH, Fjærvik E, Kjosbakken J, Weinhouse H, Ross P, Amikam D, Benziman M (1989) Cloning of a gene involved in cellulose biosynthesis in Acetobacter xylinum: complementation of cellulose-negative mutants by the UDPG pyrophosphorylase structural gene. Mol Gen Genet MGG 217(1):26–30. https://doi.org/10.1007/BF00330938

Walker KT, Goosens VJ, Das A, Graham AE, Ellis T (2019) Engineered cell-to-cell signalling within growing bacterial cellulose pellicles. Microb Biotechnol 12(4):611–619. https://doi.org/10.1111/1751-7915.13340

Walker KT, Li IS, Keane J, Goosens VJ, Song W, Lee KY, Ellis T (2025) Self-pigmenting textiles grown from cellulose-producing bacteria with engineered tyrosinase expression. Nat Biotechnol 43(3):345–354. https://doi.org/10.1038/s41587-024-02194-3

Wang Z, Li S, Zhao X, Liu Z, Shi R, Hao M (2025) Applications of bacterial cellulose in the food industry and its health-promoting potential. Food Chem 464:141763. https://doi.org/10.1016/j.foodchem.2024.141763

Yadav V, Paniliatis BJ, Shi H, Lee K, Cebe P, Kaplan DL (2010) Novel in vivo-degradable cellulose-chitin copolymer from metabolically engineered gluconacetobacter xylinus. Appl Environ Microbiol 76(18):6257–6265. https://doi.org/10.1128/AEM.00698-10

Yang F, Cao Z, Li C, Chen L, Wu G, Zhou X, Hong FF (2023) A recombinant strain of Komagataeibacter xylinus ATCC 23770 for production of bacterial cellulose from mannose-rich resources. New Biotechnol 76:72–81. https://doi.org/10.1016/j.nbt.2023.05.002

Zheng Z-J, Ye H, Guo Z-P (2024) Bacterial cellulose applications in electrochemical energy storage devices. Adv Mater 37(22):2412908. https://doi.org/10.1002/adma.202412908

Zhong C (2020) Industrial-scale production and applications of bacterial cellulose. Front Bioeng Biotechnol 8. Frontiers Media S.A. https://doi.org/10.3389/fbioe.2020.605374

Chapter 2
Biology Tools for Optimizing Bacterial Cellulose Production

Miguel Pacheco, Pedro Montenegro-Silva, Fernando Dourado, Miguel Gama, Lucília Domingues, and Óscar Dias

Abstract Metabolic systems biology integrates computational and experimental approaches to understand the complex interactions within biological networks. This subject has revolutionized biotechnological applications, including drug discovery, biofuel production, and metabolic engineering. Functional genomics, through high-throughput omics technologies, enables the characterization of gene functions, regulatory mechanisms, and metabolic pathways into valuable datasets. Genome-scale metabolic models (GEMs) are primary candidates for using such datasets by offering a mathematical framework to study the metabolic capabilities of organisms. These models allow the simulation of cellular metabolism to identify metabolic pathways, predict the effects of genetic modifications, and optimize metabolites production. The integration of multiomics data into GEMs further enhances their predictive power, leading to the development of more robust and context-specific models tailored for specific applications.

A few GEMs have been developed for specific BC producers and have demonstrated to be powerful tools to optimize substrate utilization, modify specific pathway, and perform strain improvement for cellulose biosynthesis. However, the integration of omics data and experimental validation of the identified targets remains limited. The continuous development of GEMs coupled with the integration of multiomics data, and the potential of machine learning approaches can drive further advancements in BC biotechnology, paving the way for more efficient and scalable production systems.

M. Pacheco · P. Montenegro-Silva
CEB – Centre of Biological Engineering, University of Minho, Braga, Portugal
e-mail: pmontenegro@ceb.uminho.pt

F. Dourado · M. Gama · L. Domingues · Ó. Dias (✉)
CEB – Centre of Biological Engineering, University of Minho, Braga, Portugal

LABBELS – Associate Laboratory in Biotechnology and Bioengineering and Microelectromechanical Systems, Braga, Guimarães, Portugal
e-mail: fdourado@ceb.uminho.pt; fmgama@deb.uminho.pt; luciliad@deb.uminho.pt; odias@ceb.uminho.pt

F. Dourado et al. (eds.), *Advances in Bacterial Cellulose*,
https://doi.org/10.1007/978-3-032-07042-5_2

Keywords Systems biology · Genome-scale metabolic models · Omics data integration · Bacterial cellulose

1 Introduction

Bacterial cellulose (BC) is a unique exopolysaccharide synthesized by certain bacteria, most notably *Komagataeibacter* species, forming a three-dimensional nanofibrillar network (Ryngajłło et al. 2020). The exceptional characteristics of BC have positioned it as a highly sought-after biomaterial with a vast and expanding range of potential applications across diverse sectors (Li et al. 2022). However, despite its promise, BC has yet to reach the mainstream. Many attempts have been made towards economic BC production, mainly by optimizing culture media and fermentation systems using agitated, airlift, membrane and horizontal bioreactors for strain selection. These efforts had limited success due to the low productivity rates, high capital investment, and high operating costs (Blanco Parte et al. 2020; Taokaew 2024). Another significant hurdle is the inherent complexity of the microbial metabolism involved in BC production. Bacterial cells producing cellulose must efficiently channel carbon from the growth substrate towards the synthesis of UDP-glucose, the direct precursor for cellulose polymerization, while also allocating resources for cell growth, maintenance, and the synthesis of other essential cellular components (Taokaew 2024). This intricate network of biochemical reactions involves numerous enzymes and regulatory pathways (Wang et al. 2018b). Competition for carbon flux can occur between cellulose synthesis and other metabolic pathways, such as those leading to biomass formation or the production of by-products like gluconic acid or acetic acid, which can lower the overall BC yield and process efficiency (Wu et al. 2023).

Consequently, more rational and systematic approaches are needed, involving an integrated understanding of the entire cellular system, to optimize BC production. A holistic view of the cell's metabolism is required to identify bottlenecks in precursor supply, competing pathways that divert carbon away from cellulose, and regulatory elements that could be targeted to enhance flux towards the desired product (Kim et al. 2016). This is where the power of systems biology and metabolic engineering, guided by computational tools, becomes indispensable (Lee et al. 2012). The advent of genome-scale metabolic models (GEMs) has revolutionized metabolic engineering by enabling a holistic view of cellular metabolism (Gu et al. 2019). GEMs are mathematical representations of an organism's entire metabolic network, integrating genomic, biochemical, and physiological data to simulate metabolic fluxes under different conditions (Bi et al. 2022). By leveraging these models, researchers can identify key metabolic bottlenecks, predict genetic modifications to enhance product yield, and optimize fermentation strategies in silico before experimental validation (Gu et al. 2019). It is precisely this need for a systems-level understanding and rational design that brings genome-scale metabolic models (GEMs) to the forefront as powerful tools for optimizing BC production. In the past 5 years, a few synthetic biology toolkits have begun to be developed for BC producers (Ryngajłło et al. 2020; Singh et al. 2020; Teh et al. 2019), and the endeavor towards the targeted control of gene expression to generate industrially relevant mutant strains with improved cellulose yield is just beginning.

This chapter aims to provide a comprehensive review of genome-scale metabolic modeling and its application to bacterial cellulose production. It explores the construction and validation of GEMs, their integration with multiomics data, and their role in optimizing BC biosynthesis in *Komagataeibacter* species. By synthesizing current research and case studies, the chapter evaluates the potential of GEMs to enhance BC yield and sustainability, particularly for nanotechnology applications. Additionally, it discusses prospects, including the incorporation of advanced omics and machine learning approaches, to guide rational metabolic engineering strategies for industrial-scale BC production.

2 Metabolic Systems Biology

All life forms have an inherent metabolic ability to convert nutrients into chemicals that can serve as building blocks for well-known macromolecules such as proteins, lipids, carbohydrates, and finally, DNA and RNA (Nielsen 2017). With the help of computational/mathematical models, metabolic systems biology aims to better understand the interactions that make this conversion possible and ultimately predict a determined behavior (Palsson 2006). The use of multiomics techniques, such as genomics (study of organism's genome), proteomics (study of organism's proteins), and metabolomics (study of organism's metabolome), greatly benefits and contributes to a more accurate analysis or understanding of cardiovascular diseases (Joshi et al. 2020), cancer cell signaling pathways (Janes and Lazzara 2024), improving the yield of crops for industrial agriculture (Pazhamala et al. 2021), engineering of prokaryotes for optimized production of secondary metabolites (Kim et al. 2016; Malcı et al. 2024), among many other examples and applications. The further generation, accessibility, and integration of biological data as well as other technological advancements (discussed ahead) have also allowed metabolic systems biology to cover a wider range of applications and to improve the quality of analysis (Dias and Rocha 2016).

2.1 *Functional Genomics*

Functional genomics aims to understand the functions, interactions, and regulation of genes and their products on a genome-wide scale within an organism. One of the technological advancements that enabled functional genomics was the methodology for DNA sequencing, which appeared in 1975 (Sanger et al. 1977). However, the first whole-genome sequencing and assembly was only achieved 20 years later by Fleischmann, et al. for the species *Haemophilus influenzae* Rd. (Fleischmann

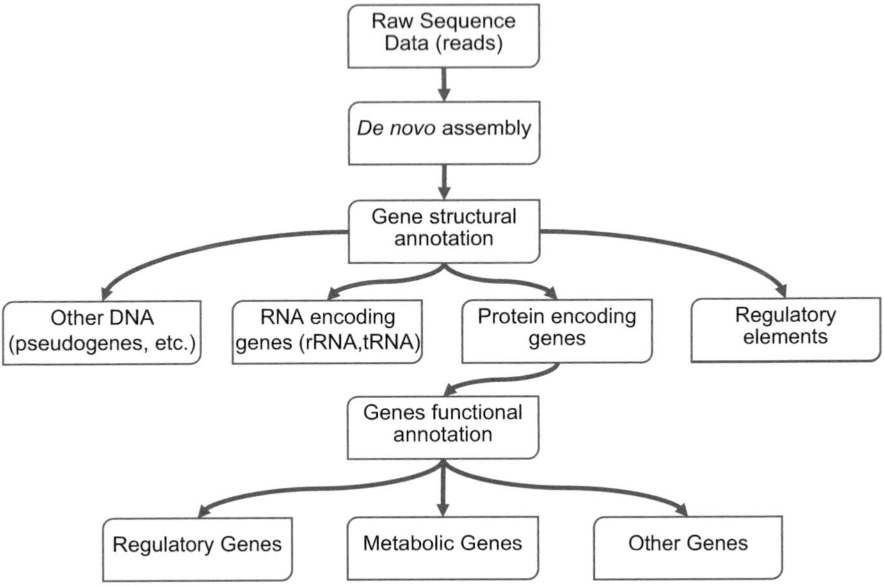

Fig. 2.1 General pipeline for obtaining functional annotations from the raw sequence data. Adapted with permission from Taylor & Francis Group LLC—Books, from Dias and Rocha (2016), ©2015. Permission conveyed through Copyright Clearance Center, Inc.

et al. 1995), a widely known threatening species that can cause infections in the human respiratory system. As of August 2022, the National Center for Biotechnology Information (NCBI[1]) database supports over one million of partially or completely assembled genomes from different species (Sayers et al. 2021), thus confirming the importance of high-throughput data for systems biology. However, from sequencing to the assignment of specific roles such as identifying metabolic genes, there are several complex steps (Fig. 2.1) (Dias and Rocha 2016).

The first step in DNA sequencing is to perform a de novo assembly that consists of the assembly of several DNA sequences from a fragment, called reads, the output of sequencing machines. Several bioinformatics tools such as SPAdes[2] (St. Petersburg genome assembler) (Bankevich et al. 2012), Velvet[3] (Zerbino and Birney 2008), aBySS[4] (Assembly by Short Sequences) (Simpson et al. 2009), and MEGAHIT[5] (Li et al. 2015), among others, are specialized in this task through developed algorithms.

The next step, called genome structural annotation, is to identify the genomic regions in the assembled genome. All genomes have several components with

[1] https://www.ncbi.nlm.nih.gov/

[2] https://github.com/ablab/spades

[3] https://github.com/dzerbino/velvet

[4] https://github.com/bcgsc/abyss

[5] https://github.com/voutcn/megahit

different purposes and structural elements (Chen et al. 2024a), such as protein-encoding genes, which are sequences that are transcribed and afterwards translated into proteins; RNA encoding genes, sequences that are transcribed into several types of RNA genes, such as transfer RNA (tRNA) and ribosomal RNA (rRNA), among others; sequences that control gene expression (regulatory elements); gene copies that no longer encode proteins (pseudogenes); and sequences that appear multiple times in the genome (repeat elements), among other types of DNA sequences. The genome structural annotation step mainly focuses on identifying the protein-encoding genes using bioinformatics tools and resources (Delcher et al. 1999; Majoros et al. 2004), to locate open reading frames (ORFs) in the assembled genome.

Once the protein-encoding genes have been identified, one can perform the final task, the gene functional annotation. This step is intended to associate biological meaning with the protein-encoding genes, either by sequence similarity, experimental data, and/or by functional prediction based on the previously mentioned methods (Gabaldón 2008; Klimke et al. 2011; Lin et al. 2024). Typically, through BLAST[6] (Basic Local Alignment Search Tool) (Altschul et al. 1990), HMMER[7] (hidden Markov models) (Eddy 2011), or other similar algorithms, it is possible to identify the protein function of a given protein-encoding gene sequence, whether it is involved in an enzymatic reaction in the carbon metabolism (metabolic gene), the activation/repression of signaling molecules that control gene expression (regulatory gene), the transport of certain metabolites into the cell (transporters), or proteins involved in the structure of the cell/movement, for example.

Understanding the biological role of genes has significant real-world application in several areas, allowing to improve the overall quality of life (Ungerer et al. 2008), such as disease gene discovery and drug targets (Roy et al. 2011), improving crop yield and stress resistance (Anuradha Singh et al. 2017), engineering microorganisms for biofuels (Wang et al. 2024) and other bioproducts such as cellulose, as a more sustainable solution (Fernandes et al. 2019; Leitão et al. 2013; Silva et al. 2020; Singh et al. 2020; Soares da Silva et al. 2019).

The experimental validation of the genes' functions predicted during annotation can prove to be challenging, expensive, and time-consuming laboratory work. For this reason, one may consider the help of genome-scale metabolic models (GEMs) which can provide powerful insights into metabolism, even more so, through the incorporation of omics data (Gu et al. 2019) and regulatory mechanisms (Sertbas and Ulgen 2020). Besides, it can also enable researchers to identify and validate potential metabolic interventions, thus guiding experimental designs with much more efficiency (O'Brien et al. 2015).

[6] https://blast.ncbi.nlm.nih.gov/Blast.cgi

[7] https://github.com/EddyRivasLab/hmmer

2.2 Genome-Scale Metabolic Models

Before genomic data became widely available, metabolic reconstructions were primarily based on biochemical literature and enzyme characterization (Dias and Rocha 2016). However, with the availability of whole-genome sequences and their consecutive downstream processing, as mentioned above (Fig. 2.1), as well as documented biochemical reactions from various biological databases (Palsson 2009; Rocha et al. 2008), it became possible to reconstruct metabolic networks on a genome-wide scale, even for less-characterized organisms (Feist et al. 2009). Likewise, with the first organism to have its genome completely sequenced in 1995 (Fleischmann et al. 1995), the first GEM was constructed in 1999 for *Haemophilus influenzae Rd* (Edwards and Palsson 1999), and ever since, advancements in high-throughput technologies and systems biology have facilitated the reconstruction of GEMs for a wide range of organisms, including bacteria, fungi, animals, plants, and humans.

Genome-scale metabolic networks (GSMNs) represent the biochemical reactions associated with enzymes encoded in an organism's genome. To construct these networks, it is essential to identify the substrates and products of the reactions, their stoichiometry, reversibility, and localization within the cell (Feist et al. 2009). GSMNs provide insights into the biochemical properties of cells. However, only GEMs allow for the predictive simulations of metabolic capabilities. Along with the metabolic network, these models further incorporate enzyme kinetics (Chen et al. 2024b) and regulatory mechanisms (Cruz et al. 2020), although such data are currently limited to well-studied organisms (Rocha et al. 2008). GEMs also contain information on biomass composition, energy demands, and additional constraints, enabling the in silico prediction of microbial responses to perturbations (Rocha et al. 2008). The reconstruction process for GEM applies to both prokaryotic and eukaryotic models, but eukaryotic models present additional challenges due to their typically larger genomes, complex cellular compartments, and broader knowledge requirements (Thiele and Palsson 2010).

To streamline the process of reconstructing a GEM, researchers have established a standardized approach consisting of four main steps (Feist et al. 2009; Thiele and Palsson 2010). Initially, genome annotation supplies the necessary information to create a draft metabolic network. Subsequently, biochemical knowledge from literature and databases refines this network, which is then converted into a GEM through the formulation of biomass production equations and organism-specific constraints. The model is then subsequently validated by comparing its predictions with experimental data, where the annotation and formulation stages are iteratively adjusted until the predictions match the observed results (Thiele and Palsson 2010) as observed in Fig. 2.2.

Given the complexity of GEM reconstruction, several software tools have been developed to facilitate and automate various tasks; however, manual curation remains essential. Examples of such tools include metabolic model reconstruction

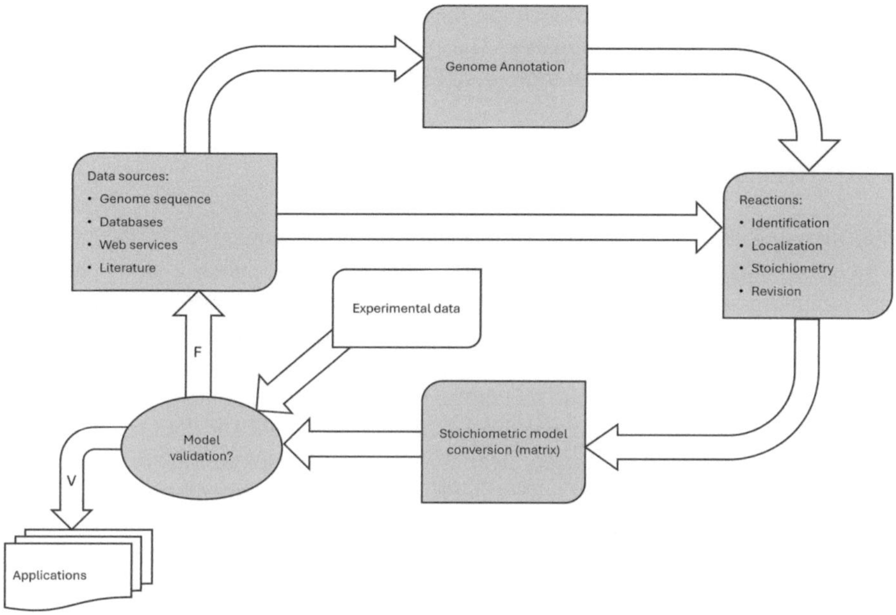

Fig. 2.2 Representation of the iterative process of reconstructing a genome-scale metabolic model. The workflow involves the main phases: 1: data collection and integration from several sources (e.g., the genome raw sequence data and current databases), genome annotation, and the identification of all possible reactions to form a metabolic network; 2: transformation of the metabolic network into a stoichiometric model; 3: validation through experimental data and finally applications of the model predictions. Adapted with permission from Taylor & Francis Group LLC—Books, from Dias and Rocha (2016), ©2015. Permission conveyed through Copyright Clearance Center, Inc.

using genome-scale information (*merlin*[8]) (Capela et al. 2022; Dias et al. 2015), CarveMe[9] (Machado et al. 2018), the Constraint-Based Reconstruction and Analysis (COBRA[10]) toolbox (Schellenberger et al. 2011), and ModelSEED[11] (Henry et al. 2010), which is implemented in the Department of Energy Systems Biology Knowledgebase (KBase[12]) (Arkin et al. 2016). Among these, *merlin* offers a user-friendly framework designed to annotate genomes and generate compartmentalized draft GEMs while assisting manual curation efforts (Dias et al. 2015).

Throughout the reconstruction process, various biochemical databases are consulted to gather updated organism-specific information, including whole-genome sequences, genome annotations, metabolic reaction data, and enzymatic functions.

---

[8] https://merlin.bio.di.uminho.pt

[9] https://github.com/cdanielmachado/carveme

[10] https://github.com/opencobra/cobratoolbox

[11] https://modelseed.org/

[12] https://www.kbase.us/

These resources are crucial for refining metabolic networks and ensuring the accuracy and applicability of the GEMs. The primary databases utilized for the GEM reconstruction are described in more detail in the following section.

2.2.1 Genome Annotation

The first step in GEM reconstruction is the functional annotation of the genome, which consists of identifying the metabolic genes within the genomic DNA sequence and assigning specific biochemical roles to their protein products. Acquiring high quality annotations such as cellular functions, gene or Open Reading Frame (ORF) names, and Enzyme Commission (EC) and Transport Classification (TC) numbers is essential to obtain an accurate metabolic network (Thiele and Palsson 2010); this information can be retrieved from databases such as NCBI and KEGG.[13] In cases where genome annotations are not available or not reliable, the functional annotation may have to be carried out with sequence similarity alignment tools, such as the Basic Local Alignment Search Tool (BLAST[14]) (Altschul et al. 1990), DIAMOND[15] (Buchfink et al. 2014), or HMMER[16] (Eddy 2011). These tools search for annotated genes of other organisms that are homologous to the unannotated gene under analysis and assign a statistical score to each match, representing the confidence level in such similarity. According to the best score, the unannotated gene should be assigned cellular function(s), gene or ORF, EC, and TC numbers, when available. Both EC and TC are particularly useful for automatic network reconstruction.

2.2.2 Metabolic Network Assembling

The second step in the reconstruction of a GEM is to assemble a metabolic network based on biochemical reactions that involve the protein-encoding genes annotated in the previous step. For example, glycolysis is a well-known biological process that converts glucose into pyruvate while producing NADH and ATP. This process occurs through a series of reactions catalyzed by enzymes (Fig. 2.3) that are encoded in the organism's genome. By assembling all the biochemical reactions of an organism, including the transport systems that mediate its interactions with the environment, the fundamental framework for GEM reconstruction can be established.

[13] https://www.kegg.jp/kegg/kegg1.html

[14] https://blast.ncbi.nlm.nih.gov/Blast.cgi

[15] https://github.com/bbuchfink/diamond

[16] https://github.com/EddyRivasLab/hmmer

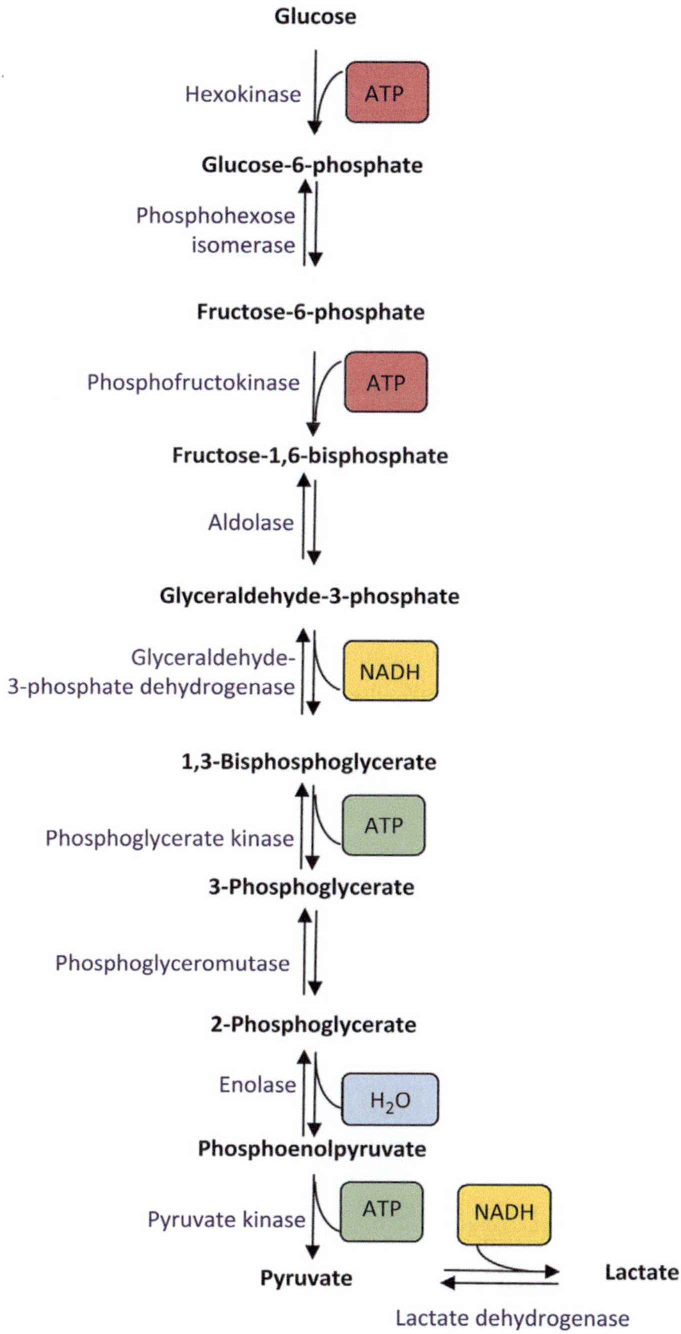

Fig. 2.3 Example of the glycolysis metabolic network, containing all the metabolites involved, as well as the enzymes responsible for the reactions. Adapted from Figure 4.1: Reaction steps of fatty acid oxidation (β-oxidation) in Werner et al. (2015), © 2016 Elsevier Inc. Reproduced with permission under license number 6075430927588

Genes, Proteins, and Reactions

The association of annotated genes, proteins, and reactions (GPR associations) is typically carried out by searching biological databases using protein names, EC numbers, or other identifiers, such as KEGG reaction numbers (Francke et al. 2005). Transporter proteins, which facilitate the movement of metabolites, are classified using Transporter Classification (TC) numbers, in the Transporter Classification Database (TCDB) (Saier et al. 2016) which then provides details on the metabolites and transport mechanisms associated with specific carrier proteins that should also be integrated into the draft network.

Defining GPR associations requires particular attention in cases involving isoenzymes—different proteins capable of catalyzing the same reaction—and promiscuous enzymes—which can act on multiple substrates. Additionally, when dealing with protein complexes and lumped reactions, it is essential to ensure that each gene subunit is correctly linked to the corresponding catalytic reaction (Thiele and Palsson 2010).

The reconstruction of a draft metabolic network involves linking enzymes and transporters to their corresponding substrates and products using EC and TC numbers, with databases such as TCDB,[17] BRENDA,[18] MetaCyc,[19] and KEGG[20] playing a crucial role in this process.

Afterwards, non-enzymatic and spontaneous reactions are also added to the network. These reactions can be found in published literature or in a few online data sources, such as KEGG. To streamline this process, several tools such as *merlin*[21] (Capela et al. 2022), RAVEN 2.0[22] (Wang et al. 2018a), gapseq[23] (Zimmermann et al. 2021), AutoKEGGRec[24] (Karlsen et al. 2018), MicrobesFlux[25] (Feng et al. 2012), among others, can be used for the automatic retrieval of relevant data. However, despite being associated with EC numbers, certain proteins, including those involved in DNA methylation and RNA modification, are generally not incorporated into metabolic models, due to their low metabolic flux interference and metabolite consumption. Finally, all the GPR associations should also be validated using the above-mentioned databases.

[17] https://www.tcdb.org/

[18] https://www.brenda-enzymes.org/

[19] https://metacyc.org/

[20] https://www.genome.jp/kegg/

[21] https://merlin.bio.di.uminho.pt

[22] https://github.com/SysBioChalmers/RAVEN

[23] https://github.com/jotech/gapseq

[24] https://github.com/AlmaasLab/AutoKEGGRec

[25] https://github.com/microbesflux/MicrobesFluxSource

Stoichiometry and Localization

After collecting the set of reactions, their stoichiometry should also be revised using databases and literature. Stoichiometry is paramount and can be used to leverage the conversion of the metabolic model into a mathematical model.

The next step, localization, is to assign the reactions to specific cellular compartments. Determining enzyme localization is essential for the development of an accurate GSMM. In prokaryotes, compartments are typically restricted to the cytosol, the periplasmic space, and the extracellular space. In eukaryotes, however, additional organelles may be considered if they are relevant to the GEM. It is crucial to differentiate between similar reactions that involve the same metabolites and stoichiometry but occur in distinct compartments, treating them as separate reactions. Likewise, metabolites should be uniquely represented for each cell compartment using distinct identifiers that consider their location. For example, glucose present outside the cell and in the cytosol can be labeled as glucose_ext and glucose_cyt, respectively. Various bioinformatics tools have been developed to predict intracellular or extracellular enzyme localization based on amino acid sequences and physiological data of the organism's physiology. The PSort[26] (Yu et al. 2010) (Protein Subcellular Localization Prediction) and TargetP[27] (Armenteros et al. 2019) are among the most widely used applications, and DeepLoc[28] (Almagro Armenteros et al. 2017) and DeepPSL (Wei et al. 2018)[29] are deep-learning alternatives that focus more on eukaryotic subcellular protein location (Gillani and Pollastri 2024).

Manual Curation

The final step involves manually curating and refining the metabolic network by reviewing the literature and databases relevant to the target organism or closely related species (Thiele and Palsson 2010). This process addresses several key aspects, including verifying the metabolic network (ensuring the correct presence or absence of reactions), replacing vague terms with specific ones, confirming substrate and cofactor specificity, ensuring accurate stoichiometry and mass balance, and determining the directionality and reversibility of the reactions. Additionally, gap-filling is an essential step to resolve network inconsistencies caused by missing reactions, which would otherwise lead to the unintended buildup of metabolites. Several databases can ensure that the drafted metabolic network is as close as possible to the organism, as listed below (Table 2.1).

[26] https://psort.org/

[27] https://services.healthtech.dtu.dk/services/TargetP-2.0/

[28] https://services.healthtech.dtu.dk/services/DeepLoc-2.0/

[29] https://arxiv.org/abs/2109.13662

Table 2.1 Common and relevant databases for metabolic data curation. Reproduced from Sampaio et al. (2022), under the terms of the Creative Commons Attribution 4.0 International License (CC BY 4.0): https://creativecommons.org/licenses/by/4.0/

Database (Ref.)	Description	Data
KEGG[a] (Kanehisa et al. 2017)	Generic resource that comprises genomes, metabolic pathways, chemical compounds, diseases, and drugs	Metabolic data
MetaCyc[b] (Caspi et al. 2019)	Comprehensive database of extensively curated metabolic pathways, containing information on reactions, enzymes, genes, and compounds for several organisms	Curated metabolic data
BioCyc[c] (Karp et al. 2019)	Collection of Pathway/Genome databases (PGDBS), each containing the complete genome and predicted metabolic network of an organism	Organism-specific predicted metabolic data
NCBI[d] (Agarwala et al. 2018)	Online repository containing several databases for genomics and biomedical information and tools for extracting and analyzing the data	Reviewed and unreviewed sequence data
UniProt[e] (Bateman 2019)	Resource for protein sequence and related information, including manually reviewed data (Swiss-Prot) and automatic, non-reviewed protein annotations (TrEMBL)	Curated and predicted protein data
BRENDA[f] (Chang et al. 2021)	Main database of manually annotated enzyme functional data, which uses the EC classification system	Curated enzyme data
TCDB[g] (Saier et al. 2016)	Curated database contains information on transport systems from several organisms, including sequence, structure, and function, and uses the TC system to classify transport proteins.	Curated transport data
BiGG[h] (Schellenberger et al. 2010)	Online repository containing large-scale metabolic models from various organisms. It provides reactions and metabolites of all the present models	Reviewed and unreviewed models data
KBase[i] (Cottingham 2015)	Open-source platform that contains several genomes, gene expression profiles, and metabolic networks	Reviewed and unreviewed data

[a]https://www.genome.jp/kegg/
[b]https://metacyc.org/
[c]https://biocyc.org/
[d]https://www.ncbi.nlm.nih.gov/
[e]https://www.uniprot.org/
[f]https://www.brenda-enzymes.org/
[g]https://www.tcdb.org/
[h]http://bigg.ucsd.edu/
[i]https://www.kbase.us/

2.2.3 From Metabolic Network to Stoichiometric Model

After assembling the metabolic network, a pseudo-reaction representing the requirements for generating a new unit of biomass must be added to the model. This reaction represents the macromolecular composition of the cell, and the building blocks used to generate those molecules. Typically, biomass components include proteins (amino acids), DNA, RNA (nucleotides), cofactors, lipids, and carbohydrates, among other metabolites. The following equation represents the biomass formulation (Dias and Rocha 2016):

$$\sum_{k=1}^{p} C_k . X_k \rightarrow \text{biomass} \quad (2.1)$$

where C_k represents the coefficient of a certain metabolite and X_k each metabolite. Biomass is then normalized to grams, as it expresses the specific growth rate of the organism (h^{-1}), i.e., the amount of biomass (in grams) produced per gram of existing biomass per hour (Rocha et al. 2008). If the biomass composition of an organism is unavailable, both experimentally and in the literature, the composition of a closely related organism can be used. This equation should also account for energy requirements in the form of ATP, for the growth-associated energy demand, representing the number of ATP molecules required per gram of biomass produced (Rocha et al. 2008). The energy requirements can be determined from literature or experimental data, as aforementioned (Thiele and Palsson 2010). The final phase of the GEM reconstruction involves the conversion of the network into the stoichiometric matrix. Based on the fundamental concepts of chemical engineering, variations in metabolite concentrations over time can be described using ordinary differential equations, as demonstrated in Eq. (2.2) for a specific metabolite (Dias and Rocha 2016):

$$\frac{dX_i}{d_t} = \sum_{j=1}^{N} S_{ij} * v_j + \mu X, i = 1, \ldots, M \quad (2.2)$$

where

- X_i—is the concentration of metabolite i
- v_j—is the metabolic flux of reaction j
- S_{ij}—is the stoichiometric coefficient of the metabolite i in reaction j
- μX—is the system growth rate

Predicting the concentration profiles over time relies on knowing the kinetic rates of all reactions. However, the limited availability of kinetic data remains a challenge in the development of dynamic models for most organisms. To address this, a pseudo-steady-state assumption is applied, where metabolite concentrations are considered constant over time (Thiele and Palsson 2010). This simplifies the dynamic equations, enabling their transformation into a stoichiometric matrix, as illustrated in Eq. (2.3) (Orth et al. 2010).

$$S * v = 0 \quad (2.3)$$

where S is the stoichiometric matrix in which the rows represent metabolites, columns, and reactions, and v is the flux vector.

In metabolic networks, the number of reactions typically exceeds the number of metabolites, leading to an underdetermined system with numerous possible solutions that satisfy the given constraints. To limit this extensive solution space to a feasible range, known as the flux cone of solutions, additional constraints are applied (Rocha et al. 2008). These constraints account for the reversibility of the reactions and establish the limits for the minimum and maximum fluxes for each reaction. Mathematically, these constraints are expressed using inequalities, as illustrated in Eq. (2.4) (Dias and Rocha 2016).

$$v\min_j \leq v_j \leq v\max_j, j = 1,\ldots,N \tag{2.4}$$

where

- v_j is the flux vector
- $v\min_j$ is the lower bound
- $v\max_j$ is the upper bound

Reversible reactions are usually assigned with flux boundaries that range from negative to positive infinity, whereas irreversible reactions are constrained by setting one boundary to zero, depending on their direction (Dias and Rocha 2016). Figure 2.4 illustrates the construction of a metabolic network and its transformation into a stoichiometric model.

2.2.4 Model Validation

After the conversion of a metabolic network into a stoichiometric matrix, simulations are now possible provided that an objective function is defined. For clarity, an objective function is a mathematical function that represents the biological goal that the model aims to achieve, for instance, maximizing growth, producing a specific metabolite, or any other relevant to the study. The most commonly used objective function is the maximization of biomass growth rate (biomass reaction), as it was demonstrated that the metabolic networks of most microorganisms have evolved to maximize growth rates, under carbon-limiting conditions (Rocha et al. 2008); however, other products of interest might be considered, provided that the simulations are in agreement with experimental data and/or literature, especially for more complex organisms or models of tissues.

If discrepancies are found, the genome annotation, gap-filling, and manual curation steps are repeatedly refined until the model produces accurate predictions. Once validated, the model should be stored in SBML format (Hucka et al. 2003), allowing for in silico phenotype simulations using methods such as Flux Balance Analysis (FBA) (Orth et al. 2010).

Flux Balance Analysis (FBA) is the most widely used constraint-based approach for simulating GEMs (Orth et al. 2010; Varma and Palsson 1994). It employs linear programming to determine the optimal flux distribution, regarding the objective function, while satisfying the mass balance and flux constraints. FBA can simulate phenotypic behavior under various conditions, such as different carbon or nitrogen sources, enhancing or decreasing metabolite production, evaluating gene

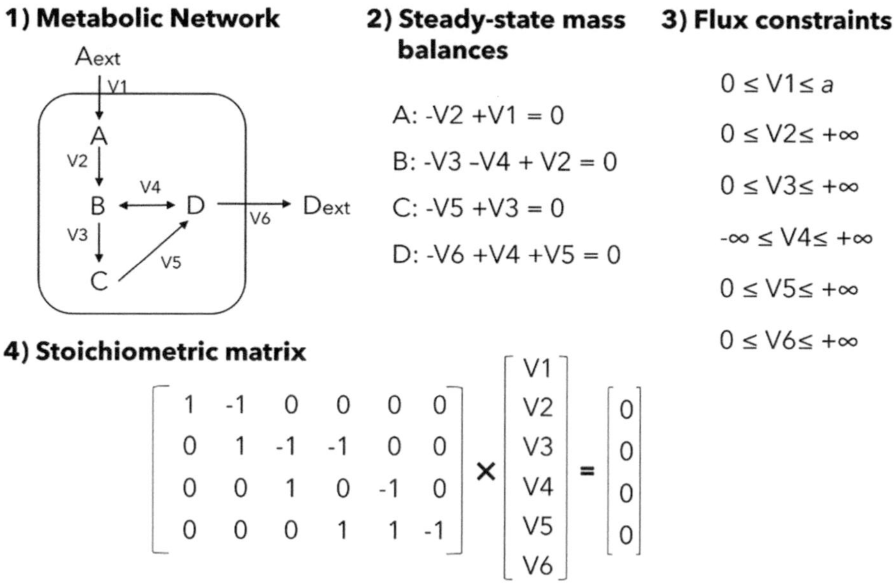

Fig. 2.4 Assembly of a metabolic network with four metabolites (A to D) and six fluxes (V1 to v6) into a stoichiometric matrix. The metabolic network and boundaries of a system (e.g., cell) are shown in (1). Fluxes V1 and V6 represent exchange fluxes with the environment. Reversible reactions are indicated by double arrows, and irreversible reactions are indicated by a forward arrow. (2) Represents the steady-state mass balances, and panel (3) shows the constraints for flux values, in which ("a") represents the uptake rate of the metabolite A. Finally, (4) shows the representation of the mass balances in the stoichiometric matrix format, where the rows correspond to the metabolites A, B, C, and D and columns to fluxes from V1 to V6. Reproduced from Orth et al. (2010), under the terms of the Creative Commons Attribution 4.0 International License (CC BY 4.0): https://creativecommons.org/licenses/by/4.0/

knockouts, and preventing toxic metabolite synthesis. Additionally, FBA helps to identify essential reactions by systematically removing them and assessing their impact on biomass formation (Rocha et al. 2008; Thiele and Palsson 2010).

To engineer metabolic phenotypes, strain optimization algorithms like OptKnock (Burgard et al. 2003) and OptRAM (Shen et al. 2019) use evolutionary approaches to systematically manipulate gene expression or perform knockouts in a genome-scale metabolic model. However, FBA solutions are not always unique, implying that multiple flux distributions can satisfy the same constraints and objective functions. This limitation can be addressed with Parsimonious Flux Balance Analysis (pFBA), which selects the flux distribution with the lowest overall flux while maximizing growth (Lewis et al. 2010). Flux Variability Analysis (FVA) is another approach that evaluates the robustness of a model by determining the minimum and maximum flux values for each reaction under given constraints (Mahadevan and Schilling 2003). These methods and optimizations have also been implemented using several bioinformatics tools, including the COBRA[30] toolbox (Schellenberger et al. 2011), COBRApy[31] (Ebrahim et al. 2013), and ReFramed[32] (Kulyashov et al. 2023).

Dynamic Flux Balance Analysis (dFBA) extends the traditional FBA by incorporating temporal changes in metabolite concentrations (Radhakrishnan Mahadevan et al. 2002). While intracellular metabolism is assumed to remain in a steady state, extracellular components such as exchange metabolites and biomass are modeled using dynamic equations that capture uptake and secretion rates. This hybrid framework enables the simulation of both intracellular metabolic processes and dynamic environmental changes.

Beyond FBA, alternative approaches such as flux sampling methods including Markov chain Monte Carlo (MCMC) offer a way to explore metabolic networks without requiring a predefined objective function. These methods generate a sequence of feasible flux distributions that adhere to network constraints, providing insights into the range and probability distribution of metabolic fluxes (Herrmann et al. 2019; Schellenberger and Palsson 2009).

3 Beyond Genome-Scaled Metabolic Models

As mentioned in Sect. 2.2, GEMs can also integrate other omics data to further refine the models, making them more representative of specific cellular conditions and improving their predictive accuracy. This is particularly achieved by incorporating transcriptomics, proteomics, and metabolomics data. These methods essentially follow one of two approaches: either narrowing the solution space to improve metabolic flux predictions or developing context-specific models that capture metabolic variations across different environmental conditions, cell types, or tissues (Machado and Herrgård 2014; Robaina Estévez and Nikoloski 2014). Furthermore, most integration methods can be categorized into three different algorithmic classes: GIMME-like, iMAT-like, and MBA-like (Robaina Estévez and Nikoloski 2014).

3.1 GIMME-Like Methods

GIMME-like methods use omics data to define active and inactive genes while ensuring a Required Metabolic Function (RMF), for example, to guarantee ATP production under specific conditions, or any other metabolites that are required for

[30] https://opencobra.github.io/cobratoolbox/stable/index.html

[31] https://opencobra.github.io/cobrapy/

[32] https://bio.tools/reframed

the accuracy of the model predictions (Robaina Estévez and Nikoloski 2014). These methods are further divided into switch-based and valve-based approaches.

Switch-based methods can disable reactions linked to inactive genes in the omics data by setting their flux bounds to zero, thus ensuring consistency between gene expression and flux predictions. GIMME (Becker and Palsson 2008) focuses on the integration of transcriptomics data into GEMs, whereas GIM3E (Schmidt et al. 2013) extends GIMME by incorporating metabolomics data, forcing flux through reactions associated with experimentally detected metabolites.

In contrast, valve-based methods adjust the reaction flux bounds proportionally to the normalized expression levels of the corresponding genes (Vivek-Ananth and Samal 2016). Although initially developed for transcriptomics, GIMME-like methods have been adapted to incorporate proteomics data, such as in GIMMEp (Bordbar et al. 2012). These approaches are widely used to refine metabolic flux predictions and create-specific metabolic models.

3.2 iMAT-Like Methods

Similar to the GIMME-like methods, iMAT-like approaches classify reactions based on gene or protein expression levels but without using the RMF. This makes them useful for modeling complex multi-cellular systems (Robaina Estévez and Nikoloski 2014), such as human tissues, where identifying the necessary set of reactions and functions can be challenging. The iMAT algorithm, based on Shlomi et al. (2008), classifies genes into high, moderate, and low expression groups, maximizing reactions linked to highly expressed genes while minimizing those associated with low-expression genes. iNIT (Agren et al. 2012) expands on this by assigning positive weights to highly expressed reactions and negative weights to lowly expressed ones. An extension, tINIT (Agren et al. 2014), builds on iNIT by allowing the definition of RMF to tailor to context-specific models. As most iMAT-like methods do not impose predefined metabolic functions, they are particularly effective for generating tissue-specific models.

3.3 MBA-Like Methods

MBA-like methods focus on context-specific model reconstruction, rather than flux prediction. They classify reactions into core and non-core sets, where core reactions are supported by omics data or curated information, while non-core reactions have little or no evidence of activity. The algorithm iteratively removes non-core reactions while ensuring that core reactions remain functional (Robaina Estévez and Nikoloski 2014).

MBA (Jerby et al. 2010) randomly eliminates non-core reactions while maintaining model consistency and generates multiple candidate models before selecting a

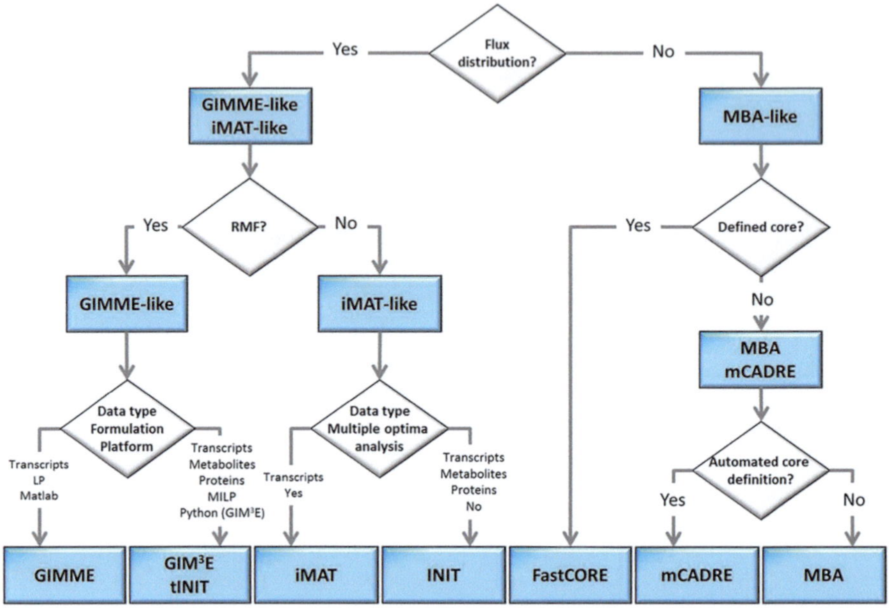

Fig. 2.5 Schematic representation of the three methods, GIMME-like, iMAT-like, and MBA-like, and several bioinformatic tools, for omics integration in GEMs based on flux distribution, the definition of an RMF or core, and types of omics data. Reproduced from Robaina Estévez and Nikoloski (2014), under the terms of the Creative Commons Attribution License (CC BY)

final model. Meanwhile, mCADRE (Wang et al. 2012) improves upon this by ranking non-core reactions based on expression data and connectivity within the metabolic network, removing the lowest-ranked reactions sequentially to construct a single, computationally efficient model. FastCORE (Vlassis et al. 2014) further optimizes model reconstruction by using linear programming to iteratively refine a consistent subnetwork containing all defined core reactions and a minimal set of non-core reactions. Particularly, these methods are more suitable for context-specific models, such as tissue-specific metabolic reconstructions or even for the identification of active pathways under certain conditions (Robaina Estévez and Nikoloski 2014).

Integrating omics data into metabolic modeling allows the refining of flux predictions and enables the reconstruction of biologically relevant, context-specific models. The three major algorithmic families: GIMME-like, iMAT-like, and MBA-like also offer different strategies for integrating transcriptomics, proteomics, and metabolomics data (Fig. 2.5), depending on the model's main purpose.

4 GEMs for Bacterial Cellulose Production

As mentioned at the beginning of this chapter, among their diverse applications, GEMs enable the rational design of metabolic engineering strategies to optimize bacterial cellulose (BC) production. GEM models have several applications, including enhancing bacterial cellulose (BC) production to meet the growing demand for this polysaccharide in a more environmentally sustainable way. Currently, genomes from over 60 BC producing species have been sequenced, including both complete and draft genome assemblies, with *Komagataeibacter xylinus*, arguably the most efficient producer, having 13 sequenced genomes including both complete and draft genome assemblies (Table 2.2), according to the taxonomy database[33] (Schoch et al. 2020).

Furthermore, the metabolic pathways that allow these species to produce BC are well documented (Abidi et al. 2022; Römling and Galperin 2015). For instance, *K. xylinus* is a widely studied model organism for BC production, which has been investigated using various carbon sources, including ethanol, monosaccharides (glucose and fructose), and disaccharides (sucrose and maltose), among others (Fig. 2.6) (Jacek et al. 2021; Ryngajłło et al. 2019; Wang et al. 2018b).

Several studies have also provided several omics data regarding cellulose production in several species. For instance, Ryngajłło et al. (2019) investigated the global gene expression profile of *K. xylinus* E25 using RNA sequencing and provided a better understanding of the effects of ethanol on cellulose biosynthesis. Similarly, Fei et al. (2023) uncovered similar effects of ethanol on *K. nataicola* Y49 through proteomics. Besides omics insights, other studies have also engineered known strains to successfully improve BC production. For example, Gwon et al. (2019) introduced the *E. coli pfkA* gene into the chromosome of the wild-type strain *Komagataeibacter xylinus* DSM 2325 to enhance intracellular ATP levels, boosting growth and cellulose production.

Despite the abundance of omics data on BC production across multiple species, there are few known GEMs specifically designed to enhance or optimize BC overproduction for improved industrial-scale performance (Table 2.3). The GEM KxyMBEL1810 (Jang et al. 2019) created through the sequenced genome of *K. xylinus* DSM 2325 contains 686 metabolic genes, 1712 different metabolites, and 1810 reactions, attributed to several categories, such as transporters, amino acid metabolism, glycan metabolism, cofactor and vitamin metabolism, and carbohydrate metabolism, among others. Following the identification of gene overexpression targets from the glycolysis and pentose phosphate pathway, the model predicted that overexpression of the genes for the enzymes glucose-6-phosphate isomerase (*pgi*) and phosphogluconate dehydrogenase (*gnd*) would increase the BC production yield. This prediction was validated experimentally and an increase by nearly 116% in BC yield was obtained compared to the control strain (Jang et al. 2019).

---

[33] https://www.ncbi.nlm.nih.gov/taxonomy

Table 2.2 Current published sequenced genomes of the most well-known species involved in BC production

Organism (Ref.)	Examples of strain identifiers	Number of datasets Genomes
Komagataeibacter xylinus (Brown, 1886)	CGMCC:17276, DSM:6513, CGMCC 21569, BCCM/LMG:1515, NBRC:15237, JCM:7644, E25, NBRC: 13693, DSM:2325, K2G30, CIP103107T	13
Novacetimonas hansenii (Gosselé et al. 1983)	NRIC:0243, HUM-1, SC-3B, Ay201, SI1, NQ5, ATCC:53582, JCM:7643, NBRC:14820, LMG:23726, ATCC:23769, C110	20
Komagataeibacter rhaeticus (Dellaglio et al. 2005)	DSM:16663, iGEM, BCCM/LMG:22126, CGMCC:2955, ENS9b, ENS9a1a	12
Komagataeibacter saccharivorans (Lisdiyanti et al. 2006)	SRCM101450, NRIC:0614, LMG:1582, CV1, JH1	11
Komagataeibacter sucrofermentans (Yamada et al. 2012)	DSM:15973, JCM:9730, BCCM/LMG:18788, SMEG01, JML:2321, JML:KO23, ATCC 700178, CECT 7291, BPR 2001	6
Komagataeibacter nataicola (Lisdiyanti et al. 2006)	DS12, BCCM/LMG:1536, NRIC:0616, RZS01, FWP-2023	5

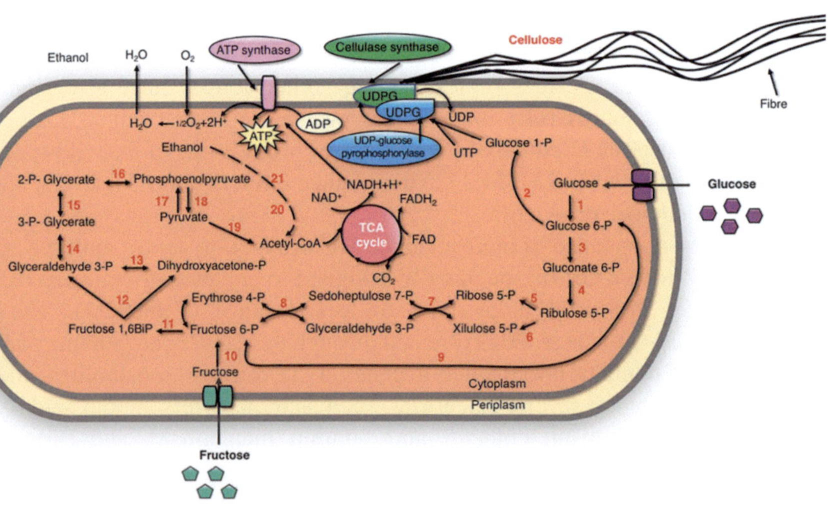

Fig. 2.6 Metabolic network regarding cellulose production from two different carbon sources (fructose and glucose) in *Komagataeibacter xylinus*: (1) glucokinase-ATP, (2) phosphoglucomutase, (3) glucose-6-phosphate dehydrogenase, (4) 6-phosphogluconate dehydrogenase, (5) phosphoribulose isomerase, (6) phosphor ribulose epimerase, (7) transketolase, (8) transaldolase, (9) phosphoglucoisomerase, (10) fructokinase, (11) fructokinase ATP, (12) aldolase, (13) triosephosphate isomerase, (14) glyceraldehyde 3-phosphate dehydrogenase, (15) phosphoglycerate mutase, (16) enolase, (17) pyruvate kinase (18) pyruvate biphosphate kinase, (19) pyruvate dehydrogenase, (20) alcohol dehydrogenase, and (21) aldehyde dehydrogenase. Reproduced from Jacek et al. (2021), published under the Creative Commons Attribution License (CC BY 4.0): https://creativecommons.org/licenses/by/4.0/

Table 2.3 Overview of published GEMs for BC overproduction

Model (Ref)	Organism (strain)	Number of genes	Number of reactions	Number of metabolites	Main predictions
KxyMBEL1810 (Jang et al. 2019)	*Komagataeibacter xylinus* DSM 2325	686 (metabolic)	1810	1712	pgi and gnd overexpression increases BC yield
iMR640 (Rezazadeh et al. 2020)	*Komagataeibacter xylinus*	640	865	783	Anaerobic metabolism and ethanol uptake
*i*HZ771 (Zhang et al. 2017)	*Komagataeibacter nataicola* RZS01	771	2014	2035	Glycerol preferred carbon source; 8 gene targets for overproduction

The GEM iMR640 (Rezazadeh et al. 2020) model, containing 640 genes, 865 metabolites, and 783 reactions was created from the model microorganism *K. xylinus* E25. The model predicted the lack of growth under anaerobic conditions and successfully predicted the fluxes of oxygen and carbon dioxide during growth with ethanol, with an accuracy of 93.7%, thus allowing for the better understanding of the species dependence on oxygen availability (Rezazadeh et al. 2020).

Lastly, the model *i*HZ771 (Zhang et al. 2017), developed for *K. nataicola* RZS01, includes 771 genes, 2035 metabolites, and 2014 reactions. It successfully identified glycerol as the preferred carbon source for BC production, determined the essential genes for cellular growth in both minimal and complex media, and pinpointed eight genes as potential targets for BC overproduction. Despite the lack of experimental data for gene validation, the model provided valuable insights into the metabolic framework underlying cell growth.

5 Conclusion and Prospects

The use of GEMs to enhance BC production still holds great promise, as these models allow for rational metabolic engineering strategies, including substrate optimization, pathway modifications, and strain improvement. By incorporating constraints based on experimental data, GEMs enable the prediction of metabolic flux distributions, helping to identify genetic targets for increased cellulose yield. Additionally, computational approaches such as flux balance analysis (FBA) and dynamic simulations provide valuable predictions on how different environmental conditions and genetic modifications influence BC production.

Looking ahead, one of the most promising prospects for improving BC production through GEMs is the integration of multiomics data, which is yet to be seen in the existing models. Including transcriptomics, proteomics, and metabolomics allows the refinement of metabolic models with omics-derived constraints, and thus, the predictive accuracy of GEMs can be significantly enhanced, leading to more precise metabolic engineering strategies. Furthermore, the development of context-specific GEMs tailored to different bacterial strains or environmental conditions will allow for strain-specific optimization, which can improve industrial-scale production efficiency. Additionally, machine learning and artificial intelligence-driven modeling approaches may further refine metabolic predictions and uncover novel engineering targets for enhanced cellulose biosynthesis.

In conclusion, the combination of GEMs with advanced omics data integration presents a powerful framework for optimizing bacterial cellulose production. As computational and experimental methodologies continue to evolve, the design of highly efficient bacterial strains for sustainable and high-yield BC production is becoming increasingly feasible. These advancements will not only contribute to fundamental metabolic research but also pave the way for the industrial-scale production of bacterial cellulose with enhanced properties and further applications.

References

Abidi W, Torres-Sánchez L, Siroy A, Krasteva PV (2022) Weaving of bacterial cellulose by the Bcs secretion systems. FEMS Microbiol Rev 46(2):1–35. https://doi.org/10.1093/FEMSRE/FUAB051

Agarwala R et al (2018) Database resources of the National Center for biotechnology information. Nucleic Acids Res 46(D1):D8–D13. https://doi.org/10.1093/NAR/GKX1095

Agren R, Bordel S, Mardinoglu A, Pornputtapong N, Nookaew I, Nielsen J (2012) Reconstruction of genome-scale active metabolic networks for 69 human cell types and 16 cancer types using INIT. PLoS Comput Biol 8(5):e1002518. https://doi.org/10.1371/JOURNAL.PCBI.1002518

Agren R, Mardinoglu A, Asplund A, Kampf C, Uhlen M, Nielsen J (2014) Identification of anti-cancer drugs for hepatocellular carcinoma through personalized genome-scale metabolic modeling. Mol Syst Biol 10(3). https://doi.org/10.1002/MSB.145122

Almagro Armenteros JJ, Sønderby CK, Sønderby SK, Nielsen H, Winther O (2017) DeepLoc: prediction of protein subcellular localization using deep learning. Bioinf (Oxf) 33(21):3387–3395. https://doi.org/10.1093/BIOINFORMATICS/BTX431

Altschul SF, Gish W, Miller W, Myers EW, Lipman DJ (1990) Basic local alignment search tool. J Mol Biol 215(3):403–410. https://doi.org/10.1016/S0022-2836(05)80360-2

Arkin AP et al (2016) The DOE systems biology knowledgebase (KBase). BioRxiv. https://doi.org/10.1101/096354

Armenteros JJA, Salvatore M, Emanuelsson O, Winther O, Von Heijne G, Elofsson A, Nielsen H (2019) Detecting sequence signals in targeting peptides using deep learning. Life Sci Alliance 2(5):e201900429. https://doi.org/10.26508/LSA.201900429

Bankevich A, Nurk S, Antipov D, Gurevich AA, Dvorkin M, Kulikov AS, Lesin VM, Nikolenko SI, Pham S, Prjibelski AD, Pyshkin AV, Sirotkin AV, Vyahhi N, Tesler G, Alekseyev MA, Pevzner PA (2012) SPAdes: a new genome assembly algorithm and its applications to single-cell sequencing. J Comput Biol 19(5):455–477. https://doi.org/10.1089/CMB.2012.0021

Bateman A (2019) UniProt: a worldwide hub of protein knowledge. Nucleic Acids Res 47(D1):D506–D515. https://doi.org/10.1093/NAR/GKY1049

Becker SA, Palsson BO (2008) Context-specific metabolic networks are consistent with experiments. PLoS Comput Biol 4(5). https://doi.org/10.1371/JOURNAL.PCBI.1000082

Bi X, Liu Y, Li J, Du G, Lv X, Liu L (2022) Construction of multiscale genome-scale metabolic models: frameworks and challenges. Biomolecules 12(5):721. https://doi.org/10.3390/BIOM12050721

Blanco Parte FG, Santoso SP, Chou CC, Verma V, Wang HT, Ismadji S, Cheng KC (2020) Current progress on the production, modification, and applications of bacterial cellulose. Crit Rev Biotechnol 40(3):397–414. https://doi.org/10.1080/07388551.2020.1713721

Bordbar A, Mo ML, Nakayasu ES, Schrimpe-Rutledge AC, Kim YM, Metz TO, Jones MB, Frank BC, Smith RD, Peterson SN, Hyduke DR, Adkins JN, Palsson BO (2012) Model-driven multi-omic data analysis elucidates metabolic immunomodulators of macrophage activation. Mol Syst Biol 8:558. https://doi.org/10.1038/MSB.2012.21

Brown AJ (1886) XLIII.—On an acetic ferment which forms cellulose. Journal of the Chemical Society, Transactions, 49(0), 432–439. https://doi.org/10.1039/CT8864900432

Buchfink B, Xie C, Huson DH (2014) Fast and sensitive protein alignment using DIAMOND. Nat Methods 12(1):59–60. https://doi.org/10.1038/nmeth.3176

Burgard AP, Pharkya P, Maranas CD (2003) Optknock: a bilevel programming framework for identifying gene knockout strategies for microbial strain optimization. Biotechnol Bioeng 84(6):647–657. https://doi.org/10.1002/BIT.10803

Capela J, Lagoa D, Rodrigues R, Cunha E, Cruz F, Barbosa A, Bastos J, Lima D, Ferreira EC, Rocha M, Dias O (2022) Merlin, an improved framework for the reconstruction of high-quality genome-scale metabolic models. Nucleic Acids Res 50(11):6052–6066. https://doi.org/10.1093/NAR/GKAC459

Caspi R, Billington R, Keseler IM, Kothari A, Krummenacker M, Midford PE, Ong WK, Paley S, Subhraveti P, Karp PD (2019) The MetaCyc database of metabolic pathways and enzymes - a 2019 update. Nucleic Acids Res 48(D1):D445. https://doi.org/10.1093/NAR/GKZ862

Chang A, Jeske L, Ulbrich S, Hofmann J, Koblitz J, Schomburg I, Neumann-Schaal M, Jahn D, Schomburg D (2021) BRENDA, the ELIXIR core data resource in 2021: new developments and updates. Nucleic Acids Res 49(D1):D498–D508. https://doi.org/10.1093/NAR/GKAA1025

Chen Y, Gustafsson J, Tafur Rangel A, Anton M, Domenzain I, Kittikunapong C, Li F, Yuan L, Nielsen J, Kerkhoven EJ (2024b) Reconstruction, simulation and analysis of enzyme-constrained metabolic models using GECKO Toolbox 3.0. Nat Protoc 19(3):629–667. https://doi.org/10.1038/s41596-023-00931-7

Chen Z, Ain NU, Zhao Q, Zhang X (2024a) From tradition to innovation: conventional and deep learning frameworks in genome annotation. Brief Bioinform 25(3):138. https://doi.org/10.1093/BIB/BBAE138

Cottingham RW (2015) The DOE systems biology knowledgebase (KBase). In: BCB '15: Proceedings of the 6th ACM Conference on bioinformatics, computational biology and health informatics. ACM, p 510. https://doi.org/10.1145/2808719.2811433

Cruz F, Faria JP, Rocha M, Rocha I, Dias O (2020) A review of methods for the reconstruction and analysis of integrated genome-scale models of metabolism and regulation. Biochem Soc Trans 48(5):1889–1903. https://doi.org/10.1042/BST20190840

Delcher AL, Harmon D, Kasif S, White O, Salzberg SL (1999) Improved microbial gene identification with GLIMMER. Nucleic Acids Res 27(23):4636–4641. https://doi.org/10.1093/NAR/27.23.4636

Dellaglio F, Cleenwerck I, Felis GE, Engelbeen K, Janssens D, Marzotto M (2005) Description of *Gluconacetobacter swingsii* sp. nov. and *Gluconacetobacter rhaeticus* sp. nov., isolated from Italian apple fruit. Int J Syst Evol Microbiol 55(Pt 6):2365–2370. https://doi.org/10.1099/IJS.0.63301-0

Dias O, Rocha I (2016) Chapter 6 - Systems biology in fungi. In: Paterson R, Lima N (eds) Molecular biology of food and water borne mycotoxigenic and mycotic fungi. CRC Press, New York, pp 69–92

Dias O, Rocha M, Ferreira EC, Rocha I (2015) Reconstructing genome-scale metabolic models with merlin. Nucleic Acids Res 43(8):3899–3910. https://doi.org/10.1093/NAR/GKV294

Ebrahim A, Lerman JA, Palsson BO, Hyduke DR (2013) COBRApy: COnstraints-based reconstruction and analysis for python. BMC Syst Biol 7(1):1–6. https://doi.org/10.1186/1752-0509-7-74/FIGURES/2

Eddy SR (2011) Accelerated profile HMM searches. PLoS Comput Biol 7(10). https://doi.org/10.1371/JOURNAL.PCBI.1002195

Edwards JS, Palsson BO (1999) Systems properties of the *Haemophilus influenzae*Rd metabolic genotype. J Biol Chem 274(25):17410–17416. https://doi.org/10.1074/JBC.274.25.17410
Fei S, Yang X, Xu W, Zhang J, Li J, Chen H, Lin X, Liu S, Li C (2023) Insights into proteomics reveal mechanisms of ethanol-enhanced bacterial cellulose biosynthesis by *Komagataeibacter nataicola*. Fermentation 9(6):575. https://doi.org/10.3390/FERMENTATION9060575/S1
Feist AM, Herrgård MJ, Thiele I, Reed JL, Palsson B (2009) Reconstruction of biochemical networks in microorganisms. Nat Rev Microbiol 7(2):129–143. https://doi.org/10.1038/nrmicro1949
Feng X, Xu Y, Chen Y, Tang YJ (2012) MicrobesFlux: a web platform for drafting metabolic models from the KEGG database. BMC Syst Biol 6(1):1–9. https://doi.org/10.1186/1752-0509-6-94/FIGURES/5
Fernandes M, Souto AP, Gama M, Dourado F (2019) Bacterial cellulose and emulsified AESO biocomposites as an ecological alternative to leather. Nanomaterials (Basel) 9(12). https://doi.org/10.3390/NANO9121710
Fleischmann RD et al (1995) Whole-genome random sequencing and assembly of *Haemophilus influenzae* Rd. Science 269(5223):496–512. https://doi.org/10.1126/SCIENCE.7542800
Francke C, Siezen RJ, Teusink B (2005) Reconstructing the metabolic network of a bacterium from its genome. Trends Microbiol 13(11):550–558. https://doi.org/10.1016/J.TIM.2005.09.001
Gabaldón T (2008) Comparative genomics-based prediction of protein function. Methods Mol Biol 439:387–401. https://doi.org/10.1007/978-1-59745-188-8_26
Gillani M, Pollastri G (2024) Protein subcellular localization prediction tools. Comput Struct Biotechnol J 23:1796–1807. https://doi.org/10.1016/J.CSBJ.2024.04.032
Gosselé F, Swings J, Kersters K, Pauwels P, De Ley J (1983) Numerical analysis of phenotypic features and protein gel electrophoregrams of a wide variety of Acetobacter strains. Proposal for the improvement of the taxonomy of the genus Acetobacter Beijerinck 1898, 215. Syst Appl Microbiol 4(3):338–368. https://doi.org/10.1016/S0723-2020(83)80020-4
Gu C, Kim GB, Kim WJ, Kim HU, Lee SY (2019) Current status and applications of genome-scale metabolic models. Genome Biol 20(1):1–18. https://doi.org/10.1186/S13059-019-1730-3
Gwon H et al (2019) A safe and sustainable bacterial cellulose nanofiber separator for lithium rechargeable batteries. Proc Natl Acad Sci USA 116(39):19288–19293. https://doi.org/10.1073/PNAS.1905527116/-/DCSUPPLEMENTAL
Palsson B (2006) Chapter 1.1 The Need for Systems Analysis in Biology. Systems Biology: Properties of Reconstructed Networks. Cambridge University Press: New York pp.1–12 https://books.google.com/books/about/Systems_Biology.html?hl=pt-PT&id=8MlPngEACAAJ
Henry CS, Dejongh M, Best AA, Frybarger PM, Linsay B, Stevens RL (2010) High-throughput generation, optimization and analysis of genome-scale metabolic models. Nat Biotechnol 28(9):977–982. https://doi.org/10.1038/NBT.1672
Herrmann HA, Dyson BC, Vass L, Johnson GN, Schwartz JM (2019) Flux sampling is a powerful tool to study metabolism under changing environmental conditions. Npj Syst Biol Appl 5(1):1–8. https://doi.org/10.1038/s41540-019-0109-0
Hucka M et al (2003) The systems biology markup language (SBML): a medium for representation and exchange of biochemical network models. Bioinf (Oxf) 19(4):524–531. https://doi.org/10.1093/BIOINFORMATICS/BTG015
Jacek P, Dourado F, Gama M, Bielecki S (2021) Molecular aspects of bacterial nanocellulose biosynthesis. Microb Biotechnol. https://doi.org/10.1111/1751-7915.13386
Janes KA, Lazzara MJ (2024) Systems biology of the cancer cell. Annu Rev Biomed Eng. https://doi.org/10.1146/ANNUREV-BIOENG-103122-030552
Jang WD, Kim TY, Kim HU, Shim WY, Ryu JY, Park JH, Lee SY (2019) Genomic and metabolic analysis of *Komagataeibacter xylinus* DSM 2325 producing bacterial cellulose nanofiber. Biotechnol Bioeng 116(12):3372–3381. https://doi.org/10.1002/BIT.27150
Jerby L, Shlomi T, Ruppin E (2010) Computational reconstruction of tissue-specific metabolic models: application to human liver metabolism. Mol Syst Biol 6:401. https://doi.org/10.1038/MSB.2010.56/SUPPL_FILE/MSB201056-SUP-0005.ZIP

Joshi A, Rienks M, Theofilatos K, Mayr M (2020) Systems biology in cardiovascular disease: a multiomics approach. Nat Rev Cardiol 18(5):313–330. https://doi.org/10.1038/s41569-020-00477-1

Kanehisa M, Furumichi M, Tanabe M, Sato Y, Morishima K (2017) KEGG: new perspectives on genomes, pathways, diseases and drugs. Nucleic Acids Res 45(D1):D353–D361. https://doi.org/10.1093/NAR/GKW1092

Karlsen E, Schulz C, Almaas E (2018) Automated generation of genome-scale metabolic draft reconstructions based on KEGG. BMC Bioinform 19(1):1–11. https://doi.org/10.1186/S12859-018-2472-Z/TABLES/3

Karp PD, Billington R, Caspi R, Fulcher CA, Latendresse M, Kothari A, Keseler IM, Krummenacker M, Midford PE, Ong Q, Ong WK, Paley SM, Subhraveti P (2019) The BioCyc collection of microbial genomes and metabolic pathways. Brief Bioinform 20(4):1085–1093. https://doi.org/10.1093/BIB/BBX085

Kim HU, Charusanti P, Lee SY, Weber T (2016) Metabolic engineering with systems biology tools to optimize production of prokaryotic secondary metabolites. Nat Prod Rep 33(8):933–941. https://doi.org/10.1039/C6NP00019C

Klimke W, O'Donovan C, White O, Brister JR, Clark K, Fedorov B, Mizrachi I, Pruitt KD, Tatusova T (2011) Solving the problem: genome annotation standards before the data deluge. Stand Genomic Sci 5(1):168–193. https://doi.org/10.4056/SIGS.2084864

Kulyashov MA, Kolmykov SK, Khlebodarova TM, Akberdin IR (2023) State-of the-art constraint-based modeling of microbial metabolism: from basics to context-specific models with a focus on methanotrophs. Microorganisms 11(12):2987. https://doi.org/10.3390/MICROORGANISMS11122987

Lee SY, Sohn SB, Kim HU, Park JM, Kim TY, Orth JD, Palsson BO (2012) Genome-scale network modeling. Syst Metab Eng. https://doi.org/10.1007/978-94-007-4534-6_1

Leitão AF, Silva JP, Dourado F, Gama M (2013) Production and characterization of a new bacterial cellulose/poly(vinyl alcohol) nanocomposite. Materials (Basel) 6(5):1956–1966. https://doi.org/10.3390/MA6051956

Lewis NE, Hixson KK, Conrad TM, Lerman JA, Charusanti P, Polpitiya AD, Adkins JN, Schramm G, Purvine SO, Lopez-Ferrer D, Weitz KK, Eils R, König R, Smith RD, Palsson B (2010) Omic data from evolved E. coli are consistent with computed optimal growth from genome-scale models. Mol Syst Biol 6. https://doi.org/10.1038/MSB.2010.47

Li D, Liu CM, Luo R, Sadakane K, Lam TW (2015) MEGAHIT: an ultra-fast single-node solution for large and complex metagenomics assembly via succinct de Bruijn graph. Bioinf (Oxf) 31(10):1674–1676. https://doi.org/10.1093/BIOINFORMATICS/BTV033

Li G, Wang L, Deng Y, Wei Q (2022) Research progress of the biosynthetic strains and pathways of bacterial cellulose. J Ind Microbiol Biotechnol 49(1):71. https://doi.org/10.1093/JIMB/KUAB071

Lin B, Luo X, Liu Y, Jin X (2024) A comprehensive review and comparison of existing computational methods for protein function prediction. Brief Bioinform 25(4):289. https://doi.org/10.1093/BIB/BBAE289

Lisdiyanti P, Navarro RR, Uchimura T, Komagata K (2006) Reclassification of *Gluconacetobacter hansenii* strains and proposals of *Gluconacetobacter saccharivorans* sp. nov. and *Gluconacetobacter nataicola* sp. nov. Int J Syst Evol Microbiol 56(Pt 9):2101–2111. https://doi.org/10.1099/IJS.0.63252-0

Machado D, Andrejev S, Tramontano M, Patil KR (2018) Fast automated reconstruction of genome-scale metabolic models for microbial species and communities. Nucleic Acids Res 46(15):7542–7553. https://doi.org/10.1093/NAR/GKY537

Machado D, Herrgård M (2014) Systematic evaluation of methods for integration of transcriptomic data into constraint-based models of metabolism. PLoS Comput Biol 10(4):e1003580. https://doi.org/10.1371/JOURNAL.PCBI.1003580

Mahadevan R, Schilling CH (2003) The effects of alternate optimal solutions in constraint-based genome-scale metabolic models. Metab Eng 5(4):264–276. https://doi.org/10.1016/J.YMBEN.2003.09.002

Mahadevan R, Edwards JS, Doyle FJ (2002) Dynamic flux balance analysis of diauxic growth in *Escherichia coli*. Biophys J 83(3):1331–1340. https://doi.org/10.1016/S0006-3495(02)73903-9

Majoros WH, Pertea M, Salzberg SL (2004) TigrScan and GlimmerHMM: two open source ab initio eukaryotic gene-finders. Bioinf (Oxf) 20(16):2878–2879. https://doi.org/10.1093/BIOINFORMATICS/BTH315

Malcı K, Li IS, Kisseroudis N, Ellis T (2024) Modulating microbial materials - engineering bacterial cellulose with synthetic biology. ACS Synth Biol 13:3875. https://doi.org/10.1021/ACSSYNBIO.4C00615/ASSET/IMAGES/LARGE/SB4C00615_0006.JPEG

Nielsen J (2017) Systems biology of metabolism. Annual Review of Biochemistry 86:245–275. https://doi.org/10.1146/annurev-biochem-061516-044757

O'Brien EJ, Monk JM, Palsson BO (2015) Using genome-scale models to predict biological capabilities. Cell 161(5):971. https://doi.org/10.1016/J.CELL.2015.05.019

Orth JD, Thiele I, Palsson BO (2010) What is flux balance analysis? Nat Biotechnol 28(3):245–248. https://doi.org/10.1038/NBT.1614

Palsson B (2009) Metabolic systems biology. FEBS Lett 583(24):3900–3904. https://doi.org/10.1016/J.FEBSLET.2009.09.031

Pazhamala LT, Kudapa H, Weckwerth W, Millar AH, Varshney RK (2021) Systems biology for crop improvement. Plant Genome 14(2):e20098. https://doi.org/10.1002/TPG2.20098

Rezazadeh M, Babaeipour V, Motamedian E (2020) Reconstruction, verification and in-silico analysis of a genome-scale metabolic model of bacterial cellulose producing *Komagataeibacter xylinus*. Bioprocess Biosyst Eng 43(6):1017–1026. https://doi.org/10.1007/S00449-020-02299-4/FIGURES/3

Robaina Estévez S, Nikoloski Z (2014) Generalized framework for context-specific metabolic model extraction methods. Front Plant Sci 5. https://doi.org/10.3389/FPLS.2014.00491

Rocha I, Förster J, Nielsen J (2008) Chapter 29 - Design and application of genome-scale reconstructed metabolic models. In: Osterman AL, Gerdes SY (eds) Methods in molecular biology. Humana Press, Totowa, NJ, pp 409–431. https://doi.org/10.1007/978-1-59745-321-9_29

Römling U, Galperin MY (2015) Bacterial cellulose biosynthesis: diversity of operons, subunits, products and functions. Trends Microbiol 23(9):545. https://doi.org/10.1016/J.TIM.2015.05.005

Roy R, Chun J, Powell SN (2011) BRCA1 and BRCA2: different roles in a common pathway of genome protection. Nat Rev Cancer 12(1):68. https://doi.org/10.1038/NRC3181

Ryngajłło M, Jacek P, Cielecka I, Kalinowska H, Bielecki S (2019) Effect of ethanol supplementation on the transcriptional landscape of bionanocellulose producer *Komagataeibacter xylinus* E25. Appl Microbiol Biotechnol 103(16):6673–6688. https://doi.org/10.1007/S00253-019-09904-X

Ryngajłło M, Jędrzejczak-Krzepkowska M, Kubiak K, Ludwicka K, Bielecki S (2020) Towards control of cellulose biosynthesis by *Komagataeibacter* using systems-level and strain engineering strategies: current progress and perspectives. Appl Microbiol Biotechnol 104(15):6565–6585. https://doi.org/10.1007/S00253-020-10671-3

Saier MH, Reddy VS, Tsu BV, Ahmed MS, Li C, Moreno-Hagelsieb G (2016) The transporter classification database (TCDB): recent advances. Nucleic Acids Res 44(D1):D372–D379. https://doi.org/10.1093/NAR/GKV1103

Sampaio M, Rocha M, Dias O (2022) Exploring synergies between plant metabolic modelling and machine learning. Comput Struct Biotechnol J 20:1885–1900. https://doi.org/10.1016/J.CSBJ.2022.04.016

Sanger F, Nicklen S, Coulson AR (1977) DNA sequencing with chain-terminating inhibitors. Proc Natl Acad Sci USA 74(12):5463–5467. https://doi.org/10.1073/PNAS.74.12.5463

Sayers EW et al (2021) Database resources of the National Center for Biotechnology Information. Nucleic Acids Res 50(D1):D20. https://doi.org/10.1093/NAR/GKAB1112

Schellenberger J, Palsson B (2009) Use of randomized sampling for analysis of metabolic networks. J Biol Chem 284(9):5457–5461. https://doi.org/10.1074/JBC.R800048200
Schellenberger J, Park JO, Conrad TM, Palsson BT (2010) BiGG: a biochemical genetic and genomic knowledgebase of large scale metabolic reconstructions. BMC Bioinform 11. https://doi.org/10.1186/1471-2105-11-213
Schellenberger J, Que R, Fleming RMT, Thiele I, Orth JD, Feist AM, Zielinski DC, Bordbar A, Lewis NE, Rahmanian S, Kang J, Hyduke DR, Palsson B (2011) Quantitative prediction of cellular metabolism with constraint-based models: the COBRA Toolbox v2.0. Nat Protoc 6(9):1290–1307. https://doi.org/10.1038/nprot.2011.308
Schmidt BJ, Ebrahim A, Metz TO, Adkins JN, Palsson B, Hyduke DR (2013) GIM3E: condition-specific models of cellular metabolism developed from metabolomics and expression data. Bioinformatics 29(22):2900–2908. https://doi.org/10.1093/BIOINFORMATICS/BTT493
Schoch CL, Ciufo S, Domrachev M, Hotton CL, Kannan S, Khovanskaya R, Leipe D, McVeigh R, O'Neill K, Robbertse B, Sharma S, Soussov V, Sullivan JP, Sun L, Turner S, Karsch-Mizrachi I (2020) NCBI taxonomy: a comprehensive update on curation, resources and tools. Database 2020. https://doi.org/10.1093/DATABASE/BAAA062
Sertbas M, Ulgen KO (2020) Genome-scale metabolic modeling for unraveling molecular mechanisms of high threat pathogens. Front Cell Dev Biol 8:566702. https://doi.org/10.3389/FCELL.2020.566702/PDF
Shen F, Sun R, Yao J, Li J, Liu Q, Price ND, Liu C, Wang Z (2019) OptRAM: in-silico strain design via integrative regulatory-metabolic network modeling. PLoS Comput Biol 15(3):e1006835. https://doi.org/10.1371/JOURNAL.PCBI.1006835
Shlomi T, Cabili MN, Herrgård MJ, Palsson B, Ruppin E (2008) Network-based prediction of human tissue-specific metabolism. Nat Biotechnol 26(9):1003–1010. https://doi.org/10.1038/nbt.1487
Silva FAGS, Dourado F, Gama M, Poças F (2020) Nanocellulose bio-based composites for food packaging. Nanomaterials (Basel) 10(10):1–29. https://doi.org/10.3390/NANO10102041
Simpson JT, Wong K, Jackman SD, Schein JE, Jones SJM, Birol I (2009) ABySS: a parallel assembler for short read sequence data. Genome Res 19(6):1117–1123. https://doi.org/10.1101/GR.089532.108
Singh A, Walker KT, Ledesma-Amaro R, Ellis T (2020) Engineering bacterial cellulose by synthetic biology. Int J Mol Sci 21(23):9185. https://doi.org/10.3390/IJMS21239185
Singh A, Septiningsih EM, Balyan HS, Singh NK, Rai V (2017) Genetics, physiological mechanisms and breeding of flood-tolerant rice (*Oryza sativa* L.). Plant Cell Physiol 58(2):185–197. https://doi.org/10.1093/PCP/PCW206
Soares da Silva FAG, Fernandes M, Souto AP, Ferreira EC, Dourado F, Gama M (2019) Optimization of bacterial nanocellulose fermentation using recycled paper sludge and development of novel composites. Appl Microbiol Biotechnol 103(21–22):9143–9154. https://doi.org/10.1007/S00253-019-10124-6
Taokaew S (2024) Bacterial nanocellulose produced by cost-effective and sustainable methods and its applications: a review. Fermentation 10(6):316. https://doi.org/10.3390/FERMENTATION10060316
Teh MY, Ooi KH, Danny Teo SX, Bin Mansoor ME, Shaun Lim WZ, Tan MH (2019) An expanded synthetic biology toolkit for gene expression control in *Acetobacteraceae*. ACS Synth Biol 8(4):708–723. https://doi.org/10.1021/ACSSYNBIO.8B00168/SUPPL_FILE/SB8B00168_SI_001.PDF
Thiele I, Palsson B (2010) A protocol for generating a high-quality genome-scale metabolic reconstruction. Nat Protoc 5(1):93–121. https://doi.org/10.1038/NPROT.2009.203
Ungerer MC, Johnson LC, Herman MA (2008) Ecological genomics: understanding gene and genome function in the natural environment. Heredity 100(2):178–183. https://doi.org/10.1038/sj.hdy.6800992
Varma A, Palsson BO (1994) Metabolic flux balancing: basic concepts, scientific and practical use. Bio/Technology 12(10):994–998. https://doi.org/10.1038/NBT1094-994

Vivek-Ananth RP, Samal A (2016) Advances in the integration of transcriptional regulatory information into genome-scale metabolic models. Biosystems 147:1–10. https://doi.org/10.1016/J.BIOSYSTEMS.2016.06.001

Vlassis N, Pacheco MP, Sauter T (2014) Fast reconstruction of compact context-specific metabolic network models. PLoS Comput Biol 10(1):e1003424. https://doi.org/10.1371/JOURNAL.PCBI.1003424

Wang C, Chang D, Zhang Q, Yu Z (2024) Enhanced bioethanol production by evolved Escherichia coli LGE2-H in a microbial electrolysis cell system. Bioresour Bioprocess 11(1):1–10. https://doi.org/10.1186/S40643-023-00717-5/FIGURES/5

Wang H, Marcišauskas S, Sánchez BJ, Domenzain I, Hermansson D, Agren R, Nielsen J, Kerkhoven EJ (2018a) RAVEN 2.0: a versatile toolbox for metabolic network reconstruction and a case study on *Streptomyces coelicolor*. PLoS Comput Biol 14(10):e1006541. https://doi.org/10.1371/JOURNAL.PCBI.1006541

Wang SS, Han YH, Chen JL, Zhang DC, Shi XX, Ye YX, Chen DL, Li M (2018b) Insights into bacterial cellulose biosynthesis from different carbon sources and the associated biochemical transformation pathways in *Komagataeibacter* sp. W1. Polymers 10(9):963. https://doi.org/10.3390/POLYM10090963

Wang Y, Eddy JA, Price ND (2012) Reconstruction of genome-scale metabolic models for 126 human tissues using mCADRE. BMC Syst Biol 6(1):1–16. https://doi.org/10.1186/1752-0509-6-153/FIGURES/5

Wei L, Ding Y, Su R, Tang J, Zou Q (2018) Prediction of human protein subcellular localization using deep learning. J Parallel Distrib Comput 117:212–217. https://doi.org/10.1016/J.JPDC.2017.08.009

Werner C, Doenst T, Schwarzer M (2015) Chapter 4 - Metabolic pathways and cycles. In: Werner C, Doenst T, Schwarzer M (eds) The scientist's guide to cardiac metabolism. Elsevier, Jena, pp 39–55. https://doi.org/10.1016/B978-0-12-802394-5.00004-2

Wu Z, Liang X, Li M, Ma M, Zheng Q, Li D, An T, Wang G (2023) Advances in the optimization of central carbon metabolism in metabolic engineering. Microb Cell Factories 22(1):1–11. https://doi.org/10.1186/S12934-023-02090-6/TABLES/2

Yamada Y, Yukphan P, Vu HTL, Muramatsu Y, Ochaikul D, Tanasupawat S, Nakagawa Y (2012) Description of *Komagataeibacter* gen. nov., with proposals of new combinations (*Acetobacteraceae*). J Gen Appl Microbiol 58(5):397–404. https://doi.org/10.2323/JGAM.58.397

Yu NY, Wagner JR, Laird MR, Melli G, Rey S, Lo R, Dao P, Cenk Sahinalp S, Ester M, Foster LJ, Brinkman FSL (2010) PSORTb 3.0: improved protein subcellular localization prediction with refined localization subcategories and predictive capabilities for all prokaryotes. Bioinformatics 26(13):1608–1615. https://doi.org/10.1093/BIOINFORMATICS/BTQ249

Zerbino DR, Birney E (2008) Velvet: algorithms for de novo short read assembly using de Bruijn graphs. Genome Res 18(5):821–829. https://doi.org/10.1101/GR.074492.107

Zhang H, Ye C, Xu N, Chen C, Chen X, Yuan F, Xu Y, Yang J, Sun D (2017) Reconstruction of a genome-scale metabolic network of *Komagataeibacter nataicola* RZS01 for cellulose production. Sci Rep 7(1):1–9. https://doi.org/10.1038/s41598-017-06918-1

Zimmermann J, Kaleta C, Waschina S (2021) gapseq: informed prediction of bacterial metabolic pathways and reconstruction of accurate metabolic models. Genome Biol 22(1):1–35. https://doi.org/10.1186/S13059-021-02295-1

Chapter 3
Bacterial Cellulose Production from Non-conventional Strains

Miguel Pacheco, Pedro Montenegro-Silva, Miguel Gama, Lucília Domingues, Fernando Dourado, and Óscar Dias

Abstract The research and exploitation of the industrial production and applications of bacterial cellulose (BC) have been predominantly driven by *Komagataeibacter*, *Acetobacter*, and *Gluconacetobacter* species, as these genera are generally regarded as high cellulose producers. However, non-conventional genera, such as *Enterobacter*, *Rhodococcus*, *Bacillus*, and *Leifsonia*, have shown promising potential to overcome key production limitations, including oxygen dependence, low yield under stress, and limited substrate utilization. A deeper understanding of the biology, metabolism, and cultivation conditions of these alternative producers can thus prove to be crucial for the development of new BC production methods. Expanding research on these genera could lead to significant advances in the efficiency, scalability, and sustainability of BC production, ultimately accelerating its industrial and commercial applications.

Keywords Bacterial cellulose (BC) · Non-conventional strains · Anaerobic production · Genomic analysis · Cellulose synthase operon

M. Pacheco · P. Montenegro-Silva · M. Gama · L. Domingues · F. Dourado
CEB – Centre of Biological Engineering, University of Minho, Campus de Gualtar, Braga, Portugal
e-mail: fmgama@deb.uminho.pt; luciliad@deb.uminho.pt; fdourado@ceb.uminho.pt

Ó. Dias (✉)
CEB – Centre of Biological Engineering, University of Minho, Campus de Gualtar, Braga, Portugal

LABBELS – Associate Laboratory in Biotechnology and Bioengineering and Microelectromechanical Systems, Braga, Guimarães, Portugal
e-mail: odias@ceb.uminho.pt

F. Dourado et al. (eds.), *Advances in Bacterial Cellulose*,
https://doi.org/10.1007/978-3-032-07042-5_3

1 Introduction

Cellulose has been present in many everyday life products across the human history; however, new presentations and derivatives such as nanocelluloses keep widening the scope of applications for cellulose fibers, reinventing an already dynamic market (Cruz et al. 2024; Dang et al. 2024; Oesef et al. 2024). Bacterial cellulose (BC) is a highly pure, nanostructured form of cellulose synthesized by certain bacterial species, exhibiting unique properties such as high crystallinity, remarkable tensile strength, and excellent biocompatibility (Gullo et al. 2018). Due to its distinct physicochemical features, BC has garnered significant attention for applications across multiple fields, including biomedical engineering, textile, food industry, and electronics (Cruz et al. 2024; Portela et al. 2019; Shi et al. 2014; Thomas et al. 2018; Ullah et al. 2019). In the biomedical sector, BC membranes are used clinically for wound dressings and the replacement of the dura mater, owing to its excellent water retention capacity and biocompatibility (Ullah et al. 2019). Meanwhile, in the food industry, BC may be used as a stabilizer, thickener, and low-calorie fiber (Shi et al. 2014). Recent advancements have also highlighted BC's potential in developing flexible electronic devices and bio-based composites, given its superior mechanical properties and ability to form transparent films (Thomas et al. 2018).

Among the microorganisms capable of producing BC, species belonging to the genera *Komagataeibacter*, *Acetobacter,* and *Gluconacetobacter*, particularly *Komagataeibacter xylinus*, is recognized as the most efficient and widely studied producer (Gullo et al. 2018; Iguchi et al. 2000). *Komagataeibacter hansenii*, *Komagataeibacter rhaeticus*, and *Komagataeibacter europaeus* are also notable for their robust BC production under various culture conditions (Fang and Catchmark 2014; Keshk 2014). The preference for *Komagataeibacter* species in both academic research and potential industrial applications stems from their ability to efficiently synthesize BC using a wide range of carbon sources, along with their relatively well-characterized biosynthetic pathways (Gullo et al. 2018; Jacek et al. 2019; Montenegro-Silva et al. 2024).

At the cellular level, BC production is catalyzed by a membrane-embedded cellulose synthase complex that polymerizes UDP-glucose into β-1,4-glucan chains, which subsequently crystallize into microfibrils (Römling and Galperin 2015; Ross et al. 1991). The cellulose synthase operon, typically composed of *bcsA*, *bcsB*, *bcsC*, and *bcsD* genes, orchestrates the polymerization, secretion, and organization of the cellulose fibers (Standal et al. 1994). Among these, the *bcsA* gene encodes the catalytic subunit responsible for glucose polymerization and binding of cyclic-di-GMP (Gullo et al. 2018), while *bcsB* facilitates the process by stabilizing the cellulose synthase complex and the binding of UDP-glucose to the BcsA protein (Saxena et al. 1991). The crystallization into highly ordered structures results in the formation of ribbon-like microfibrils that are extruded through pores in the outer membrane (Fig. 3.1) (Gullo et al. 2018; Morgan et al. 2013). Environmental factors such as pH, oxygen availability, and nutrient concentration can regulate the activity

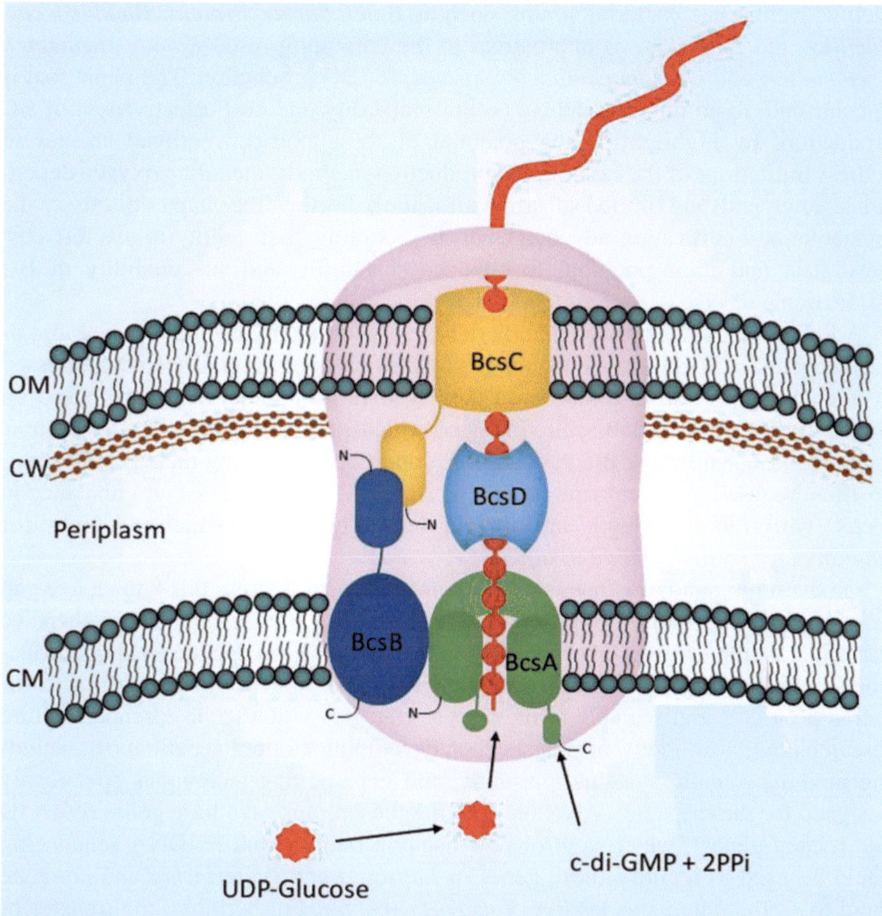

Fig. 3.1 Cellulose synthase complex present in *Komagataeibacter* membrane. *OM* Outer membrane, *CW* Cell wall, *CM* cytoplasmic membrane. Reprinted from Gullo et al. (2018), Copyright (2018), with permission from Springer-Verlag GmbH Germany, part of Springer Nature

of the cellulose synthase complex and influence the rate, efficiency, and morphology of the BC fibers (Bielecki et al. 2002; Jacek et al. 2019).

However, aside from the above-mentioned more conventional cellulose producers, there are still a wide range of other bacteria with unique properties that can be explored (Buldum et al. 2017; Hungund and Gupta 2010; Ji et al. 2016; Tanskul et al. 2013). In the next section of this chapter, several limitations of the current and standard species used for BC production will be approached, as well as other non-conventional strains that can potentially overcome some of these limitations and provide a breakthrough in the general production of BC.

This chapter aims to address a gap in BC research, where *Komagataeibacter* dominates, by investigating for the first time the potential of lesser-known (non-conventional) bacterial strains, such as *Enterobacter, Proteus, Rhodococcus, Bacillus*, and *Leifsonia*, as alternatives to the commonly used *Komagataeibacter, Acetobacter*, and *Gluconacetobacter* species, for BC production. The main goal is to contribute to an ongoing debate on the scalability and cost-effectiveness of BC production, by highlighting the potential of these non-conventional species to address limitations of the existing BC production methods, including oxygen dependence, low yield, and limited substrate utilization. Further, the chapter discusses the metabolic and cultivation advantages of these strains, their ability to use low-cost substrates, and their potential to enhance scalability and sustainability in BC production.

A literature survey showed that there is a very limited discussion on the biological mechanisms of cellulose biosynthesis by the non-conventional species addressed in this chapter. Existing research on these genera is, for the most part, based on single studies, done at lab scale, many of which lack in-depth follow-up on how these bacteria synthesize BC in anaerobic conditions, including their genetic composition, regulation, or enzyme functions. Admittedly, this creates an imbalance in focus, with more in-depth information currently being available mainly for *Enterobacter* strains relative to others.

On the other hand, this opens a great opportunity to address this knowledge gap in expanding the range of bacteria suitable for BC production. Comprehensive experiments may allow a better understanding of their growth patterns, BC production stability, and physiological behaviors. The development of standardized cultivation protocols and scalable fermentation strategies will also be essential. Future research could also focus, for instance, on developing tailored genetic tools, including modular genetic systems, plasmids, and gene-editing techniques specifically designed for these species. Also, on studying the cellulose synthase genes (bcsA–D and related genes), which control how bacteria build cellulose. DNA sequencing could be used to identify these genes in *Leifsonia* and *Beijerinckia* and compare them to well-studied species like *Komagataeibacter*. Furthermore, the transfer of these genes into easier-to-study bacteria, such as *Escherichia coli* or *Komagataeibacter rhaeticus*, could help to confirm their role in BC production (*E. coli* is a well-understood model bacterium with simple growth requirements and many genetic tools (Idalia and Bernardo 2017) while *K. rhaeticus* is a known BC producer with established methods for studying cellulose synthesis (Uğurel and Öğüt 2024)). Advanced computational tools to predict gene functions or model how cellulose synthase enzymes work could also guide these efforts by identifying promising bacteria for further study. These approaches, among others, could help uncover new BC producing strains and improve their potential for industrial use.

2 Limitations of Conventional Species for BC Production

Komagataeibacter spp. are the most efficient BC producers as compared to other genera (Cruz et al. 2024). However, the large-scale and cost-effective production are still hindered by their low production rates (Anguluri et al. 2022), sensitivity to environmental factors (Montenegro-Silva et al. 2024; Quijano et al. 2024), lack of genetic engineering tools capable of strain improvement (Malcı et al. 2024; Ryngajłło et al. 2020), and technical difficulties in scaling up the fermentation process (Cruz et al. 2024).

The Hestrin-Schramm (HS) is the most used medium in laboratory experiments (Costa et al. 2017; Cruz et al. 2024). However, BC production with this medium is generally achieved at low rate and yield and is not cost-effective for industrial production (Quijano et al. 2024). A considerable fraction of the glucose—the main carbon source in the traditional recipe of the HS media—is converted to gluconic acids (Gwon et al. 2019; L.-P. Liu et al. 2020; Montenegro-Silva et al. 2024). This by-product diverts the carbon flow away from the BC biosynthesis while significantly decreasing the culture's pH, resulting in slower growth and BC production (Gwon et al. 2019; L.-P. Liu et al. 2020; Montenegro-Silva et al. 2024). Other carbon sources and feedstocks have been employed, for instance, glycerol, fructose, sucrose, and mannitol (Mikkelsen et al. 2009; S. S. Wang et al. 2018; Zikmanis et al. 2021), with varying degrees of success in increasing BC production, and the results are highly strain dependent. Recently, Quijano et al. have published a systematic review that included "299 alternative substrates from 12 industries and 101 BC-producing strains" (Quijano et al. 2024). Most of the identified substrates are by-products from the food industry (Quijano et al. 2024). The use of cheaper carbon sources and feedstocks are one of the most sought strategies to increase the cost-effectiveness of BC production and enabling the economic viability of BC-based products (El-Gendi et al. 2022; Margarita et al. 2016).

A tight control of the environmental conditions is required for optimal BC production, with a significant influence of factors such as pH (Greser and Avcioglu 2022; Z. Li et al. 2021; Molina-Ramírez et al. 2017), temperature (Greser and Avcioglu 2022), dissolved oxygen levels (Drosos et al. 2024; Li et al. 2019; Żywicka et al. 2021), and nutrient availability (Akintunde et al. 2022; Żywicka et al. 2021). For instance, *Komagataeibacter* spp. thrive best at slightly acidic conditions (pH 4.0–6.0) and require an adequate amount of oxygen for robust BC production (Liu et al. 2020; Montenegro-Silva et al. 2024; Żywicka et al. 2021). Maintaining these strict conditions on an industrial scale presents a major challenge, requiring sophisticated monitoring and control systems (Singhania et al. 2022). In static cultures, where BC is produced at the air–liquid interface, the increase in pellicle thickness can gradually restrict cells in the upper region of the pellicle from accessing nutrients in the culture media, limiting bacterial growth and further cellulose synthesis (Hornung et al. 2006; Shavyrkina et al. 2021). Likewise, the bacterial growth and BC production are limited by the surface area of the recipient, which severely restricts the production rate. In addition, the biotechnological scale-up of static fermentation systems is complex (Cruz et al. 2024). Agitated cultures, on the other hand, offer improved oxygenation and nutrient distribution, but, while allowing higher production rates as compared to static culture, they come with their own set of challenges (Somerville and Proctor 2013; Wang et al. 2019). The constant mechanical forces in these systems disrupt the natural self-assembly of BC fibrils and alter the properties of BC (Krystynowicz et al. 2002; Wang et al. 2019). Additionally, agitation can induce shear stress, which may trigger the emergence of non-cellulose-producing mutants, further compromising production efficiency (Krystynowicz et al. 2002; J. Wang et al. 2019). Beyond oxygenation, the increasing viscosity of the BC-containing culture broths creates difficulties in mixing and mass transfer, requiring specialized reactor designs to maintain homogeneity (Lahiri et al. 2021; Putra et al. 2009). Conventional stirred-tank reactors, which are commonly used for industrial fermentation, often struggle with these issues due to the fibrous nature of BC (Krystynowicz et al. 2002; Lahiri et al. 2021; Singhania et al. 2022; Wang et al. 2019). Alternative reactor designs, such as air-lift bioreactors or rotating drum fermenters, have been explored to improve process efficiency, but these approaches introduce additional operational complexities and costs (Chao et al. 2000; Páez et al. 2024; Shoda and Sugano 2005). Moreover, the need for sterile conditions and controlled environmental parameters further complicates the scale-up process, making BC production an energy-intensive and economically challenging endeavor (Cruz et al. 2024; Singhania et al. 2022).

In conclusion, the scalability and economic feasibility of BC production remain challenging due to substrate costs, strict environmental requirements, and process limitations related to oxygenation and fermentation mode (Martirani-VonAbercron and Pacheco-Sánchez 2023; Quijano et al. 2024; Singhania et al. 2022). Optimizing culture conditions, including the choice of carbon sources and fermentation strategies, is crucial for improving BC yields while maintaining process efficiency (Quijano et al. 2024; Shoda and Sugano 2005; Singhania et al. 2022; Wang et al. 2019). Nevertheless, modifications of these parameters can directly influence the structural and mechanical properties of the final material (Krystynowicz et al. 2002). For instance, while certain alternative carbon sources may enhance BC yield, they can also affect fiber morphology, degree of polymerization, and crystallinity (MAS Keshk and Sameshima 2020; Mohammadkazemi et al. 2015). Similarly, shifting from static to agitated cultures may increase oxygenation and homogeneity but often leads to dispersed BC fibrils rather than structured pellicles, altering the material's characteristics (Chao et al. 2000; Lahiri et al. 2021; Páez et al. 2024; Singhania et al. 2022). Thus, an increase in BC production may come at the expense of its desirable physicochemical properties (Krystynowicz et al. 2002; Lahiri et al. 2021; Singhania et al. 2022; Wang et al. 2019). Addressing these trade-offs requires a balance between bioprocess engineering and material science approaches. Advances in strain engineering, metabolic optimization, and reactor design hold promise for overcoming current limitations, enabling more cost-effective and scalable BC production (Blanco Parte et al. 2020; Ryngajłło et al. 2020; Singhania et al. 2021).

These limitations exploited with conventional BC-producing strains highlight the need to explore alternative microbial sources, which may offer improved substrate flexibility, higher productivity, better genetic tractability, and enhanced process scalability, as discussed in the following sections.

3 Non-conventional Species in BC Production

Aside from the genus *Komagataeibacter*, *Acetobacter,* and *Gluconacetobacter* (among others), there are other promising species that can be further investigated and optimized to produce more BC. This section explores bacterial species beyond the above-mentioned ones that have demonstrated the ability to produce BC, with a focus on their unique metabolic properties and industrial potential.

3.1 Gram-Negative BC Producers

As mentioned, oxygen availability is one of the major limitations for BC production in most of the conventional species (Bi et al. 2014; Cruz et al. 2024; Hornung et al. 2006; Montenegro-Silva et al. 2024). Therefore, the ability to produce BC in anaerobic environments, or facultative anaerobic environments, may prove to be a reliable strategy in terms of spatial distribution and nutrient availability to overcome such limitation (Buldum et al. 2017; Ji et al. 2016; Morgan et al. 2014; Saxena and Brown 2005; Zimmer 2020). *Enterobacter* is a widely known genus that consists of mostly common gram-negative, facultatively anaerobic and non-spore-forming bacteria, known to form BC (Adeolu et al. 2016).

3.1.1 *Enterobacter* Genus

The *Enterobacter* sp. FY-07 seems to be one of the most efficient BC producers under anaerobic conditions (Gao et al. 2021; Ma et al. 2012) due to its nitrate respiration mechanism (Fig. 3.2) (Ji et al. 2016). Ji et al. (2016) investigated the cellulose biosynthesis mechanism of *Enterobacter* sp. FY-07 in relation to *Gluconacetobacter xylinum* ATCC 53582, through a comparative genomic analysis. Through genome sequencing, they identified three distinct cellulose synthase operons in FY-07 (bcs I, bcs II, and bcs III) and confirmed via gene knockout and functional reconstitution that bcs III is the essential operon for cellulose production. Gene arrangement and composition of bcs III were compared with the cellulose synthesis operon of *G. xylinum* ATCC 53582, finding that the core genes (bcsA, bcsB, bcsC, and bcsD) are structurally identical. However, the flanking sequences differ significantly, with FY-07 lacking key regulatory genes such as ccpAx and cmcAx found in *G. xylinum.* Despite these differences, the conserved operon structure and functional validation suggest that FY-07 uses a cellulose synthesis mechanism that is fundamentally like that of *G. xylinum* ATCC 53582.

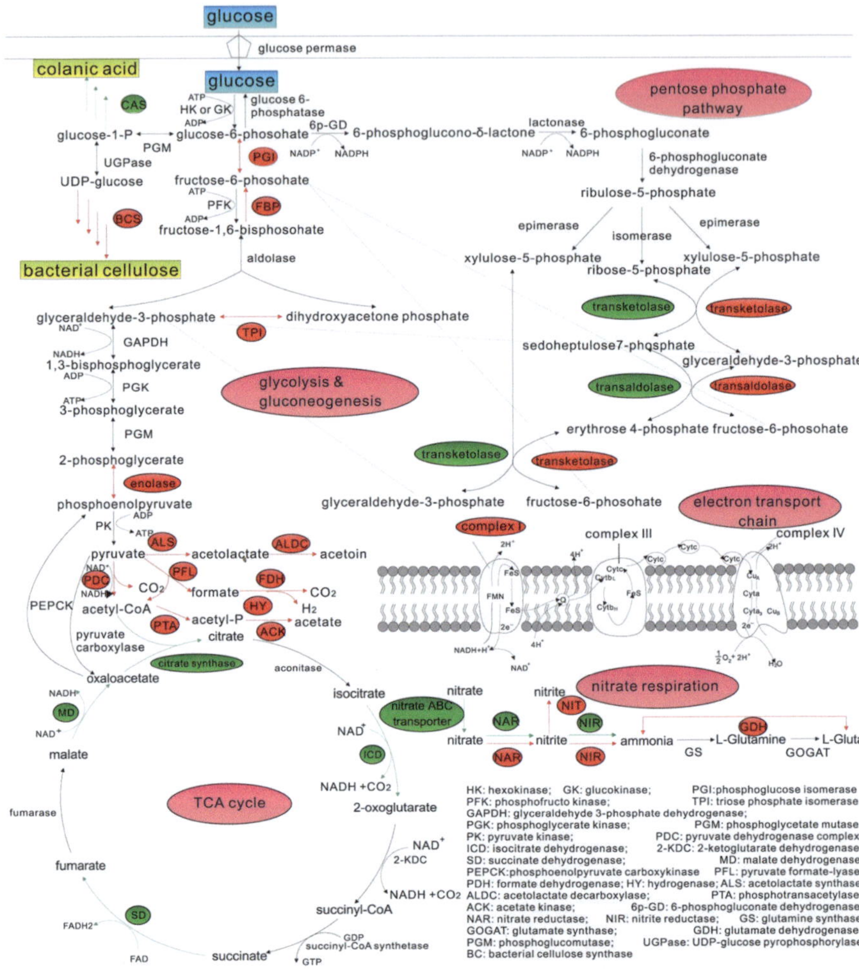

Fig. 3.2 Metabolic pathway for BC production in *Enterobacter* sp. FY-07 under both anaerobic/aerobic conditions. Green represents reactions that are downregulated in anaerobic conditions, whereas red represents upregulated in the same condition. Reproduced from Ji et al. (2016), under the terms of the Creative Commons Attribution 4.0 International License (CC BY 4.0). https://creativecommons.org/licenses/by/4.0/

Ji et al. (2016) performed a comparative kinetic study of this strain to evaluate its cellulose production capacity, cell growth, and substrate consumption under aerobic and anaerobic conditions. While the overall maximum cell density in anaerobic conditions was comparable to that of the aerobic one, the rate of growth during the exponential phase was marginally slower. Also, the transition to the stationary phase was somewhat delayed. A lower BC titer of 2.7 g/L was recorded in anaerobic conditions, compared to 3.8 g/L in aerobic conditions (Ji et al. 2016). This discrepancy can be attributed to several interrelated metabolic and genetic factors.

BC production requires ATP, which is essential for activating UDP-Glucose and synthesizing c-di-GMP. This energy requirement is mostly satisfied via the electron transport chain, with oxygen being the final acceptor under aerobic conditions (Cruz et al. 2024). However, under anaerobic conditions, *Enterobacter* sp. FY-07 seems to fulfil this energy requirement by shifting its energy metabolism to nitrate respiration (utilizing nitrate as an alternative electron acceptor). This metabolic flexibility allows the bacterium to generate sufficient ATP to support cellular growth and produce BC. However, ATP yield per glucose molecule is significantly lower under anaerobic metabolism, which limits the availability of UDP-glucose and c-di-GMP, both being crucial to cellulose biosynthesis (Ji et al. 2016).

Transcriptomic data from the work of Ji et al. showed that the UDP-glucose biosynthesis pathway and the bcsIII operon (which encodes the cellulose synthase complex) are expressed at similar levels under both aerobic and anaerobic conditions. However, the lower availability of ATP under anaerobic conditions likely restricts the rate of UDP-glucose formation and c-di-GMP signaling, thus limiting cellulose production (Morgan et al. 2014; Saxena and Brown 2005; Zimmer 2020). Also, under anaerobic conditions, the accumulation of acidic fermentation products (e.g., formic and acetic acid) may cause pH stress, potentially inhibiting enzyme activity and reducing cellular viability (Mishra et al. 2022; Sumini et al. 2024). The absence of ethanol and lactate fermentation pathways in *Enterobacter* sp. FY-07 also means that this bacterium cannot efficiently re-oxidize NADH, consequently affecting the redox balance and energy metabolism (Fei et al. 2023; Li et al. 2012; Montenegro-Silva et al. 2024).

Aside from *Enterobacter* sp. FY-07, *Enterobacter amnigenus* GH1 also proves to be a viable contender for BC production achieving a titer of 4.1 g/L in static culture using an improved medium containing fructose as the main carbon source (Hungund and Gupta 2010). Although its performance under anaerobic environment has not been evaluated, this species demonstrated high adaptability to different media compositions. Optimal BC production was achieved at a pH of 6.5 (slightly higher than *K. xylinus* strains) and at 28 °C (slightly lower than *K. xylinus* strains) (Hungund and Gupta 2010). Also, BC production was improved with the addition of metal ions (calcium, magnesium, and zinc), as observed in other studies (Hungund and Gupta 2010; Jia et al. 2004). This bacterium was also capable of using low-cost substrates such as molasses, starch hydrolysates, sugarcane, and fruit juices (orange, pineapple, and pomegranate) (Hungund and Gupta 2010), thus improving the potential for better cost-effective BC production at an industrial level, despite a lower yield than conventional species.

Regarding the use of low-cost substrates, a recent study by Al-Hasabe et al. (2024) demonstrated that *Enterobacter chuandaensis* can also produce BC using fruit extracts from *Phoenix dactylifera* (palm tree species) and *Musa acuminata* (banana plant species) extracts as main carbon sources. The study reported BC titers of 2.8 and 2.5 g/L after 7 days of aerobic static culture, for palm tree and banana plant extracts, respectively (Al-Hasabe et al. 2024), being lower than top conventional species. These authors performed the whole-genome sequencing of this species. The study showed the presence of a complete bcsABZC operon as well as multiple accessories and regulatory genes associated with BC biosynthesis (bcsA, bcsB, bcsC, bcsD, bcsE, bcsF, bcsG, bcsQ, bcsR, bcsZ, acsA, and csgD). The functional annotation of the whole genome revealed a strong representation of genes involved in carbohydrate transport, metabolism, and energy production. However, all fermentations were done under aerobic static conditions; moreover, the genomic analysis did not explore alternative energy-generating pathways or regulatory responses to environmental stress. While the genome sequencing (the first report of cellulose production by this species) of this species represented an important milestone towards the better understanding of this species genome, the *E.* sp. FY-07 dataset from Ji et al. (2016) work is a more advanced model for the purpose of energy metabolism-linked regulation of cellulose production research.

3.1.2 *Proteus* Genus

Patent WO2007062502A1 (Delaquis 2006) describes BC production from two gram-negative facultative anaerobes *Proteus myxofaciens* strains (IDAC 071005-01, isolated from carrot surfaces, and ATCC 19692, isolated from gypsy moth larvae). Under static culture, cellulose formation occurred throughout the bulk of the vessels, suggesting oxygen limitation wasn't an obstacle. Also, no agitation or aeration was required, which might imply anaerobic or microaerophilic settings, especially since static setups can naturally create oxygen gradients. The highest BC titer was obtained using *P. myxofaciens* IDAC 071005-01, static culture, reaching a modest value of about 0.3 g/L, when compared with conventional strains. The document does not explicitly state that the static culture was done under aerobic conditions. Also, no quantitative titer data for anaerobic conditions were reported.

Nevertheless, the unique production under anaerobic conditions offers an interesting opportunity to produce a unique kind of BC membranes. While in static culture, the aerobic production occurs layer-by-layer at the air–liquid interface, generating a membrane with anisotropic morphology and mechanical properties (the fibers are organized mostly in the horizontal plane), under anaerobic conditions it may be expected that the fibers will distribute isotropically. Also, the porosity may be rather distinct from the commonly obtained BC membranes, as the cellulose is biosynthesized in the bulk of the liquid medium, potentially offering novel technological possibilities.

3.2 *Gram-Positive BC Producers*

3.2.1 *Rhodococcus* Genus

Most of the conventional species for BC production are known to be gram-negative (Moradi et al. 2021), the *Rhodococcus* sp. MI2, an obligate aerobe, is one of the first to be reported with high BC production efficiency while being a gram-positive aerobic bacterium (Tanskul et al. 2013). Tanskul et al. reported a BC titer of 3.91 g/L (and 7.4 g/L in 14 days) with *Rhodococcus* sp. MI2 under static culture using a modified HS medium (containing sucrose as a carbon source and calcium carbonate), surpassing that of *Acetobacter xylinum* 988 (1.17 g/L in 6 days) under the same conditions (Tanskul et al. 2013). In a later study, using a medium supplemented with vegetable extracts from Cha-poo (*Piper sarmentosum*), the BC titer reached 10 g/L after 14 days of fermentation (Tanskul et al. 2018), thus proving to be a viable alternative to more conventional strains.

3.2.2 *Bacillus* Genus

Following the first publications of gram-positive genera producing BC, the marine-derived *Bacillus velezensis* SMR, a facultative aerobe, also proved to be a very efficient BC producer, achieving 5.2 g/L after 10 days in static culture with HS medium, surpassing *Gluconacetobacter xylinus* ATCC 10245 (2.6 g/L) in the same conditions (Abouelkheir et al. 2020). Using Ulva seaweed extract as an alternative growth medium further improved the BC titer to 13 g/L, within the same time, thus making the strain promising for applications using industrial and agricultural wastes (Abouelkheir et al. 2020).

A more recent study regarding *Bacillus tequilensis* SEE-12 using cantaloupe juice as the main carbon source and small-scale static cultures, managed to reach a BC titer of 22.8 g/L in 6 days (El-Naggar et al. 2022), thus, surpassing the previous known gram-positive BC producers, such as, *Bacillus licheniformis* ZBT2 that reached 7.19 g/L in 8 days using glucose as the main carbon source (Bagewadi et al. 2020).

3.2.3 *Leifsonia* Genus

Another genus that belongs to the group of gram-positive bacteria able to produce BC is the *Leifsonia* (Rastogi and Banerjee 2019; Velmurugan et al. 2015). Despite the few reports of BC production, a *Leifsonia* sp. isolated from low-cost fruit wastes achieved a titer of 5.97 g/L within 7 days of static culture in optimal conditions, using glucose as the main carbon source (Rastogi and Banerjee 2019). Furthermore, the low degree of crystallinity and high porosity in the BC produced also suggest that it could be applied in food packaging, biomedical industries, and cosmetics, among others (Rastogi and Banerjee 2019).

Another *Leifsonia* species was isolated from the earthworm, *Eisenia fetida*, named *Leifsonia* sp. CBNU-EW3, reached 4 g/L BC titer in a glucose-rich medium under an incubation period of 15 days (Velmurugan et al. 2015). The strain also produces a web-like structure with large pores and cellulose fibril aggregation, like other gram-positive BC producers (Rastogi and Banerjee 2019; Tanskul et al. 2013; Velmurugan et al. 2015).

Table 3.1 Comparison between *Komagataeibacter xylinus* and other non-conventional strains in BC titer

Species	Strain name	Carbon source	Day	Titer (g/L)	Reference
Komagataeibacter xylinus	BPR 2001	Molasses	9	7.5	(Rodrigues et al. 2019)
Komagataeibacter rhaeticus	K3	Eucalyptus biomass hydrolysate	10	5.46	(Jacek et al. 2021)
Rhodococcus sp.	MI 2	Sweet potato extract	14	10.42	(Tanskul et al. 2018)
	MI 2	Coconut juice and sucrose	14	2.54	(Tanskul et al. 2013)
Enterobacter amnigenus	GH-1	Orange juice	14	3.74	(Hungund and Gupta 2010)
Leifsonia sp.	CBNU-EW3	Glucose	7	5.97	(Velmurugan et al. 2015)
Bacillus velezensis	SMR	Dried Ulva sp. (algae)	10	13	(Abouelkheir et al. 2020)
Beijerinckia fluminensis	WAUPM53	Sago hydrolysate	14	0.47	(Voon et al. 2019)

3.3 Benchmarking Bacterial Strains for BC Production

While *Komagataeibacter xylinus* continues to be the standard for bacterial cellulose production due to its high yield (Dourado et al. 2016) and extensive characterization, several non-conventional genera such as *Enterobacter, Rhodococcus, Bacillus*, and *Leifsonia* are emerging as promising alternatives (Abouelkheir et al. 2020; Ji et al. 2016; Rastogi and Banerjee 2019; Tanskul et al. 2018). These species may show advantages including the ability to produce cellulose under anaerobic or facultative anaerobic conditions, utilization of diverse and low-cost substrates, and in some cases, competitive or even superior yields under certain conditions (Abouelkheir et al. 2020; Gao et al. 2021; Rastogi and Banerjee 2019; Tanskul et al. 2018). However, challenges such as lower consistency, high condition requirements, and potential biosafety concerns, particularly with bacteria like *Enterobacter* (González 2022; Zogaj et al. 2003) still limit their industrial adoption. Expanding our understanding of these alternative species could open new avenues for more flexible, cost-effective, and more importantly, application-specific bacterial cellulose production (Table 3.1).

3.4 Constraints of Non-conventional BC Producers

The integration of non-conventional BC producing strains into research or industrial settings poses significant challenges that must be carefully evaluated and overcome. The *Enterobacter* genus includes several species that are recognized as opportunistic pathogens, with well-documented associations to human infections, including gastrointestinal, nosocomial, and opportunistic conditions (González 2022; Zogaj et al. 2003). A case in point is *Enterobacter sakazakii* (currently reclassified as *Cronobacter sakazakii*), which has been implicated in neonatal meningitis and sepsis (Zogaj et al. 2003) and is now routinely monitored in infant formula production, through strict handling protocols[1] from international guidelines (e.g., WHO, FAO). These pathogenic associations present considerable biosafety concerns, especially when BC products are intended for applications involving direct or indirect human contact, such as food packaging, wound dressings, or implantable medical devices. From a regulatory perspective, these strains are not currently classified as GRAS[2] by the U.S. Food and Drug Administration (FDA) and are also excluded from the Qualified Presumption of Safety (QPS)[3] list maintained by the European Food Safety Authority (EFSA).[4] Their use in large-scale bioprocesses would first require extensive regulatory scrutiny, especially regarding containment, endotoxin levels, downstream purification protocols, and many other safety assessment demonstrations and procedures. Besides biosafety, working with these bacteria species poses significant challenges given their low genetic amenability, making it difficult to modify cellulose production pathways, optimize yields, tune material properties, or implement biocontainment strategies.

To address these specific limitations, possible strategies could be the genetic modification of the pathogenic species to attenuate/eliminate their virulence (for instance, through targeted disruption of virulence genes) or the use of contained bioreactor systems. Further research into lesser-known, non-pathogenic species occupying the same metabolic niche may provide safer production platforms, while still retaining the distinctive features that make non-conventional species valuable.

4 Discovering Novel BC Producers via Genomic Analysis

Besides the species and strains mentioned above, there is still the wide margin of possibility that there are unknown species that can also produce BC. This section explores the use of bioinformatics tools to identify potential new producers. Specifically, sequence similarity searches were carried out using Basic Local Alignment Search Tool (Altschul et al. 1990) for nucleotides (BLASTn[5]) to compare the cellulose synthase operon (bcsA–D) from *Komagataeibacter xylinus* using the core nucleotide database (core_nt) and a Basic Local Alignment Search Tool for proteins (BLASTp) using the BcsA protein of *Salmonella typhi* against the NCBI

[1] https://www.fao.org/food/food-safety-quality/a-z-index/enterobacter/en/

[2] *Enterobacter* species do not appear in the FDA GRAS Inventory for food or biomedical use

[3] https://www.efsa.europa.eu/en/topics/topic/qualified-presumption-safety-qps

[4] https://efsa.onlinelibrary.wiley.com/doi/pdf/10.2903/j.efsa.2025.9510

[5] BLAST: Basic Local Alignment Search Tool

Table 3.2 Top 100 sequences from the BLAST results of the *Komagataeibacter xylinus* operon excluding all acetic acid bacteria species (Altschul et al. 1990)

Species	Max score	Total score	E-value	% Identity
Roseomonas haemaphysalidis	287	287	8.00E-71	74.38%
Roseomonas gilardii	278	278	5.00E-68	71.84%
Roseomonas mucosa	267	267	1.00E-64	71.66%
Reyranella sp.	272	272	2.00E-66	73.99%
Pseudoxanthomonas suwonensis 11-1	243	243	2.00E-57	73.31%
Erwinia aphidicola	209	209	2.00E-47	76.02%
Cupriavidus necator	211	211	5.00E-48	73.87%

non-redundant protein database[6] (accessed July 2025) (Wong et al. 1990). The BcsA protein from *S. typhi* was selected due to its well-documented role in cellulose production, as well as its conserved domains shared with other cellulose-producing bacteria (Gill et al. 2018; Upadhyay et al. 2023). *Komagataeibacter xylinus* was used as a well-referenced high-yield BC producer (Sumini et al. 2024; Wong et al. 1990). The top hits were selected based on the lowest E-value (Expect value) to the highest, first 100 sequences. This metric estimates the expected number of random alignments with a particular score or higher in a given database, allowing a better understanding of whether the result is by chance (higher E-value) or by a true biological relationship between the sequences (lower E-value). The results were filtered to exclude species from known BC-producing families, such as *Acetobacteraceae* and *Enterobacteriaceae*, to focus on novel genera. For the phylogenetic tree generation, phyloT (PhyloT 2023) was used based on the given species from the BLAST analysis using the NCBI taxonomy database (Schoch et al. 2020). After the tree generation, no nodes were collapsed, guaranteeing the full visualization of distance between the species observed. An initial BLAST analysis of the *Komagataeibacter xylinus* cellulose synthase operon revealed several non-conventional bacterial species with moderate to high sequence similarity to core cellulose synthase genes, suggesting the potential for BC production. As observed in Table 3.2, species like *Roseomonas haemaphysalidis* showed 74.38% identity to one of such bcs operon genes; this metric refers to the percentage of identical matches between the query and the subject over the aligned region, allowing a numerical understanding of the similarity of the sequence relative to the query. *Reyranella* and *Pseudoxanthomonas* exhibited high sequence similarity, as indicated by elevated percent identity values and low E-values (expectation values). These results suggest that, aside from the well-known BC producers, homologous bcs-like sequences may be present in many other bacteria. However, sequence similarity alone does not confirm functional cellulose synthase activity. As such, these results are to be considered as in silico predictions of the BC production potential of these bacteria, in terms of operon completeness. Therefore, regulatory elements such as c-di-GMP signaling and expression under physiological conditions must be investigated experimentally.

[6] https://blast.ncbi.nlm.nih.gov/Blast.cgi?PAGE=Proteins

Figure 3.3 is a phylogenetic tree generated through phyloT with the species presented in Table 3.2, identified through BLAST analysis of the *Komagataeibacter xylinus* cellulose synthase operon as the query. The *Roseomonas* and *Reyranella* genera appear in well-supported clades, clustering closely with known cellulose-producing species. This suggests that there is some sequence conservation in the regions of alignment of the operons. However, as previously observed, while sequence similarity provides insight into the genetic diversity and evolutionary relatedness of these candidates, it is, per se, insufficient to confirm the cellulose-producing capability of these species. Experimental validation (e.g., by transcriptomic profiling or functional gene expression studies) is needed to determine whether these species are indeed capable of biosynthesizing cellulose.

Comparing the data from Fig. 3.4 (with all *Enterobacteriaceae* species omitted to emphasize distantly related taxa) with that of Fig. 3.3, the dataset shows a broader taxonomic range, as it includes bacteria belonging to other classes and families of bacteria, such as *Pseudomonas, Acinetobacter, Zobellella*, and *Oceanisphaera* genera. Notably, despite significant phylogenetic divergence, many of these species display high sequence identity and scores with the reference BcsA protein (Table 3.3), suggesting that conserved cellulose synthase-like domains are distributed over a wider phylogenetic spectrum than previously appreciated.

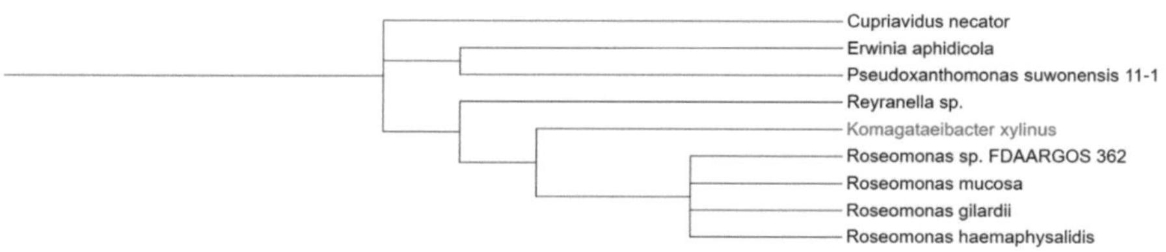

Fig. 3.3 Phylogenetic tree of the BLAST results for the *Komagataeibacter xylinus* cellulose synthase operon. In green the reference for the query. Generated from https://phylot.biobyte.de

Table 3.3 Top 100 sequences from the BLAST results of the *Salmonella typhi* BscA protein from the cellulose synthase complex, excluding all species from the *Enterobacteriaceae* family (Altschul et al. 1990)

Species	Max score	Total score	E-value	% Identity
Pseudomonas aeruginosa	1590	1590	0	86.80%
Mesorhizobium sp.	1586	1586	0	86.45%
Pseudomonadota bacterium	1512	1512	0	85.28%
Acinetobacter baumannii	1236	1236	0	69.28%
Zobellella aerophila	1118	1118	0	63.22%
Candidatus Oceanisphaera merdipullorum	1116	1116	0	61.21%
Zobellella maritima	1108	1108	0	64.94%
Oceanisphaera arctica	1098	1098	0	62.95%
Aeromonas finlandensis	1057	1057	0	60.21%
Martelella alba	1051	1051	0	61.01%
Aeromonas rivipollensis	1048	1048	0	60.42%
Oceanisphaera psychrotolerans	1033	1033	0	69.73%

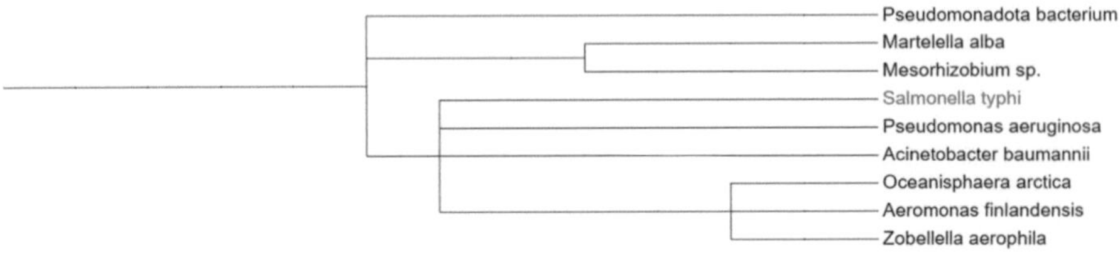

Fig. 3.4 Phylogenetic tree of the BLAST results for the *Salmonella typhi* BscA protein from the cellulose synthase complex, excluding all species from the Enterobacteriaceae family. In green the reference for the query. Generated from https://phylot.biobyte.de

5 Conclusion and Prospects

Despite *Komagataeibacter, Acetobacter*, and *Gluconacetobacter* species dominating the academic and potential industrial use of the produced BC, due to their high yields and well-characterized pathways (Gullo et al. 2018; Jacek et al. 2019), exploration of non-conventional genera may reveal promising alternatives that could overcome current production limitations such as slow growth rates, oxygen dependency, and substrate specificity (Anguluri et al. 2022; Montenegro-Silva et al. 2024; Quijano et al. 2024). In addition, particularly in the case of the facultative anaerobic producers, it may be possible to obtain BC membranes with novel properties.

Recent studies have identified *Enterobacter* species, such as *Enterobacter* sp. FY-07, as capable of efficient BC production under both aerobic and anaerobic conditions, potentially reducing production costs and solving oxygen dependency (Ji et al. 2016). *Rhodococcus* sp. MI 2 has demonstrated an ability to synthesize cellulose under a broader range of culture conditions, offering flexibility for industrial applications (Tanskul et al. 2013, 2018). Additionally, *Bacillus* species, like *Bacillus velezensis*, have been reported to produce BC with properties that can be applied to anti-bacterial gels, as well as other medicine applications (Abouelkheir et al. 2020). Other promising bacteria include *Leifsonia* sp., a genus traditionally associated with plant environments, recently shown to harbor cellulose-producing strains with potential for biotechnological exploitation (Rastogi and Banerjee 2019). Altogether, these non-conventional BC producers not only diversify the phylogenetic background of BC synthesis but also provide functional advantages that could surpass traditional species in titer, yield or productivity, environmental tolerance, and functionalization potential, thereby broadening the industrial applicability of bacterial cellulose.

Furthermore, several presented studies regarding most of the non-conventional strains for BC production are recent, thus, to improve the current state of BC industrial applications, more information regarding genetics, as well as optimization processes are required in future studies.

The discovery of new BC-producing strains undoubtedly presents unique industrial opportunities. However, the biosafety concerns, especially for species belonging to the *Enterobacter* genus, must be addressed. Existing regulatory constraints, including the absence of GRAS or QPS status, highlight the need for prudent strain selection, potential genetic attenuation approaches, and containment policies. Also, comparative genomics and sequence alignment software provide a potential route to increase the inventory of BC producing species. In silico predictions relying on bcs operon similarity should now be accompanied by transcriptomic verification and heterologous expression studies to confirm functionality in new hosts.

References

Abouelkheir SS, Kamara MS, Atia SM, Amer SA, Youssef MI, Abdelkawy RS, Khattab SN, Sabry SA (2020) Novel research on nanocellulose production by a marine Bacillus velezensis strain SMR: a comparative study. Sci Rep 10(1):1–14. https://doi.org/10.1038/s41598-020-70857-7

Adeolu M, Alnajar S, Naushad S, Gupta RS (2016) Genome-based phylogeny and taxonomy of the 'Enterobacteriales': proposal for Enterobacterales ord. nov. divided into the families Enterobacteriaceae, Erwiniaceae fam. nov., Pectobacteriaceae fam. nov., Yersiniaceae fam. nov., Hafniaceae fam. nov., Morganellaceae fam. nov., and Budviciaceae fam. nov. Int J Syst Evol Microbiol 66(12):5575–5599. https://doi.org/10.1099/IJSEM.0.001485

Akintunde MO, Adebayo-Tayo BC, Ishola MM, Zamani A, Horváth IS (2022) Bacterial cellulose production from agricultural residues by two *Komagataeibacter* sp. strains. Bioengineered 13(4):10010–10025. https://doi.org/10.1080/21655979.2022.2062970

Al-Hasabe ASH, Razis A, Bin AF, Baharum NAB, Yu CY, Mat Isa N (2024) Production and characterization of bacterial cellulose synthesized by *Enterobacter chuandaensis* strain AEC using *Phoenix dactylifera* and *Musa acuminata*. Arch Microbiol 206(11):1–19. https://doi.org/10.1007/S00203-024-04182-2

Altschul SF, Gish W, Miller W, Myers EW, Lipman DJ (1990) Basic local alignment search tool. J Mol Biol 215(3):403–410. https://doi.org/10.1016/S0022-2836(05)80360-2

Anguluri K, La China S, Brugnoli M, Cassanelli S, Gullo M (2022) Better under stress: improving bacterial cellulose production by *Komagataeibacter xylinus* K2G30 (UMCC 2756) using adaptive laboratory evolution. Front Microbiol 13. https://doi.org/10.3389/fmicb.2022.994097

Bagewadi ZK, Bhavikatti JS, Muddapur UM, Yaraguppi DA, Mulla SI (2020) Statistical optimization and characterization of bacterial cellulose produced by isolated thermophilic *Bacillus licheniformis* strain ZBT2. Carbohydr Res 491:107979. https://doi.org/10.1016/J.CARRES.2020.107979

Bi J-C, Liu S-X, Li C-F, Li J, Liu L-X, Deng J, Yang Y-C (2014) Morphology and structure characterization of bacterial celluloses produced by different strains in agitated culture. J Appl Microbiol 117(5):1305–1311. https://doi.org/10.1111/jam.12619

Bielecki S, Krystynowicz A, Turkiewicz M, Kalinowska H (2002) Bacterial Cellulose. Biopolym Online. https://doi.org/10.1002/3527600035.BPOL5003

Blanco Parte FG, Santoso SP, Chou CC, Verma V, Wang HT, Ismadji S, Cheng KC (2020) Current progress on the production, modification, and applications of bacterial cellulose. Crit Rev Biotechnol 40(3):397–414. https://doi.org/10.1080/07388551.2020.1713721

Buldum G, Bismarck A, Mantalaris A (2017) Recombinant biosynthesis of bacterial cellulose in genetically modified *Escherichia coli*. Bioprocess Biosyst Eng 41(2):265. https://doi.org/10.1007/S00449-017-1864-1

Chao Y, Ishida T, Sugano Y, Shoda M (2000) Bacterial cellulose production by *Acetobacter xylinum* in a 50-L internal-loop airlift reactor. Biotechnol Bioeng 68(3):345–352. https://doi.org/10.1002/(SICI)1097-0290(20000505)68:3<345::AID-BIT13>3.0.CO;2-M

Costa AFS, Almeida FCG, Vinhas GM, Sarubbo LA (2017) Production of bacterial cellulose by *Gluconacetobacter hansenii* using corn steep liquor as nutrient sources. Front Microbiol 8(OCT). https://doi.org/10.3389/fmicb.2017.02027

Cruz MA, Flor-Unda O, Avila A, Garcia MD, Cerda-Mejía L (2024) Advances in bacterial cellulose production: a scoping review. Coatings 14(11). https://doi.org/10.3390/coatings14111401

Dang X, Li N, Yu Z, Ji X, Yang M, Wang X (2024) Advances in the preparation and application of cellulose-based antimicrobial materials: A review. Carbohydr Polym 342:122385. https://doi.org/10.1016/J.CARBPOL.2024.122385

Delaquis P (2006) Method of bacterial cellulose production using proteus strains. WO2007062502A1 WIPO (PCT)

Dourado F, Fontão A, Leal M, Cristina Rodrigues A, Gama M (2016) Process modeling and techno-economic evaluation of an industrial bacterial nanocellulose fermentation process. In: Bacterial nanocellulose: from biotechnology to bio-economy. Elsevier, pp 199–214. https://doi.org/10.1016/B978-0-444-63458-0.00012-3

Drosos A, Kordopati GG, Anastasopoulos C, Zafeiropoulos J, Koutinas AA, Kanellaki M (2024) Comparative study and characterization of water-treated bacterial cellulose produced by solid or liquid inoculum of *Komagateibacter sucrofermentans*. Cellulose 31(9):5545–5573. https://doi.org/10.1007/s10570-024-05919-4

El-Gendi H, Taha TH, Ray JB, Saleh AK (2022) Recent advances in bacterial cellulose: a low-cost effective production media, optimization strategies and applications. Cellulose 29(14):7495–7533. https://doi.org/10.1007/s10570-022-04697-1

El-Naggar NEA, El-Malkey SE, Abu-Saied MA, Mohammed ABA (2022) Exploration of a novel and efficient source for production of bacterial nanocellulose, bioprocess optimization and characterization. Sci Rep 12(1):1–22. https://doi.org/10.1038/s41598-022-22240-x

Fang L, Catchmark JM (2014) Structure characterization of native cellulose during dehydration and rehydration. Cellulose 21(6):3951–3963. https://doi.org/10.1007/S10570-014-0435-8

Fei S, Yang X, Xu W, Zhang J, Li J, Chen H, Lin X, Liu S, Li C (2023) Insights into proteomics reveal mechanisms of ethanol-enhanced bacterial cellulose biosynthesis by *Komagataeibacter nataicola*. Fermentation 9(6):575. https://doi.org/10.3390/FERMENTATION9060575/S1

Gao G, Liao Z, Cao Y, Zhang Y, Zhang Y, Wu M, Li G, Ma T (2021) Highly efficient production of bacterial cellulose from corn stover total hydrolysate by *Enterobacter* sp. FY-07. Bioresour Technol 341:125781. https://doi.org/10.1016/J.BIORTECH.2021.125781

Gill MA, Rafique MW, Manan T, Slaeem S, Römling U, Matin A, Ahmad I (2018) The cellulose synthase BcsA plays a role in interactions of *Salmonella typhimurium* with *Acanthamoeba castellanii* genotype T4. Parasitol Res 117(7):2283–2289. https://doi.org/10.1007/S00436-018-5917-4/METRICS

González LV (2022) Human pathogenic Enterobacterales. Encycl Infect Immun 1:628–636. https://doi.org/10.1016/B978-0-12-818731-9.00160-9

Greser AB, Avcioglu NH (2022) Optimization and physicochemical characterization of bacterial cellulose by *Komagataeibacter nataicola* and *Komagataeibacter maltaceti* strains isolated from grape, thorn apple and apple vinegars. Arch Microbiol 204(8). https://doi.org/10.1007/s00203-022-03083-6

Gullo M, La China S, Falcone PM, Giudici P (2018) Biotechnological production of cellulose by acetic acid bacteria: current state and perspectives. Appl Microbiol Biotechnol 102(16):6885–6898. https://doi.org/10.1007/S00253-018-9164-5

Gwon H, Park K, Chung SC, Kim RH, Kang JK, Ji SM, Kim NJ, Lee S, Ku JH, Do EC, Park S, Kim M, Shim WY, Rhee HS, Kim JY, Kim J, Kim TY, Yamaguchi Y, Iwamuro R, Saito S, Kim G, Jung I-S, Park H, Lee C, Lee S, Jeon WS, Jang WD, Kim HU, Lee SY, Im D, Doo S-G, Lee SY, Lee HC, Park JH (2019) A safe and sustainable bacterial cellulose nanofiber separator for lithium rechargeable batteries. Proc Natl Acad Sci USA 116(39):19288–19293. https://doi.org/10.1073/pnas.1905527116

Hornung M, Ludwig M, Gerrard AM, Schmauder HP (2006) Optimizing the production of bacterial cellulose in surface culture: evaluation of substrate mass transfer influences on the bioreaction (Part 1). Eng Life Sci 6(6):537–545. https://doi.org/10.1002/elsc.200620162

Hungund BS, Gupta SG (2010) Production of bacterial cellulose from Enterobacter amnigenus GH-1 isolated from rotten apple. World J Microbiol Biotechnol 26(10):1823–1828. https://doi.org/10.1007/S11274-010-0363-1/TABLES/5

Idalia V-MN, Bernardo F (2017) *Escherichia coli* as a model organism and its application in biotechnology. In: *Escherichia coli* - recent advances on physiology, pathogenesis and biotechnological applications. IntechOpen. https://doi.org/10.5772/67306

Iguchi M, Yamanaka S, Budhiono A (2000) Bacterial cellulose - a masterpiece of nature's arts. J Mater Sci 35(2):261–270. https://doi.org/10.1023/A:1004775229149/METRICS

Jacek P, da Silva FAGS, Dourado F, Bielecki S, Gama M (2021) Optimization and characterization of bacterial nanocellulose produced by *Komagataeibacter rhaeticus* K3. Carbohydr Polymer Technol Appl 2:100022. https://doi.org/10.1016/J.CARPTA.2020.100022

Jacek P, Dourado F, Gama M, Bielecki S (2019) Molecular aspects of bacterial nanocellulose biosynthesis. Microb Biotechnol 12(4):633–649. https://doi.org/10.1111/1751-7915.13386

Ji K, Wang W, Zeng B, Chen S, Zhao Q, Chen Y, Li G, Ma T (2016) Bacterial cellulose synthesis mechanism of facultative anaerobe *Enterobacter* sp. FY-07. Sci Rep 6(1):1–12. https://doi.org/10.1038/srep21863

Jia S, Ou H, Chen G, Choi DB, Cho KA, Okabe M, Cha WS (2004) Cellulose production from *Gluconobacter oxydans* TQ-B2. Biotechnol Bioprocess Eng 9(3):166–170. https://doi.org/10.1007/BF02942287/METRICS

Keshk SM (2014) Bacterial cellulose production and its industrial applications. J Bioprocess Biotech 04(02). https://doi.org/10.4172/2155-9821.1000150

Keshk SMAS, Sameshima K (2020) Evaluation of different carbon sources for bacterial cellulose production. Int J Biochem Biotechnol 9(2) www.internationalscholarsjournals.org

Krystynowicz A, Czaja W, Wiktorowska-Jezierska A, Gonçalves-Miśkiewicz M, Turkiewicz M, Bielecki S (2002) Factors affecting the yield and properties of bacterial cellulose. J Ind Microbiol Biotechnol 29(4):189–195. https://doi.org/10.1038/sj.jim.7000303

Lahiri D, Nag M, Dutta B, Dey A, Sarkar T, Pati S, Edinur HA, Kari ZA, Noor NHM, Ray RR (2021) Bacterial cellulose: production, characterization and application as antimicrobial agent. Int J Mol Sci 22(23). https://doi.org/10.3390/ijms222312984

Li J, Chen G, Zhang R, Wu H, Zeng W, Liang Z (2019) Production of high crystallinity type-I cellulose from *Komagataeibacter hansenii* JR-02 isolated from Kombucha tea. Biotechnol Appl Biochem 66(1):108–118. https://doi.org/10.1002/bab.1703

Li Y, Tian C, Tian H, Zhang J, He X, Ping W, Lei H (2012) Improvement of bacterial cellulose production by manipulating the metabolic pathways in which ethanol and sodium citrate involved. Appl Microbiol Biotechnol 96(6):1479–1487. https://doi.org/10.1007/S00253-012-4242-6/METRICS

Li Z, Chen SQ, Cao X, Li L, Zhu J, Yu H (2021) Effect of ph buffer and carbon metabolism on the yield and mechanical properties of bacterial cellulose produced by *Komagataeibacter hansenii* atcc 53582. J Microbiol Biotechnol 31(3):429–438. https://doi.org/10.4014/JMB.2010.10054

Liu LP, Yang X, Zhao XJ, Zhang KY, Li WC, Xie YY, Jia SR, Zhong C (2020) A lambda red and FLP/FRT-mediated site-specific recombination system in *Komagataeibacter xylinus* and its application to enhance the productivity of bacterial cellulose. ACS Synth Biol 9(11):3171–3180. https://doi.org/10.1021/acssynbio.0c00450

Ma T, Ji K, Wang W, Wang J, Li Z, Ran H, Liu B, Li G (2012) Cellulose synthesized by *Enterobacter* sp. FY-07 under aerobic and anaerobic conditions. Bioresour Technol 126:18–23. https://doi.org/10.1016/J.BIORTECH.2012.09.040

Malcı K, Li IS, Kisseroudis N, Ellis T (2024) Modulating microbial materials - engineering bacterial cellulose with synthetic biology. ACS Synth Biol 13(12):3857–3875. https://doi.org/10.1021/acssynbio.4c00615

Margarita A, Gallegos A, Carrera SH, Parra R, Keshavarz T, Iqbal HMN (2016) Bacterial cellulose. BioResources 11(2)

Martirani-VonAbercron SM, Pacheco-Sánchez D (2023) Bacterial cellulose: a highly versatile nanomaterial. Microb Biotechnol 16(6):1174–1178. https://doi.org/10.1111/1751-7915.14243

Mikkelsen D, Flanagan BM, Dykes GA, Gidley MJ (2009) Influence of different carbon sources on bacterial cellulose production by *Gluconacetobacter xylinus* strain ATCC 53524. J Appl Microbiol 107(2):576–583. https://doi.org/10.1111/j.1365-2672.2009.04226.x

Mishra S, Singh PK, Pattnaik R, Kumar S, Ojha SK, Srichandan H, Parhi PK, Jyothi RK, Sarangi PK (2022) Biochemistry, synthesis, and applications of bacterial cellulose: a review. Front Bioeng Biotechnol 10:780409. https://doi.org/10.3389/FBIOE.2022.780409/BIBTEX

Mohammadkazemi F, Azin M, Ashori A (2015) Production of bacterial cellulose using different carbon sources and culture media. Carbohydr Polym 117:518–523. https://doi.org/10.1016/j.carbpol.2014.10.008

Molina-Ramírez C, Castro M, Osorio M, Torres-Taborda M, Gómez B, Zuluaga R, Gómez C, Gañán P, Rojas OJ, Castro C (2017) Effect of different carbon sources on bacterial nanocellulose production and structure using the low pH resistant strain *Komagataeibacter medellinensis*. Materials 10(6). https://doi.org/10.3390/ma10060639

Montenegro-Silva P, Ellis T, Dourado F, Gama M, Domingues L (2024) Enhanced bacterial cellulose production in *Komagataeibacter sucrofermentans*: impact of different PQQ-dependent dehydrogenase knockouts and ethanol supplementation. Biotechnol Biofuels Bioprod 17(1). https://doi.org/10.1186/s13068-024-02482-9

Moradi M, Jacek P, Farhangfar A, Guimarães JT, Forough M (2021) The role of genetic manipulation and in situ modifications on production of bacterial nanocellulose: a review. Int J Biol Macromol 183:635–650. https://doi.org/10.1016/J.IJBIOMAC.2021.04.173

Morgan JLW, McNamara JT, Zimmer J (2014) Mechanism of activation of bacterial cellulose synthase by cyclic di-GMP. Nat Struct Mol Biol 21(5):489–496. https://doi.org/10.1038/NSMB.2803

Morgan JLW, Strumillo J, Zimmer J (2013) Crystallographic snapshot of cellulose synthesis and membrane translocation. Nature 493(7431):181–186. https://doi.org/10.1038/NATURE11744

Oesef K, Cranston ED, Abdin Y (2024) Current advances in processing and modification of cellulose nanofibrils for high-performance composite applications. Mater Des 247:113417. https://doi.org/10.1016/J.MATDES.2024.113417

Páez MA, Casa-Villegas M, Aldas M, Luna M, Cabrera-Valle D, López O, Fernández D, Cruz MA, Flor-Unda O, García MD, Cerda-Mejía L (2024) Insights into agitated bacterial cellulose production with microbial consortia and agro-industrial wastes. Fermentation 10(8). https://doi.org/10.3390/fermentation10080425

phyloT (2023) phyloT: a phylogenetic tree generator. Version 2023.2. https://phylot.biobyte.de/

Portela R, Leal CR, Almeida PL, Sobral RG (2019) Bacterial cellulose: a versatile biopolymer for wound dressing applications. Microb Biotechnol 12(4):586–610. https://doi.org/10.1111/1751-7915.13392

Putra A, Kakuco A, Furukawa H, Gong JP (2009) Orientated bacterial cellulose culture controlled by liquid substrate of silicone oil with different viscosity and thickness. Polym J 41(9):764–770. https://doi.org/10.1295/polymj.PJ2009023

Quijano L, Rodrigues R, Fischer D, Tovar-Castro JD, Payne A, Navone L, Hu Y, Yan H, Pinmanee P, Poon E, Yang JH, Barro E (2024) Bacterial cellulose cookbook: A systematic review on sustainable and cost-effective substrates. J Bioresour Bioprod. https://doi.org/10.1016/j.jobab.2024.05.003

Rastogi A, Banerjee R (2019) Production and characterization of cellulose from *Leifsonia* sp. Process Biochem 85:35–42. https://doi.org/10.1016/J.PROCBIO.2019.06.008

Rodrigues AC, Fontão AI, Coelho A, Leal M, Soares da Silva FAG, Wan Y, Dourado F, Gama M (2019) Response surface statistical optimization of bacterial nanocellulose fermentation in static culture using a low-cost medium. New Biotechnol 49:19–27. https://doi.org/10.1016/J.NBT.2018.12.002

Römling U, Galperin MY (2015) Bacterial cellulose biosynthesis: diversity of operons, subunits, products and functions. Trends Microbiol 23(9):545. https://doi.org/10.1016/J.TIM.2015.05.005

Ross P, Mayer R, Benziman M (1991) Cellulose biosynthesis and function in bacteria. Microbiol Rev 55(1):35. https://doi.org/10.1128/MR.55.1.35-58.1991

Ryngajłło M, Jędrzejczak-Krzepkowska M, Kubiak K, Ludwicka K, Bielecki S (2020) Towards control of cellulose biosynthesis by *Komagataeibacter* using systems-level and strain engineering strategies: current progress and perspectives. Appl Microbiol Biotechnol. https://doi.org/10.1007/s00253-020-10671-3

Saxena IM, Brown RM (2005) Cellulose biosynthesis: current views and evolving concepts. Ann Bot 96(1):9. https://doi.org/10.1093/AOB/MCI155

Saxena IM, Lin FC, Brown RM (1991) Identification of a new gene in an operon for cellulose biosynthesis in *Acetobacter xylinum*. Plant Mol Biol 16(6):947–954. https://doi.org/10.1007/BF00016067

Schoch CL, Ciufo S, Domrachev M, Hotton CL, Kannan S, Khovanskaya R, Leipe D, McVeigh R, O'Neill K, Robbertse B, Sharma S, Soussov V, Sullivan JP, Sun L, Turner S, Karsch-Mizrachi I (2020) NCBI taxonomy: a comprehensive update on curation, resources and tools. Database 2020. https://doi.org/10.1093/DATABASE/BAAA062

Shavyrkina NA, Skiba EA, Kazantseva AE, Gladysheva EK, Budaeva VV, Bychin NV, Gismatulina YA, Kashcheyeva EI, Mironova GF, Korchagina AA, Pavlov IN, Sakovich GV (2021) Static culture combined with aeration in biosynthesis of bacterial cellulose. Polymers 13(23). https://doi.org/10.3390/polym13234241

Shi Z, Zhang Y, Phillips GO, Yang G (2014) Utilization of bacterial cellulose in food. Food Hydrocoll 35:539–545. https://doi.org/10.1016/J.FOODHYD.2013.07.012

Shoda M, Sugano Y (2005) Recent advances in bacterial cellulose production. Biotechnol Bioprocess Eng 10(1):1–8. https://doi.org/10.1007/BF02931175

Singhania RR, Patel AK, Tsai ML, Chen CW, Di Dong C (2021) Genetic modification for enhancing bacterial cellulose production and its applications. Bioengineered 12(1):6793–6807. https://doi.org/10.1080/21655979.2021.1968989

Singhania RR, Patel AK, Tseng YS, Kumar V, Chen CW, Haldar D, Saini JK, Dong CD (2022) Developments in bioprocess for bacterial cellulose production. Bioresour Technol 344. https://doi.org/10.1016/j.biortech.2021.126343

Somerville GA, Proctor RA (2013) Cultivation conditions and the diffusion of oxygen into culture media: the rationale for the flask-to-medium ratio in microbiology. BMC Microbiol 13(1). https://doi.org/10.1186/1471-2180-13-9

Standal R, Iversen TG, Coucheron DH, Fjaervik E, Blatny JM, Valla S (1994) A new gene required for cellulose production and a gene encoding cellulolytic activity in *Acetobacter xylinum* are colocalized with the bcs operon. J Bacteriol 176(3):665–672. https://doi.org/10.1128/JB.176.3.665-672.1994

Sumini M, Andrade GJS, Tischer CA, Kobayashi RKT, Nakazato G (2024) Production of bacterial cellulose by *Komagataeibacter xylinus*: biochemistry, synthesis and applications. Cellulose 32(1):81–94. https://doi.org/10.1007/S10570-024-06179-Y

Tanskul S, Amornthatree K, Jaturonlak N (2013) A new cellulose-producing bacterium, *Rhodococcus* sp. MI 2: screening and optimization of culture conditions. Carbohydr Polym 92(1):421–428. https://doi.org/10.1016/J.CARBPOL.2012.09.017

Tanskul S, Damthongsen T, Jaturonlak N (2018) A medium supplemented with vegetable extract used for cellulose production of *Rhodococcus* SP. MI 2. Biosci J 34(3):666–673. https://doi.org/10.14393/BJ-V34N3A2018-36954

Thomas B, Raj MC, Athira BK, Rubiyah HM, Joy J, Moores A, Drisko GL, Sanchez C (2018) Nanocellulose, a versatile green platform: from biosources to materials and their applications. Chem Rev 118(24):11575–11625. https://doi.org/10.1021/ACS.CHEMREV.7B00627

Uğurel C, Öğüt H (2024) Optimization of bacterial cellulose production by *Komagataeibacter rhaeticus* K23. Fibers 12(3):29. https://doi.org/10.3390/FIB12030029

Ullah H, Badshah M, Correia A, Wahid F, Santos HA, Khan T (2019) Functionalized bacterial cellulose microparticles for drug delivery in biomedical applications. Curr Pharm Des 25(34):3692–3701. https://doi.org/10.2174/1381612825666191011103851

Upadhyay A, Pal D, Kumar A (2023) Deciphering target protein cascade in salmonella typhi biofilm using genomic data mining, and protein-protein interaction. Curr Genomics 24(2):100–109. https://doi.org/10.2174/1389202924666230815144126

Velmurugan P, Myung H, Govarthanan M, Yi YJ, Seo SK, Cho KM, Lovanh N, Oh BT (2015) Production and characterization of bacterial cellulose by *Leifsonia* sp. CBNU-EW3 isolated from the earthworm, *Eisenia fetida*. Biotechnol Bioprocess Eng 20(3):410–416. https://doi.org/10.1007/S12257-014-0793-y

Voon WWY, Muhialdin BJ,Yusof NL, Rukayadi Y, Meor Hussin AS (2019) Bio-cellulose Production by Beijerinckia fluminensis WAUPM53 and Gluconacetobacter xylinus 0416 in Sago By-product Medium Applied Biochemistry and Biotechnology 187(1):211–220. https://doi.org/10.1007/s12010-018-2807-2

Wang J, Tavakoli J, Tang Y (2019) Bacterial cellulose production, properties and applications with different culture methods – a review. Carbohydr Polym 219:63–76. https://doi.org/10.1016/j.carbpol.2019.05.008

Wang SS, Han YH, Chen JL, Zhang DC, Shi XX, Ye YX, Chen DL, Li M (2018) Insights into bacterial cellulose biosynthesis from different carbon sources and the associated biochemical transformation pathways in *Komagataeibacter* sp. W1. Polymers 10(9). https://doi.org/10.3390/polym10090963

Wong HC, Fear AL, Calhoon RD, Eichinger GH, Mayer R, Amikam D, Benziman M, Gelfand DH, Meade JH, Emerick AW, Bruner R, Ben-Bassat A, Tal R (1990) Genetic organization of the cellulose synthase operon in *Acetobacter xylinum*. Proc Natl Acad Sci USA 87(20):8130–8134. https://doi.org/10.1073/PNAS.87.20.8130

Zikmanis P, Kolesovs S, Ruklisha M, Semjonovs P (2021) Production of bacterial cellulose from glycerol: the current state and perspectives. Bioresour Bioprocess 8(1). https://doi.org/10.1186/s40643-021-00468-1

Zimmer J (2020) Activation of bacterial cellulose biosynthesis by cyclic di-GMP. In: Microbial cyclic di-nucleotide signaling. Springer, pp 211–221. https://doi.org/10.1007/978-3-030-33308-9_13

Zogaj X, Bokranz W, Nimtz M, Römling U (2003) Production of cellulose and curli fimbriae by members of the family *Enterobacteriaceae* isolated from the human gastrointestinal tract. Infect Immun 71(7):4151. https://doi.org/10.1128/IAI.71.7.4151-4158.2003

Żywicka A, Ciecholewska-juśko D, Drozd R, Rakoczy R, Konopacki M, Kordas M, Junka A, Migdał P, Fijałkowski K (2021) Preparation of *Komagataeibacter xylinus* inoculum for bacterial cellulose biosynthesis using magnetically assisted external-loop airlift bioreactor. Polymers 13(22). https://doi.org/10.3390/polym13223950

Chapter 4
Agroindustrial Wastes for Bacterial Cellulose Production: Challenges and Innovations

Fan Yang, Wenqiang Xu, Jiantang Chen, and Feng F. Hong

Abstract Bacterial cellulose (BC) is an important biological material with diverse applications across various sectors. However, its industrialization and large-scale commercialization encounter significant obstacles, primarily due to high fermentation costs and low productivity. To tackle the challenges in BC production, extensive research has explored the use of low-cost feedstocks and large amounts of waste biomass by-products from various sources. Agricultural, food processing, brewing, textile, and pulp and paper industry wastes are promising alternative ingredients for fermentation. This chapter discusses the use of agricultural and industrial wastes, highlighting the common pretreatment methods employed to process these raw materials. It categorizes media sources into sugar-rich, starch-based, cellulosic and lignocellulosic, and miscellaneous wastes. This chapter further introduces innovative strategies to enhance waste utilization for BC production, aiming to reduce production costs and enable wider industrial applications. The use of waste biomass as culture media not only has the potential to diminish BC production costs but also to promote the recycling of waste resources and mitigate environmental pollution and associated treatment expenses. Nevertheless, the majority of research is limited to laboratory environments, with a notable lack of large-scale production instances. The transition to large-scale BC production also requires the implementation of advanced reactors, optimization of fermentation processes, and maintenance of stable microbial strains. Therefore, realizing BC's full potential as a sustainable, cost-effective material requires collaborative efforts between academia and industry to overcome the current limitations and advance toward environmentally sustainable, commercially viable production.

F. Yang · W. Xu · J. Chen · F. F. Hong (✉)
State Key Laboratory of Advanced Fiber Materials, College of Biological Science and Medical Engineering, Donghua University, Shanghai, China

Scientific Research Base for Bacterial Nanofiber Manufacturing and Composite Technology, China Textile Engineering Society, Shanghai, China
e-mail: fhong@dhu.edu.cn

F. Dourado et al. (eds.), *Advances in Bacterial Cellulose*,
https://doi.org/10.1007/978-3-032-07042-5_4

Keywords Bacterial cellulose · Biorefining · Cost-effectiveness · Agroindustrial waste · Sustainable production

1 Introduction

Bacterial cellulose (BC), also known as bacterial nanocellulose (BNC), is a type of nanoscale cellulose fiber produced by specific bacterial strains. Similar to plant cellulose, BC consists of glucan chains formed by glucose monomers linked through β-1,4 glycosidic bonds, which contribute to its unique properties, including high crystallinity and purity. However, BC differs from plant cellulose in that it is an extracellular polymer, facilitating its purification. The purified form of BC is free from impurities like hemicellulose, lignin, or cellular debris (Czaja et al. 2007). BC is noted for its high degree of polymerization and crystallinity, as well as its distinctive three-dimensional fiber network structure at the nanoscale. Additionally, its excellent physico-chemical performance and biocompatibility support its significant potential for use in various applications, especially in the medical field (El-Gendi et al. 2022). Global business research providers estimate that the global BC market will reach $570 million by 2024 and $680 million by 2025, growing at a Compound Annual Growth Rate (CAGR) of 13.3% between 2018 and 2025 (Moradali and Rehm 2020).

The bacteria responsible for producing BC include *Acetobacter*, *Rhizobium*, *Sarcina*, *Komagataeibacter*, and *Achromobacter* (Jacek et al. 2019). Currently, *Komagataeibacter xylinus* is the primary commercial producer of BC, largely due to the presence of multiple copies of the BC synthesis gene within its genome, which contributes to its high yield of BC (Lu et al. 2020). Its yield is significantly affected by nutrient availability in the growth medium, with the carbon source being a key factor (Fernandes et al. 2020). Hestrin-Schramm (HS) remains the most commonly used medium in research (Lu et al. 2020). The high production cost of HS as a lab-developed semi-synthetic medium (about 30% of the total cost) is a major challenge for the commercial production and application of BC (Fernandes et al. 2020; Rahman et al. 2021).

In recent decades, the identification of agricultural and industrial wastes for sustainable production has emerged as a promising alternative to traditional media used in BC production (Raiszadeh-Jahromi et al. 2020; Saleh et al. 2023). Optimizing the carbon source and adjusting the production conditions can significantly boost yields and reduce costs. This chapter discusses the utilization of wastes suitable for BC production, with an emphasis on pretreatment methods prior to fermentation. Additionally, innovative strategies for enhancing waste valorization in BC production are presented.

2 Types of Biomass Waste Suitable for BC Production

2.1 *Sugar-Rich Wastes*

Sugar-rich wastes typically do not require extensive pretreatment or strict hydrolysis. Such materials encompass by-products from various industrial and agricultural processes, such as the sugar industry, fermentation residues, fruit processing, and other residuals arising from agricultural production and industrial manufacturing. Specific details of BC production from diverse sugar-rich wastes are provided in Table 4.1.

Table 4.1 Sugar-rich wastes used for BC production

Source	Pretreatment	Cultivation time (days)	BC productivity ($g{\cdot}L^{-1}{\cdot}d^{-1}$)	Strain	Reference
Sugar industry wastes					
Sweet lime pulp waste	After comminuting 250 g pulp in 500 mL aqueous medium, the mixture was subjected to 90 °C for 1 h followed by centrifugation	16	0.393	*Komagataeibacter europaeus* SGP37	(Dubey et al. 2018)
Scum of sugarcane jaggery or gur	NM	9	0.279	*Gluconacetobacter xylinum ATCC 23768*	(Khattak et al. 2015)
Molasses	Crude molasses was diluted five-fold (w/v) and centrifuged. Following pH adjustment to 3.0 with 4N H_2SO_4, the molasses solution was thermally processed at 120 °C for 20 min prior to centrifugal separation	3	1.77	*Acetobacter xylinum* BPR2001	(Bae and Shoda 2005)
Sugarcane molasses	NM	5	0.802	*Komagataeibacter rhaeticus*	(Machado et al. 2018)
Sugarcane molasses	NM	7	223% as compared to 100% in HS medium	*Acetobacter xylinum* ATCC 10245	(Premjet et al. 2007)
Extracted date syrup	NM	8	0.725	*Gluconacetobacter xylinus* 0416 MARDI	(Lotfiman et al. 2018)

(continued)

Table 4.1 (continued)

Source	Pretreatment	Cultivation time (days)	BC productivity (g.L^{-1}.d^{-1})	Strain	Reference
Black strap molasses and brewery molasses	NM	10	0.305	*Gluconacetobacter xylinum* ATCC 23768	(Khattak et al. 2015)
Treated sugar beet molasses	NM	14	0.326	*Gluconacetobacter xylinus* PTCC 1734	(Salari et al. 2019)
Sugar beet molasses	NM	40	0.256	*Gluconacetobacter xylinus* NRRL B-759	(Oz and Kalender 2023)
Sugarcane molasses	NM	15	–	*Gluconacetobacter hansenii*	(Carvalho et al. 2025)
Molasses	NM	5	0.580	*Komagataeibacter sucrofermentans* H-110	(Revin et al. 2021)
Cashew apple juice and soybean molasses	NM	7	0.643	*Gluconacetobacter xylinus*	(Souza et al. 2020)
Sugarcane molasses	Raw molasses was diluted 5 times and centrifuged. Following pH adjustment to 3.0 with 4 N H_2SO_4, supernatant was thermally processed at 120 °C for 20 min prior to centrifugal separation	7	1.800	*Gluconacetobacter intermedius* SNT-1	(Tyagi and Suresh 2016)
Diverse fermentation wastes					
Distillery effluent	NM	8	1.063	*Gluconacetobacter oboediens*	(Jahan et al. 2018)
Waste from beer fermentation broth	NM	14	0.604	*Gluconacetobacter hansenii* PJK KCTC 10505BP	(Ha et al. 2008)
Grape pomace/ corn steep liquor	NM	25	0.268	*Acetobacter xylinum* NRRL B-42	(Cerrutti et al. 2016)
Wheat thin stillage and cheese whey	NM	3	2.063	*Gluconacetobacter sucrofermentans* B-11267	(Revin et al. 2018)
Thin stillage wastewater	NM	7	0.894	*Gluconacetobacter xylinus* BCRC 12334	(Wu and Liu 2013)

(continued)

Table 4.1 (continued)

Source	Pretreatment	Cultivation time (days)	BC productivity (g.L^{-1}.d^{-1})	Strain	Reference
Makgeolli sludge filtrate	NM	7	0.229	*Gluconacetobacter xylinus* ATCC 10788	(Hyun et al. 2014)
Cheese whey	NM	14	0.254	*Gluconacetobacter xylinus* PTCC 1734	(Salari et al. 2019)
Wastewater after pullulan polysaccharide fermentation	NM	10	0.118	*Gluconacetobacter xylinum* BC-11	(Zhao et al. 2018)
Saccharogenic liquid food waste	NM	5	0.092	*Acetobacter xylinum* KJ1	(Moon et al. 2006)
Waste beer yeast	15% (w/v) waste beer yeast suspension was subjected to ultrasonication combined with mild acid hydrolysis for 40 min	14	0.501	*Gluconacetobacter hansenii* CGMCC 3917	(Lin et al. 2014)
Cacao mucilage exudate	Cacao mucilage exudate was subjected to heat-sterilization at 120 °C and 125 psi for 20 min	7	1.876	*Gluconacetobacter xylinus* ATCC® 23,768™	(Saavedra-Sanabria et al. 2021)
Wastewater from lipid fermentation	NM	10	0.066	*Gluconacetobacter xylinus*	(Huang et al. 2016)
Fruit processing wastes					
Grape pomace hydrolysate	Grounded grape pomace were dispersed in Milly-Q water containing HCl	3	0.400	*Komagataeibacter melomenusus*	(Gorgieva et al. 2023)
Sugarcane combined with apple pomace	Juice derived from apple residue and sugarcane was dissolved in water	14	0.179	*Gluconacetobacter medellinensis* ID13488	(Urbina et al. 2017)

(continued)

Table 4.1 (continued)

Source	Pretreatment	Cultivation time (days)	BC productivity ($g{\cdot}L^{-1}{\cdot}d^{-1}$)	Strain	Reference
Grape bagasse	Grape bagasse underwent homogenization using a blender Subsequently, the extract was filtered via Whatman No. 3 filter paper, and the pH was adjusted to 5.0 with sodium hydroxide	14	0.571	*Gluconacetobacter xylinus* NRRL B-42	(Vazquez et al. 2013)
Citrus peels (lemon, mandarin, orange, and grapefruit)	Hydrolysis of citrus peel was performed at 100 °C with 0.6 M H_2SO_4 for 2 h using liquid/solid ratio 10 mL/g	21	3.92 BC/100 g peel	*Komagataeibacter hansenii* GA2016	(Guzel and Akpinar 2019)
Apple residues (carbon source) + Tea mixture (nitrogen source)	Apple residues were dehydrated at 60 °C for 8 h. Then, the feedstock was boiled for 30 min with a 1:30 liquor ratio, the solution obtained was filtered to remove solid matter and debris, and oven-dried until a constant weight was achieved	7	0.034	*Kombucha*	(Amorim et al. 2023)
White grape bagasse (WGB)	Bagasse was heated at 70 °C with distilled water to attain the correct density, then subjected to blending and sieving to yield a nearly fluid mash	14	0.050	*Komagataeibacter xylinus* DSM 6513	(Ogrizek et al. 2021)
Red grape bagasse (RGB)		14	0.021	*Komagataeibacter xylinus* DSM 6513	(Ogrizek et al. 2021)
Pineapple peel waste hydrolysate	After drying and grinding pineapple peel, it was mixed into a 2.5% v/v H_2SO4 solution with a solid-to-liquid ratio of 1:10. Hydrolysis was carried out at 130 °C for 60 min	9	0.313	*Komagataeibacter xylinus* ATCC 23769	(Santoso et al. 2021)

(continued)

Table 4.1 (continued)

Source	Pretreatment	Cultivation time (days)	BC productivity ($g{\cdot}L^{-1}.d^{-1}$)	Strain	Reference
Oil palm frond juice and coconut water	NM	7	0.643	*Acetobacter xylinum*	(Sharifah et al. 2021)
Mango peel waste hydrolysate	Fruits peels were prepared as hydrolysates through nitric acid treatment	7	0.174	*Achromobacter* S33	(Hasanin et al. 2023)
Citrus orange		7	0.00026	*Achromobacter S33*	(Hasanin et al. 2023)
Banana		7	0.046	*Achromobacter* S33	(Hasanin et al. 2023)
Pomegranate		7	0.0009	*Achromobacter* S33	(Hasanin et al. 2023)
Pineapple		7	0.054	*Achromobacter* S33	(Hasanin et al. 2023)
Mango		7	0.174	*Achromobacter* S33	(Hasanin et al. 2023)
Oil palm empty fruit bunch hydrolysate	Oil palm empty fruit bunch fibers were mixed with 2 and 5% (w/v) NaOH, respectively with a solid-to-liquid ratio of 1:10. The mixture was pretreated at 121 °C, 117 kPa for 20 min using an autoclave	7	0.324	*Komagataeibacter xylinus*	(Zaki et al. 2023)
Pear peel and pomace	Pear peel and pomace were crushed and mixed with water and beaten at a ratio of 1:6 (w/v)	7	1.563	*K. rhaeticus* and *K. intermedius*	(Ma et al. 2021)
Palm date waste Molasses hydrolysate	100 g palm date waste was immersed in an equal amount of water and homogenized, then transferred to a water bath at 70 °C to maintain temperature for 1 h before filtering	7	0.557	*K. saccharivorans* MD1	(Abol-Fotouh et al. 2020)
Cashew apple juice	NM	7	0.649	*Gluconacetobacter xylinus* ATCC 53582	(Souza et al. 2020)

(continued)

Table 4.1 (continued)

Source	Pretreatment	Cultivation time (days)	BC productivity ($g \cdot L^{-1} \cdot d^{-1}$)	Strain	Reference
Orange juice Pineapple juice	NM	14	0.421 0.293	*Acetobacter xylinum* NBRC 13693	(Kurosumi et al. 2009)
Tomato residue	Tomatoes were blended at 18,000 rpm for 2 min to form a homogeneous mass and remove excess bagasse	14	0.595	*K. hansenii* UCP1619	(Cavalcanti et al. 2023)
Pineapple peel	Pineapple peel and water were mixed and filter	12	0.368	*Acetobacter xylinum* ATCC2376	(Lee et al. 2024)
Orange peel	50 g of frozen orange peels was homogenized for 5 min with 100 mL of acetate buffer	8	0.763	*Gluconacetobacter xylinus*	(Kuo et al. 2019)
Orange peel	80 g of orange peels were combined with 250 mL of deionized water, and the pH was adjusted to 1.5 using 0.5 M HCl. The resulting mixture was then heated at 90 °C for 1 h	13	0.400	*Komagataeibacter sucrofermentans* DSM 15,973	(Andritsou et al. 2018)
Durian shell	Durian shell was subjected to treatment using 2% (w/v) dilute sulfuric acid at a solid-to-liquid ratio of 1:10 (w/v) for 60 min at 120 °C	10	0.267	*Gluconacetobacter xylinus*	(Luo et al. 2017)

NM Not mentioned

2.1.1 Sugar Industry Wastes

The waste generated during sugar production mainly includes bagasse, molasses, sugar beet pulp, etc. (Szczerbowski et al. 2014). These by-products are characterized by the significant presence of organic matter, carbohydrates, proteins, and associated trace elements.

Molasses is a by-product of the sugar industry and contains abundant fermentable sugars, making it an excellent source of biomass for the synthesis of

BC. Molasses can be derived from diverse agricultural sources, such as corn, sorghum, sugarcane, soybeans, and sugar beets, offering a wide range of options (Park et al. 2010). Before use in BC fermentation, molasses may be treated to enhance metabolization and sometimes to eliminate or reduce inhibitory substances that can hinder bacterial growth. This pretreatment helps convert oligosaccharides into monosaccharides, aids in ash settling, and reduces viscosity (Álvarez-Cao et al. 2020). Tyagi et al. utilized sugarcane molasses as a carbon source to produce BC, achieving a yield of 12.6 g/L from 45.8 g/L of diluted (1:4) molasses following H_2SO_4-heat pretreatment (Tyagi and Suresh 2016). Cakar et al. examined BC production in molasses medium using a static semi-batch operation mode. They found that *G. xylinus* produced BC more efficiently in molasses via this method than in a static culture process (Cakar et al. 2014).

Research on the co-culture of bacteria and yeasts in molasses media has attracted significant attention. Devanthi and colleagues investigated the effects of *Dekkera bruxellensis* yeast on *Komagataeibacter intermedius* bacterial development and BC synthesis in molasses. They found that co-culture increased utilization of reducing sugars and number of *K. intermedius* cells but reduced BC production (Devanthi et al. 2021). Additionally, Kalashnikova et al. demonstrated a low-cost BC production using H_2SO_4-heat-treated soybean molasses with *Medusomyces gisevii*, also known as "Kombucha fungus" or "black tea fungus" (Kalashnikova et al. 2024). The symbiotic system of *Medusomyces gisevii* includes 15–30 yeast species belonging to different genera (*Pichia* sp., *Candida* sp., *Zygosaccharomyces bailii*, *Saccharomyces* sp.) and 8–10 genera of acetic acid bacteria (*Acetobacter* sp., *Acetobacter aceti*, *Gluconobacter oxydans*, *Acetobacter xylinum*, etc.). The symbiotic organisms have a specific metabolism and synthesize all of the nutrients required for its normal functioning (Gladysheva et al. 2018). Developing a stable production process can reduce environmental pollution from sugar production and lower the BC production costs. Therefore, using molasses for BC production offers promising opportunities in regions with established sugar industries.

2.1.2 Fermentation Wastes

Fermentation wastewater from beer, dairy, and other industries contains high levels of carbohydrates, organic acids, and alcohols (Table 4.1), making it a suitable substrate for BC cultivation. Huang et al. used fermentation wastewater from acetone, butanol, and ethanol (ABE) production facilities to create a culture medium, achieving a yield of 1.34 g/L of *Gluconacetobacter xylinus* CH001 after 7 days of incubation (Huang et al. 2015).

In traditional Korean rice wine distilleries, substantial quantities of Makgeolli sludge are discarded without further processing. This sludge medium has been utilized as a carbon source for the growth of *G. xylinus* in the production of BC. Analytical characterization indicates the sludge contains total nitrogen (0.81 g/L), glucose (10.24 g/L), organic acids (1.15 g/L), alcohol (0.93% v/v), and metal ions. Structural analysis of the BC produced in the Makgeolli sludge medium showed a fibrous network and a cellulose I polymorphic form (Hyun et al. 2014).

After physical and chemical treatments, the hydrolyzed product derived from waste beer yeast (WBY) functions as a growth substrate for *G. hansenii* CGMCC 3917 to generate BC in static culture. The results showed that the BC yield increased almost 6 times upon ultrasonication treatment, compared to untreated WBY. Analysis of the surface of the acquired BC revealed that the microfibrils in the membranes generated using WBY were arranged in a loose, random fashion within a network exhibiting high porosity.

Grape bagasse, a wine production residue, has also been evaluated for BC production (Lin et al. 2014). The characterization studies showed microfibrils of several micra in length, containing rectangular cross-sections with widths range of 35–70 nm and thicknesses of 13–24 nm and the presence of cellulose I polymorph (Vazquez et al. 2013).

The exudate from cacao mucilage, known as CME, is a nutrient-dense fluid that is considered a waste from cacao bean fermentation processes. Three distinct growth media formulated from CME were assessed: sterile raw cacao mucilage exudate, supplemented cacao mucilage exudate (containing sterile CME, 5 g/L peptone, 5 g/L yeast extract, and 2.7 g/L sodium citrate), and diluted/supplemented cacao mucilage exudate (prepared with a 1:2 ratio of sterile CME to water, along with 5 g/L peptone, 5 g/L yeast extract, and 2.7 g/L sodium citrate). The BC yield in the CME-based medium was 0.55 ± 0.16 g/L, while the supplemented CME-based medium yielded 14.06 ± 0.69 g/L, and the diluted/supplemented CME-based medium produced 13.13 ± 1.09 g/L (Saavedra-Sanabria et al. 2021).

Furthermore, Huang and colleagues employed lipid fermentation wastewater as a substrate for BC production with *G. xylinus*. In this research, lipid fermentation wastewater, defined as fermentation broth after separation from yeast biomass, featuring a high chemical oxygen demand (COD) value of 25.6 mg/L, was utilized for BC production. After 5 days of fermentation, the highest BC yield (0.659 g/L) was obtained (Huang et al. 2016).

2.1.3 Fruit Processing Wastes

In the early days of the coconut food industry, coconut juice (liquid endosperm) was discarded as waste due to difficulties in preservation. Coconut juice is rich in proteins, carbohydrates, and trace elements and has been proven to be an excellent culture medium for BC production, superior to pineapple juice (Kongruang 2008).

Peels of some fruits and vegetables are discarded because they are inedible. Inedible peels account for 5–40% of the total weight of fruits and vegetables and are rich sources of reducing sugars, vitamins, proteins, and organic acids (Kamal et al. 2022), thus they can be used as substrates for value-added products such as BC. Several studies have reported the feasibility of converting different fruit and vegetable peels into BC through biological processes. For example, orange peel residues derived from the juice manufacturing, characterized by 10% moisture,

30–40% sugars, 15–25% pectin, 8–10% cellulose, and 5–7% hemicellulose, serve as viable substrates for BC synthesis. Enzymatic pretreatment using cellulase and pectinase significantly enhanced the yield of fermentable sugars, achieving concentrations of 60–80 g/L. This resulted in the production of 4.2–6.32 times more BC, compared to HS medium (Kuo et al. 2019).

In addition, banana peel extract combined with glucose in various mixtures was investigated. The results suggest that banana peels can partially replace conventional nitrogen and carbon sources for BC production (Hussain et al. 2019). Güzel's research group studied the BC production potential of fruit and vegetable peels (cucumber, melon, kiwifruit, tomato, apple, quince, and pomegranate) using *Gluconacetobacter* GA2016 (Guzel and Akpinar 2020). Recently, BC production by a Kombucha microbial consortium was successfully sustained in an alternative medium incorporating apple waste (10% w/v) into a tea mixture, achieving yields comparable to those obtained using the HS medium (control), as reported by Amorim (Amorim et al. 2023). A study effectively used enzymatically hydrolyzed prickly pear peels as a promising carbon source for BC production, achieving a yield of 6.01 g/L (El-Gendi et al. 2023).

2.2 *Starch-Based Wastes*

Starch-based wastes encompass low-value crops and agricultural by-products, such as potato peels, and waste generated from baking processes. In contrast to sugar-based raw materials, starch sources exhibit greater diversity and widespread availability, positioning them as economically viable options for BC production (Table 4.2). Hong's group took the lead in the production of BC from dilute acid-hydrolyzed and detoxified konjac (glucomannan) powder. Their findings indicated that the BC yield reached 2.12 g/L (dry basis), whereas the control groups utilizing glucose or mannose as carbon sources yielded only 0.69 g/L and 0.33 g/L, respectively (Hong and Qiu 2008).

Abdelraof et al. reported the application of potato peel waste (PPW) pretreated with mild nitric acid to yield fermentable sugars in an alternative culture medium for BC production. Upon optimizing the culture conditions, the BC yield from the PPW nitric acid hydrolysate medium was found to be four times higher than that from the HS medium (Abdelraof et al. 2019).

A recent investigation employed unsterilized rice soaking wastewater (RSW) as a cultivation medium. *Novacetimonas hansenii* S-08 underwent co-fermentation with indigenous microorganisms naturally present in RSW. Compared to cultivation in HS medium alone, BC production in the RSW + HS combination exhibited a 146% increase (Zhang et al. 2024).

Bakery waste, such as bread, dough, and flour, frequently discarded by catering services owing to spoilage or expiration, can be harnessed for BC production. Tsouko et al. documented that hydrolyzed bakery waste abundant in flour, processed via batch fermentation with *Komagataeibacter sucrofermentans* DSM15973, generated 13 g/L of high-grade BC. This BC displayed a substantial water-holding capacity of up to 102–138 g of water per gram of dry material. Relative to BC cultivated in sucrose- and yeast extract-enriched media, BC from bakery waste exhibited superior mechanical attributes, including stress at break (72.3–139.5 MPa) and Young's modulus (0.97–1.64 GPa). These features match those of BC derived from a costly chemically defined medium.(Tsouko et al. 2015).

More examples are shown in Table 4.2.

Table 4.2 Starch-rich wastes used for BC production

Source	Pretreatment	Cultivation time (days)	BC productivity ($g{\cdot}L^{-1}{\cdot}d^{-1}$)	Strain	Reference
Konjac powder	Konjac powder was immersed in dilute acids, agitated to ensure mixing within a 100 °C water bath, and subsequently filtered	8	0.265	*Gluconacetobacter xylinus* ATCC 23770	(Hong and Qiu 2008)
Potato peel	A mixture of potato peel and water was treated with 2M nitric acid at 100 °C for 2 h	4	0.653	*Gluconacetobacter xylinus*	(Abdelraof et al. 2019)
Potato juice	NM	7	0.286–0.571	*Komagataeibacter xylinus*	(Ciecholewska-Jusko et al. 2021)
Carob and haricot bean medium	Carobs and haricot beans were milled or broken into small pieces independently, then hydrolyzed in five volumes of distilled water (w/v) at 121 °C for 30 min	9	0.356	*Gluconacetobacter xylinus* ATCC 700178	(Bilgi et al. 2016)

(continued)

Table 4.2 (continued)

Source	Pretreatment	Cultivation time (days)	BC productivity ($g \cdot L^{-1} \cdot d^{-1}$)	Strain	Reference
Fermentation of wine residues and surplus potatoes	Mixtures containing equal weights of wine bagasse powder and potato powder (50:50 w/w) were created by mixing with a 2% sulfuric acid solution to achieve a final solid-to-liquid ratio of 1:10. The mixture was then heated at 125 °C for 1 h	6	0.667	*Komagataeibacter xylinus*	(Cazon et al. 2022)
Sweet potato residues	Sweet potato residues (SPG) underwent pulverization The starting SPG concentration was 14% w/v, with 9% w/v SPG supplements introduced at 3 h, 9 h, and 12 h, culminating in a final SPG level of 41% w/v. A cellulase, amylase, and pectinase blend was employed at a 4:1:1 ratio	14	0.811	*Gluconacetobacter xylinus*	(Xu et al. 2022)
Sweet potato peel	Sweet potato peel was mixed with water and blended	14	–	*Acetobacter xylinum*	(Betlej et al. 2022)

(continued)

Table 4.2 (continued)

Source	Pretreatment	Cultivation time (days)	BC productivity ($g·L^{-1}·d^{-1}$)	Strain	Reference
Cassava Residue (CR)	The α-amylase was at a loading of 150 U/g-CR, glucoamylase dosage was of 20 U/g-CR, and cellulase was 18 FPU/g cellulose (substrate concentration of 2% (w/v)). The hydrolysis was performed in the acidic buffer solution of pH 4.8 at 48 °C and 150 rpm for 72 h	10	Mass Yield 40%	*Acetobacter xylinum*	(Ji et al. 2022)
Wastewater from the Arenga Starch Industry	NM	14	0.311	*Acetobacter xylinum*	(Sulaiman 2023)
Algal starch	72% (w/w) H_2SO_4 was used and kept at 30 °C, for 1 h. Then, mixture was treated with 4% (w/w) H_2SO_4 at 121 °C, for 1 h	7	0.157	*Komagataeibacter hansenii*	(Uzyol and Sacan 2017)
Starchy kitchen waste (SKW)	SKW concentration of 24.3 g/100 mL with enzyme concentration of 0.1 mg/mL for 5 h, 40 min	2	1.055	*Komagataeibacter hansenii* AS.5	(Saleh et al. 2023)

NM Not mentioned

2.3 Cellulosic and Lignocellulosic Wastes

2.3.1 Cellulosic Wastes

Utilizing cellulose waste for BC biorefinery presents a notable advantage due to the diverse availability of raw materials, with substantial potential for further development. For example, the recycling of textile waste aligns with ecological principles, promoting sustainability, low carbon emissions, and environmental friendliness (Hong et al. 2012). Aged cotton clothing and discarded regenerated cellulose fabrics, such as viscose, are primarily composed of cellulose polymers, potentially serving as an affordable source of polysaccharides. Hong's research team investigated cotton-based waste textiles for BC production. These cellulosic materials were processed using the ionic liquid 1-allyl-3-methylimidazolium chloride ([AMIM]Cl). The BC obtained from the cotton cloth hydrolysate had a yield of 10.8 g/L, which was 83% higher than that obtained from glucose-based culture medium (Hong et al. 2012). Guo et al. reported that BC was produced by using the enzymatic hydrolysate obtained from discarded dyed cotton fabric. Reactive dyes exhibited minimal inhibition of cellulase activity at 5 g/L, but reduced BC production when the concentration surpassed 1 g/L. The BC yield reached 12.8 g/L with cotton hydrolysate, which was 48% higher than that with glucose-based medium (Guo et al. 2016). Kuo et al. conducted dissolution pretreatment of waste cotton textiles using NMMO, [BMIM]Cl, and 85% phosphoric acid, to reduce the crystallinity of cellulose and enhance saccharification. Fermentation of hydrolysates from non-decolorized and decolorized textiles yielded BC of 1.59 g/L and 1.88 g/L, respectively, whereas the BC yield from the glucose control medium was only 1.14 g/L (Kuo et al. 2010).

Discarded fiber sludge, a residue of negligible value, stems from pulp mills and lignocellulosic biorefineries. Sludges derived from sulfate (SAFS) and sulfite (SIFS) methods were subjected to enzymatic hydrolysis without thermochemical pretreatment, and the resulting hydrolysates were employed for BC production. The BC yield on initial sugar in a hydrolysate-based medium reached 0.3 g/g after 7 days of cultivation (Cavka et al. 2013b).

Examples of the production of BC from cellulosic waste are shown in Table 4.3.

2.3.2 Lignocellulosic Wastes

For the realization of large-scale and cost-effective BC biorefineries, large amounts of accessible biomass resources, such as diverse crop straws, would be valuable feedstock. Agricultural wastes are a renewable resource of substantial economic importance on a global scale. Less than 10% of the vast quantity of agricultural waste generated daily is utilized as an alternative feedstock in other industries (Ul-Islam et al. 2020). The full utilization of abundant agricultural waste not only holds great significance for environmental conservation, but also has potential value

Table 4.3 Cellulosic wastes used for BC production

Source	Pretreatment	Cultivation time (days)	BC productivity ($g \cdot L^{-1} \cdot d^{-1}$)	Strain	Reference
Waste dyed cotton fabrics hydrolysate	1 g cotton sheet was agitated (500 rpm) with 10 g [AMIM]Cl in a 50 mL flask at 110 °C	10	1.28	*Gluconacetobacter xylinus*	(Guo et al. 2016)
Waste cotton textiles	0.5 g waste textile was treated with 10 mL aqueous solution (12% w/v urea, 7% w/v NaOH) at RT (30 min), cooled to −20 °C overnight, thawed at RT, and diluted with 60 mL deionized water.	14	0.134	*Gluconacetobacter xylinus*	(Kuo et al. 2010)
Waste fiber sludge sulfate (SAFS) and sulfite fiber sludges (SIFS)	SAFS (290 g) and SIFS (343 g) were blended with citrate buffer (693.8 g and 640.8 g, respectively), enzymatically treated at 1.6% w/w mixture concentration, and incubated (50 °C, 150 rpm, 48 h) under orbital shaking.	14	0.786	*Gluconacetobacter xylinus* ATCC 23770	(Cavka et al. 2013b)
Water-soluble fraction (WS) from pulping waste liquor	Crude WS was concentrated via evaporation to 1/5 volume, followed by multi-pass purification through activated carbon layers to yield purified WS	7	120% BC yield (yield was percentage of the BC production using glucose)	*Acetobacter xylinum* ATCC 10245	(Uraki et al. 2002)

for the low-cost production of products. A summary of the BC produced from lignocellulosic waste from different sources is shown in Table 4.4.

Cheng et al. utilized a hydrolysate from acetic acid-treated corn stalks as the sugar source, followed by a detoxification process to remove furfural and other inhibitors, and adjusted the pH to optimize the subsequent fermentation process. Before detoxification the corn stalk hydrolysate boasted a high sugar content and an

Table 4.4 Lignocellulosic wastes used for BC production

Source	Pretreatment	Cultivation time (days)	BC productivity ($g{\cdot}L^{-1}{\cdot}d^{-1}$)	Strain	Reference
Corn stalk	200 g corn stalk underwent overnight soaking in 1000 mL acetic acid (2 wt%), followed by 60-min hydrothermal treatment at 160 °C	7	0.409	*Acetobacter xylinum* ATCC 23767	(Cheng et al. 2017)
Wheat straw	10 g wheat straw underwent overnight impregnation with 60 mL H_2SO_4 (1.5% w/v), followed by hydrothermal treatment (190 °C, 12 × 10^5 Pa, 10 min) and filtration. The residue was subsequently re-processed with 30 mL H_2SO_4 (3% w/v) under intensified conditions (210 °C, 19 × 10^5 Pa, 10 min)	7	1.23	*Gluconacetobacter xylinus* ATCC 23770	(Hong et al. 2011)
Sugarcane straw	Sugarcane straw was mixed with distilled water (1:10 w/v). The system was heated up to 195 °C and stirred at 200 rpm for 10 min	15	0.301	*Komagataeibacter xylinus* ATCC 11142	(Dhar et al. 2019)
Sweet sorghum	Sweet sorghum and PHP (phosphoric acid and hydrogen peroxide) was mixed (1:10, w/w) at 40 °C, 160 rpm for 2 h.	6	Stalk 0.303 Leaf 0.423 Root 0.380 Juice 0.145	*Acetobacter xylinum* ATCC 23767	(Wang et al. 2021a)

(continued)

Table 4.4 (continued)

Source	Pretreatment	Cultivation time (days)	BC productivity ($g{\cdot}L^{-1}{\cdot}d^{-1}$)	Strain	Reference
Sweet sorghum bagasse	1 g sweet sorghum bagasse was pretreated in 5 g ionic liquid [EMIM]Fm. The temperature was 110 °C and lasted for 4 h	14	0.710	*Komagataeibacter xylinus* ATCC 23770	(Chen et al. 2018)
Cashew tree residues	A 10% w/v cashew tree exudate solution underwent 12-h stirring, 1 M NaOH neutralization, and subsequent filtration	7	0.857	*Komagataeibacter rhaeticus*	(Pacheco et al. 2017)
Red maple	Red maple underwent tap water extraction (liquid-to-wood ratio 4:1) in a high-pressure reactor via 50-min heating from ambient to 160 °C, followed by 25-min isothermal retention	28	0.005	*Acetobacter xylinus* 23,769	(Erbas Kiziltas et al. 2015)
Pecan nutshell	10 g pecan nutshell underwent hydration in 100 mL distilled water via 2-h heating at 80 °C with 500-rpm agitation, followed by 60-min sonication	28	0.101	*Gluconacetobacter entanii*	(Dorame-Miranda et al. 2019)
Caragana Korshinskii Kom (CKK)	20 mL EDA was mixed with 20 g CKK, and the mixture was then transferred into an autoclave reactor for 210 °C and 40 min	7	0.657	*Gluconacetobacter xylinus* CGMCC 2955	(Li et al. 2021)

(continued)

Table 4.4 (continued)

Source	Pretreatment	Cultivation time (days)	BC productivity (g·L^{-1}·d^{-1})	Strain	Reference
Oat hulls	Miscanthus and oat hulls underwent dilute nitric acid pretreatment (3 wt% HNO_3, solid-liquid ratio 1:15 w/v) through 10-h hydrothermal processing at 95 °C	7	NM	*Medusomyces gisevii* Sa-12	(Kashcheyeva et al. 2019)
Silver grass		7	NM	*Medusomyces gisevii* Sa-12	(Kashcheyeva et al. 2019)
Oat hulls	Oat hulls underwent 1-h nitric acid treatment (0.2%, 90 °C) at a 1:20 (w/v) solid-liquid ratio. Oat hulls were mixed with 0.2% nitric acid and treated at 90 °C for 1 h. Oat hull-to-solution ratio of 1:20 (w/v).	24	0.029–0.917	*Medusomyces gisevii* Sa-12	(Skiba et al. 2020)
Elephant grass	The hydrolysis of elephant grass was carried out in an 8-m^3 bioreactor with sulfuric acid (2.5% w/v) at 135 °C for 1 h.	14	0.457	*Gluconacetobacter xylinus* CH001	(Yang et al. 2013)
Coffee cherry husk (CCH)	Powdered CCH underwent 30-min aqueous boiling (1:1 w/v), followed by muslin cloth filtration of the resultant slurry.	14	0.586	*Gluconacetobacter hansenii* UAC09	(Rani and Appaiah 2013)

(continued)

Table 4.4 (continued)

Source	Pretreatment	Cultivation time (days)	BC productivity ($g{\cdot}L^{-1}{\cdot}d^{-1}$)	Strain	Reference
Mulberry leaves	Mulberry leaf protein, which was extracted by alkaline and precipitated by acid, was hydrolyzed through hydrochloric acid. And the residue of the extract process was hydrolyzed by sulfuric acid.	7	0.291	*Acetobacter xylinum* NUST 4.2	(Chen et al. 2019b)
Spruce hydrolysate	Unbarked wood chips underwent continuous SO_2 treatment (18 bar, 204 °C; 7–8 min).	14	0.586	*Gluconacetobacter xylinus* ATCC 23770	(Guo et al. 2013)
Dry olive mill residue (DOR)	DOR (50 g) was submitted to water extraction (200 mL) at 40 °C and then added 200 mL of H_2SO_4 (1 M) at 100 °C, for 2 h.	4	0.213	*Gluconacetobacter sacchari*	(Gomes et al. 2013)
Industrial hardwood	Hardwood was treated by acidic sulfite pulping	4	0.068	*Gluconacetobacter sacchari*	(Carreira et al. 2011)
Acerola by-product	The by-product was dried and milled, followed by preparation of acerola-compound-rich extracts via 15-min hydrothermal extraction (121 °C) in an autoclave	15	0.267	*Komagataeibacter rhaeticus*	(Leonarski et al. 2021)
Rice husk	Rice husk was treated with Avizyme® for 96 h.	10	Aerated ~0.09 Static ~0.13	*Acetobacter xylinum* ATCC 23769	(Goelzer et al. 2009)

NM Not mentioned

organic composition comprising 1.84 g/L of mannose, 3.87 g/L of glucose, 29.61 g/L of xylose, 2.95 g/L of furfural, 4.01 g/L of lignin, and 18.71 g/L of acetic acid. The final BC yield in the hydrolysate of corn stalk was 2.86 g/L, with the resulting fibrils ranging from 300 nm to several micrometers in length, and a width within 20–70 nm (Cheng et al. 2017).

Rice husks were subjected to pretreatment with an enzyme mixture comprising endo-1 → 4-α-xylanase, endo-1,3(4)-α-glucanase, α-amylase, subtilisin, and polygalacturonase. An enzymatic hydrolysate with glucose concentrations of up to 40 g/L was evaluated for BC production. Aerated conditions yielded 1.57 g/L BC, while static conditions achieved 2.42 g/L (Goelzer et al. 2009).

Rice straw is also a common agricultural waste product. Ionic liquid [AMIM]Cl was used at 110 °C for 1 h to pretreat the rice straw to enhance hydrolysis. The yield of reducing sugars from the hydrolysis of regenerated straw with 33 U/mL cellulase was 53.3% (Hong and Han 2011). Wheat straw, another prevalent agricultural waste, is typically disposed of through burning, causing significant air pollution. However, it can be transformed into a BC production medium via acid or enzyme hydrolysis. Hong and colleagues successfully utilized dilute acid-hydrolyzed wheat straw as the starting substrate for BC production, employing *Gluconacetobacter aceti* subsp. *xylinus*. Growth inhibitors arising from acid hydrolysis were removed by detoxifying the wheat straw hydrolysates through the application of various alkalis, such as sodium hydroxide, calcium hydroxide, and ammonia, combined with laccase or activated charcoal. The results showed about a 50% improvement in the production of BC in the detoxified hydrolysate, as compared to the chemically defined medium (Hong et al. 2011). Chen et al. used ionic liquid [AMIM]Cl to pretreat wheat straw and then hydrolyzed it using cellulose, to obtain fermentable sugar. Treating wheat straw with [AMIM]Cl at 115 °C for 1.5 h, before enzymatic hydrolysis, resulted in a 71.2% fermentable sugar yield (Chen et al. 2013). In another study, wheat straw underwent heat treatment followed by chemical treatment. Following initial chemical and thermal treatments, the wheat straw underwent enzymatic hydrolysis employing cellulase, β-glucosidase, and xylanase. This approach yielded a hydrolysate-derived medium capable of generating up to 10.6 g/L of BC (Al-Abdallah and Dahman 2013).

Oat hulls are inedible hard shells that are removed during oat ripening, constituting 28% of the grain weight and containing 45% fiber. These hulls are produced in large quantities by oat-processing companies that manufacture oat cereals, biscuits, and oatmeal (Skiba et al. 2023). Due to the inherent inability of BC-producing strains to metabolize oat hulls directly as a carbon source, pretreatment procedures become essential for converting this lignocellulosic material into fermentable sugars. A recent study detailed the pilot-scale production of BC from oat hulls through three main steps: (1) chemical pretreatment with HNO_3 at 26 wt% under atmospheric pressure; (2) enzymatic saccharification catalyzed by commercial enzyme preparations, which attained a sugar yield of 79.5% within a 100-L fermentation system, and (3) fermentation of the hydrolysate with mixed cultures by *Medusomyces gisevii*. This integrated methodology generated BC at a concentration of 3.2 g/L, exhibiting exceptional purity alongside a crystallinity index of 93% (Skiba et al. 2020).

An alternative investigation has examined grass straw, grass husk, wheat husk, and corn cob as lignocellulosic substrates for BC production under cost-effective conditions. The lignocellulosic biomass was processed by a consortium of four cellulolytic bacterial strains to produce microcrystalline cellulose and sugar hydrolysate, which served as a carbon source for BC production. The highest BC yields of 1.54 and 1.40 g/L were obtained from grass straw and wheat husk hydrolysates, respectively. This research proposed a simple and eco-friendly method for managing lignocellulosic biomass waste and converting it into valuable products (Sadalage and Pawar 2021). Collectively, these studies underscore the potential for large-scale, low-cost industrial BC production.

2.4 Glycerol and Other Wastes

In addition to sugar, starch, cellulose, and lignocellulosic wastes, other types of waste can be used (Table 4.5). Crude glycerol has been demonstrated to be easily metabolized by certain BC-producing strains. The yield of BC was 3.4 g/L, using waste glycerol from Hokusei Kigyo Co., Ltd. (Sapporo) as a raw material (Kose et al. 2013). Another study showed that 12.31 g/L BC can be produced using non-detoxified crude glycerol (Soemphol et al. 2018).

Table 4.5 Other wastes used for BC production

Source	Pretreatment	Cultivation time (days)	BC productivity ($g{\cdot}L^{-1}{\cdot}d^{-1}$)	Strain	Reference
Crude glycerol combined with flour-rich waste streams and sunflower meal hydrolysates	*Aspergillus oryzae* and *Aspergillus awamori* fungal strains were utilized for the hydrolysis of organic matter	15	0.887	*Komagataeibacter sucrofermentans* DSM 15973	(Tsouko et al. 2015)
Waste glycerol	NM	4	0.850	*Gluconacetobacter intermedius* NEDO-01	(Kose et al. 2013)
Non-detoxified crude glycerol	NM	14	0.879	*Gluconacetobacter xylinus* BNKC19	(Soemphol et al. 2018)
Tobacco extract	Tobacco extract was boiled in water and subjected to steam distillation to remove nicotine	7	0.743	*Acetobacter xylinum* ATCC 23767	(Ye et al. 2019)

NM Not mentioned

3 Treatments of Wastes Before BC Production

3.1 Pretreatment of Wastes

An overview of the pretreatment of wastes for the preparation of the culture medium for BC production is shown in Fig. 4.1. Pretreatment is a crucial phase in the development of biomass waste as a feedstock, as it modifies the size, structure, and chemical characteristics of biomass to enhance efficient hydrolysis. Typically, agricultural residues like corn stalks, rice husks, and wheat straw are primarily composed of lignocellulose (Li et al. 2023). Significant research and technologies have been developed to enhance the bioconversion of lignocellulosic biomass. However, cost-effective bioconversion remains challenging due to the complex and dense structure of lignocellulose. This structure composed of cross-linked cellulose, hemicellulose, and lignin limits its accessibility. Cellulose is intertwined with hemicellulose and is embedded in a lignin matrix (Li et al. 2023). Hemicellulose consists of various monosaccharides, including xylose, arabinose, and mannose. Lignin is a complex heteropolymer of phenylpropanoid monomers, such as p-hydroxyphenyl, guaiacyl, and syringyl (Chundawat et al. 2011). Lignin is linked to hemicellulose through ester and ether bonds and plays a significant role in the rigidity of lignocellulose (Wang et al. 2021b).

It is difficult to directly hydrolyze natural biomass to release all reducing sugars, making it essential to employ efficient pretreatment techniques that influence the structure, chemical makeup, and hydrolysis speed of the pretreated biomass. (Podgorbunskikh et al. 2019; Sun et al. 2016). As a result, pretreatment should achieve the following objectives: (1) enhance sugar yield through enzymatic hydrolysis; (2) prevent further breakdown of carbohydrates; and (3) be economically viable. Both chemical and physical methods have been employed for the standard pretreatment of agricultural waste (Abol-Fotouh et al. 2020; Li et al. 2021).

3.1.1 Chemical Pretreatment

Chemical pretreatment involves the use of various acids, bases, acid-base mixtures, or ionic liquids to process lignocellulosic biomass. Acid pretreatment is a widely used technique to break down biomass waste and enhance cellulose utilization (Song et al. 2014). For instance, Wang et al. treated sweet sorghum residues with phosphoric acid and hydrogen peroxide to produce fermentable sugars for BC production, demonstrating that sweet sorghum can be effectively utilized for this purpose (Wang et al. 2021a). Kashcheyeva and colleagues investigated hydrothermal, nitric acid, or sodium hydroxide pretreatments for *Miscanthus* and oat hulls, subsequently conducting enzymatic hydrolysis and BC synthesis. Post-pretreatment enzymatic hydrolysis demonstrated a 4–7 times increase in yields relative to untreated controls (Kashcheyeva et al. 2019).

For the biosynthesis of BC, single-stage pretreatments with either dilute HNO_3 or dilute NaOH are advised due to their simplicity and the high performance of enzymatic hydrolysis of pretreatment products (RS yield 79.7–83.4%) (Kashcheyeva et al. 2019).

Li et al. systematically evaluated hydrothermal, alkaline (NaOH), and ethylenediamine (EDA) pretreatment regimens to enhance the enzymatic saccharification efficiency of *Caragana korshinskii* Kom. The results indicated that a 210 °C pretreatment for 40 mins with EDA was the most effective, with 72.8% lignin being removed and 88.0% glucose yield being obtained (Li et al. 2021).

Skiba et al. implemented a pilot-scale HNO_3 and NaOH pretreatment to produce BC from agricultural residue oat hulls. Overall, optimum results presented a 72.4% reducing sugar yield and a BC yield of 3.2 g/L under pilot-plant conditions (Skiba et al. 2020).

Silver grass was pretreated with a range of 3–6 wt% NaOH solution at 90–95 °C for 6–8 h, yielding 85% reducing sugars (Kashcheyeva et al. 2019). In NaOH pretreatment, OH^- targets the carbon of the ester linkage between lignin and hemicellulose, resulting in the irreversible hydrolysis of the ester bond and weakening of the structural integrity of lignocellulose (Modenbach and Nokes 2014).

Imidazolium-based ionic liquids (ILs), having anions of chloride, acetate, or alkyl phosphonate as cellulose solvents, have been used for the pretreatment of wastes like switchgrass (Singh et al. 2009) and wheat straw (Chen et al. 2013). Implementing ionic liquid [AMIM]Cl pretreatment at 110 °C for 1.5 h with 3% (w/w) wheat straw loading, Chen et al. achieved 71.2% sugar yield through enzymatic saccharification, thereby facilitating BC production. The results showed that the yield of BC pellicle obtained from a lime-detoxified hydrolysate of straw was 8.3 g/L, while that of BC pellicle cultured in glucose-based medium was 3.7 g/L (without buffer, pH 5.0) and 4.5 g/L (50 mM acetate buffer, pH 5.0), respectively (Chen et al. 2013).

3.1.2 Physical Pretreatment

Physical pretreatment methods, including grinding, irradiation, and the application of temperature and pressure, are commonly employed to break down waste materials. These techniques enhance the surface area and pore size of the biomass (El-Gendi et al. 2022). Through physical pretreatment, water extracts from the waste of mixed fruit skins (such as pineapples, oranges, sweet limes, and bananas) are utilized as nutrient and carbon sources. Using *Komagataeibacter xylinus* IITR DKH20 for BC production, the BC yield reached 11.4 g/L (Khan et al. 2021). Extracts derived from citrus processing waste, alongside pomelo and lemon peels, have been reported for their application in bioconversion processes. By grinding and water extraction, the aqueous extracts from lemon peel, grapefruit peel, and orange peel yielded 5.2 g/L, 5.0 g/L, and 2.9 g/L of BC, respectively (Andritsou et al. 2018).

3.2 *Hydrolysis of Wastes*

Lignocellulosic biomass pretreatment targets structural destabilization of the lignocellulose complex, enhancing cellulose accessibility while diminishing crystallinity, thus optimizing enzymatic (or acidic) hydrolysis efficacy. At this stage, acids or enzymes break down polysaccharide chains into monosaccharides that bacteria can assimilate for BC biosynthesis (Fig. 4.1). The effectiveness of the hydrolysis in waste materials is influenced by the type of pretreatment technology and the hydrolysis method (El-Gendi et al. 2022).

3.2.1 Acid Hydrolysis

Compared to enzymatic hydrolysis, the operational procedures associated with dilute acid hydrolysis are relatively straightforward. While the cost of enzymes tends to be high, dilute acid hydrolysis is generally more economical. However, this method is associated with significant energy consumption, potential equipment corrosion, generation of inhibitory by-products, and environmental impact resulting from the use of chemicals. The effluent produced would have severe impacts on water bodies, soil, and ecosystems, if not processed. It is essential to design the process flow reasonably and treat wastewater to reduce pollution. This process effectively deconstructs complex polysaccharides into simpler sugars. Diluted sulfuric acid is the most frequently employed acid for the saccharification of waste materials in the production of BC, necessitating subsequent detoxification (Sect. 2.3) of the hydrolysate (Santoso et al. 2021). For instance, Hong et al. utilized 0.5 mol/L sulfuric acid to hydrolyze konjac powder, resulting in a sugar yield of 80% in the final hydrolysate (Hong and Qiu 2008). In another study, elephant grass was hydrolyzed using 2.5% (w/v) sulfuric acid at 135 °C for 1 h, yielding 12 g/L glucose and 20.3 g/L xylose, and 2.3 g/L arabinose (Yang et al. 2013).

3.2.2 Enzymatic Hydrolysis

Enzymatic hydrolysis serves as an effective technique for saccharifying biomass waste, owing to its eco-friendly sustainability and the lack of inhibitory by-products. Different enzymes have been used for the saccharification of waste, including cellulases for cotton cloth (Guo et al. 2016), Cellic® CTec2 for sugarcane straw (Dhar et al. 2019), Avizyme® for rice husks (Goelzer et al. 2009), amylase for starch kitchen waste (Saleh et al. 2023), and lactase for cheese whey (Salari et al. 2019). A particular investigation subjected pretreated sugarcane straw at 10% w/v to enzymatic hydrolysis with cellulase dosed at 15 FPU/g. The hydrolysis process was conducted in a citrate buffer (pH 5.0) at a stirring speed of 250 rpm and a temperature of 50 °C, resulting in the production of 49.2 g/L of glucose, 6.1 g/L of xylose, 1.8 g/L of cellobiose, and 2.9 g/L of acetic acid, after 2-day enzymatic hydrolysis (Dhar et al. 2019).

3.3 Detoxification of Hydrolysates

Multiple research findings have demonstrated that effective pretreatment techniques, such as steam explosion, acid-alkali processes, and ionic liquids, can improve the enzymatic hydrolysis of lignocellulose and support its conversion to BC. Nevertheless, these approaches face difficulties including high operational temperatures and pressures, the application of corrosive acids or alkalis, significant capital expenditures, and the production of inhibitory compounds. The acid hydrolysis of lignocellulose yields the desired sugars but also results in the formation of various by-products that can inhibit bacterial activity (Kim et al. 2021). The types and concentrations of these inhibitory compounds are closely associated with the pretreatment techniques employed. Common inhibitors include aliphatic acids, furan aldehydes, and phenolic compounds (Molina-Ramirez et al. 2018). Lignocellulosic hydrolysates contain characteristic short-chain aliphatic acids, with formic, acetic, and levulinic acids constituting the dominant fraction. Acetic acid is generated primarily from the hydrolysis of the acetyl groups of hemicellulose, while formic and levulinic acids are formed during 5-hydroxymethylfurfural (HMF) degradation (Wang et al. 2018). Furfural and HMF represent typical furan aldehydes present in lignocellulose hydrolysates, originating from the degradation of pentoses and hexoses, respectively. Acid-based and hydrothermal approaches act as key pretreatment methods responsible for generating these furan aldehydes. The concentrations of furan aldehydes in lignocellulose hydrolysates depend on the biomass species and the severity of the pretreatment (Wang et al. 2018). Phenolic compounds, including acids (ferulic, vanillic, 4-hydroxybenzoic, and syringic acids), alcohols (guaiacol, catechol, and vanillyl alcohol), and aldehydes (vanillin, syringaldehyde, and 4-hydroxybenzaldehyde), are mainly formed during the degradation of lignin in the biomass pretreatment process by acid-based, hydrothermal, alkaline, and oxidative methods (Wang et al. 2018). A deeper understanding of how the various derivatives of lignocellulosics impact bacterial growth and BC synthesis would benefit the utilization of lignocellulose resources and the optimization of fermentation process.

Zhang et al. conducted a detailed study on how four phenolic model compounds—coniferyl aldehyde, ferulic acid, vanillin, and 4-hydroxybenzoic acid—inhibited the growth of *G. xylinus* and the production of BC. Among these compounds, coniferyl aldehyde emerged as the strongest inhibitor, followed by vanillin, ferulic acid, and 4-hydroxybenzoic acid. Specifically, coniferyl aldehyde completely prevented BC production at concentrations as low as 2 mM. When the concentrations ranged from 0.5 to 1.5 mM, the volumetric BC yield was reduced to 3.4–6.4 g/L, below the reference medium yield of 6.7 g/L. Vanillin also significantly inhibited BC production at 2.5 mM, decreasing the yield to approximately 40% of that in the inhibitor-free control medium. Yields dropped below 0.4 g/L at vanillin concentrations of 5 mM or higher. In contrast, ferulic acid and 4-hydroxybenzoic acid showed negligible impact on BC yield below 2.5 mM. Interestingly, the bacterial cultures metabolized coniferyl aldehyde to ferulic acid with conversion rates reaching 81%, while vanillin was reduced to vanillyl alcohol with yields of up to 80% (Zhang et al. 2014b). Zhang also examined the resistance of *G. xylinus* to other inhibitors from lignocellulose, such as formic acid, acetic acid, levulinic acid, furfural, and 5-hydroxymethylfurfural. Formic acid at 100 mM fully blocked the metabolism of *G. xylinus*, but 250 mM concentrations of acetic acid or levulinic acid enabled BC production to proceed. Conversely, the addition of 20 mM furfural and 30 mM 5-hydroxymethylfurfural resulted in the complete suppression of glucose utilization and BC production (Zhang et al. 2014a). These findings are vital for revealing the inhibition mechanisms in lignocellulosic hydrolysates and for boosting BC production efficiency from lignocellulosic sources.

Although inhibitors can negatively affect BC synthesis, detoxification can significantly reduce their concentration in the hydrolysate (Fig. 4.1). Guo and colleagues evaluated various detoxification techniques for BC production utilizing spruce hydrolysate. The results suggest that after activated charcoal detoxification, 88% of phenolic compounds, 94% of furfural, 28% of acetic acid, and 39% of formic acid from spruce hydrolysate were removed and the BC yield by *G. xylinus* ATCC 23770 increased from non-producing to 8.2 g/L (Guo et al. 2013). Similarly, Hong et al. reported that treatment of wheat straw hydrolysates with $Ca(OH)_2$ and activated charcoal was more effective in removing inhibitors compared to other alkaline treatments when producing BC (Hong et al. 2011). While these detoxification methods can effectively remove inhibitors, they may increase the operating costs and cause the loss of sugars (Wang et al. 2018). Future research should more extensively and comprehensively investigate inhibitors, including their detoxification approaches, underlying mechanisms, and the resistance/synergistic interactions among various tolerance genes.

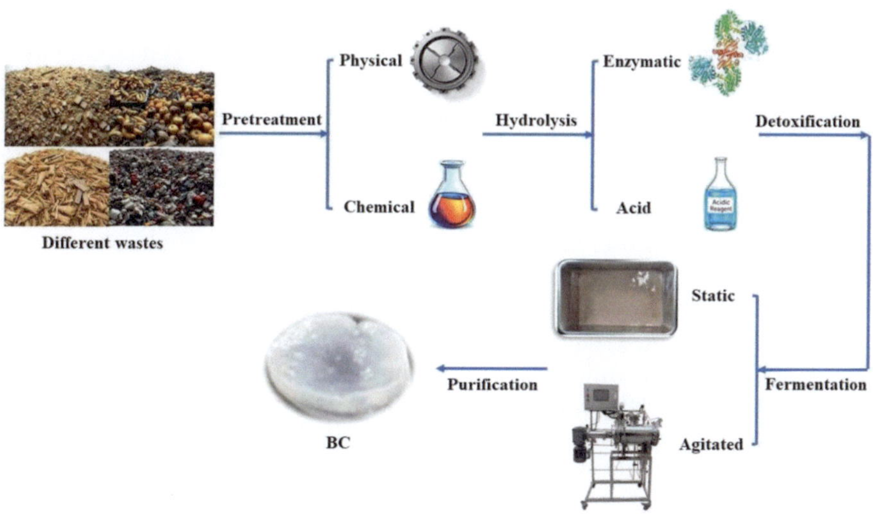

Fig. 4.1 Overview of the pretreatment of wastes for the preparation of fermentation culture medium

4 Emerging Technology for the Biorefinery of Wastes to BC

Biomass-derived culture media are complex and contain both easily usable carbon and nitrogen sources for BC-producing bacteria, as well as inhibitors and other monosaccharides such as mannose and xylose, which are poorly utilized by microorganisms (Chen et al. 2019a). These substances inhibit bacterial growth and BC production. While detoxification improves fermentability, it increases production costs due to complex processes. An effective method to boost microbial tolerance to these inhibitors is the adaptive evolution strategy, which involves acclimating BC fermentation strains to lignocellulosic hydrolysate (Joensson and Martin 2016). Research has identified *Gluconacetobacter xylinus*, a significant bacterial strain for cellulose fermentation, as having substantial tolerance to organic acids found in lignocellulosic hydrolysate (Zou et al. 2017; Zhang et al. 2014a, 2014b). For example, it can tolerate 25 mM of formic acid, metabolize 175 mM of acetic acid, and enhance BC production when exposed to 100 mM of acetylpropionic acid. Moreover, the strain can reduce low concentrations of furan and phenolic compounds to less toxic forms, such as converting furfural to furoic acid, 5-hydroxymethylfurfural to 5-hydroxymethyl-2-furoic acid, cedrol to ferulic acid, and vanillin to vanillic acid and vanillyl alcohol. By leveraging these metabolic abilities, BC-producing strains can be adapted through batch or continuous cultivation, by exposing them to hydrolysates containing sub-lethal inhibitor concentrations. Such short- or long-term adaptations may further boost the capacity of microorganisms to convert inhibitors (Vasconcellos et al. 2019). Therefore, adaptive evolution strategies can generate BC fermentation strains with high tolerance to fermentation inhibitors, representing a simple and effective method to increase the lignocellulosic raw material content in synchronous saccharification and fermentation processes to boost BC production.

Gene modification directly alters the characteristics of strains, enhancing their ability to utilize a specific biomass medium. For example, to improve the use of sucrose for BC production, the sucrose phosphorylase gene from *Leuconostoc mesenteroides* was overexpressed in an *Acetobacter* strain, resulting in a 27% increase in BC production (Tonouchi et al. 1998), In another study, a sucrose synthase from mung beans was expressed in *K. xylinus*, which directed carbon from sucrose to cellulose via UDP-glucose, achieving a 100% increase in BC production (Nakai et al. 1999). To improve the efficiency of mannose (derived from hemicellulose) utilization, Yang et al. overexpressed mannose kinase and phosphate mannose isomerase from *E. coli* in *K. xylinus*, increasing BC production from mannose by 84% (Yang et al. 2023). Furthermore, the random insertion of the *E. coli* β-galactosidase gene into *K. xylinus* ITDI via transposon-mediated integration enabled this strain to convert lactose into cellulose, achieving BC concentrations ranging from 0.78 g/Lto 1.82 g/L (Battad-Bernardo et al. 2004).

Agricultural and industrial wastes, despite being low-cost, require additional processing costs for hydrolysis. Enzymatic hydrolysis produces minimal inhibitors during the process and meets environmental protection requirements, but the high cost of hydrolytic enzymes is one of the factors hindering its development. Consequently, an innovative method with considerable promise is arising, enabling the cost-effective transformation of waste into two high-value microbial products. Cavka et al. employed fiber sludge as a raw material, integrating BC production with the synthesis of hydrolytic enzymes for lignocellulose degradation. Specifically, *Gluconacetobacter xylinus* generated BC from enzymatic hydrolysates of fiber sludge, while the filamentous fungus *Trichoderma reesei* produced enzymes from spent fermentation broth after BC harvest. Continuous production of BC and enzymes may yield two high-value products from extremely low-value or valueless residual streams. The maximum volumetric yield of BC from fiber sludge obtained from sulfate (SAFS) and sulfite (SIFS) processes was 11.1 g/L and 10 g/L (dry weight), respectively. The cellulase activity (CMCase activity) in spent SAFS and SIFS hydrolysates reached 5.2 U/mL (87 nkat/mL) (Cavka et al. 2013a). Adopting an approach to harness waste for BC production and repurpose spent fermentation broth for enzyme synthesis will not only lower BC manufacturing expenses but also mitigate environmental contamination from broth discharge, propelling BC fermentation toward a circular economic system. Interestingly, another work from the same group found that the xylanase activities obtained with the spent fermentation broth were 3–4 times higher than those obtained with glucose medium (Guo et al. 2016). The finding indicates that there may be some factors in the spent fermentation broth that promote the synthesis of xylanase, such as oligosaccharides, which can be used by fungi as nutrients to increase xylanase production or function as inducers for the synthesis of xylanase.

5 Conclusions

Optimizing BC culture media using waste biomass is crucial for addressing the cost challenges of BC production. Numerous studies have shown that waste biomass holds significant potential as a BC medium. Cost reduction can enhance the widespread application of BC in diverse fields, including food, packaging, papermaking, and textiles. Culture medium prepared from waste biomass not only decreases BC synthesis costs but also effectively utilizes waste biomass, thereby reducing environmental pollution and treatment expenses. Despite their aforementioned advantages, hydrolysis and detoxification processes incur additional costs. Furthermore, the organic load of fermentation waste requires wastewater treatment. These expenditures may partially offset the economic benefits of using waste biomass. The preparation process for waste-based media requires a comprehensive consideration of various costs. Therefore, new technologies are required to promote BC production from agroindustrial wastes. For example, the spent fermentation broth after BC harvest can be used to produce enzymes such as cellulase and xylanase, reducing overall production costs and achieving combined economic benefits. Producing both nanofiber materials and enzymes from waste enhances the economic viability of biorefinery factories, thus lowering the overall cost of biorefinery using waste as raw material. Nevertheless, most research efforts remain confined to laboratory settings, with a notable absence of large-scale production. Moreover, large-scale commercial manufacture of BC requires the support of reactors, optimized fermentation processes, and utilization of stable strains. Despite BC's immense potential as a green material across numerous fields, translating this potential into practical production applications still requires further exploration and sustained development.

References

Abdelraof M, Hasanin MS, El-Saied H (2019) Ecofriendly green conversion of potato peel wastes to high productivity bacterial cellulose. Carbohydr Polym 211:75–83

Abol-Fotouh D, Hassan MA, Shokry H, Roig A, Azab MS, Kashyout AE-HB (2020) Bacterial nanocellulose from agro-industrial wastes: low-cost and enhanced production by Komagataeibacter saccharivorans MD1. Sci Rep 10:3491

Al-Abdallah W, Dahman Y (2013) Production of green biocellulose nanofibers by *Gluconacetobacter xylinus* through utilizing the renewable resources of agriculture residues. Bioprocess Biosyst Eng 36:1735–1743

Álvarez-Cao M-E, Becerra M, González-Siso M-I (2020) Biovalorization of cheese whey and molasses wastes to galactosidases by recombinant yeasts. In: Biovalorisation of wastes to renewable chemicals and biofuels. Elsevier

Amorim LFA, Li L, Gomes AP, Fangueiro R, Gouveia IC (2023) Sustainable bacterial cellulose production by low cost feedstock: evaluation of apple and tea by-products as alternative sources of nutrients. Cellulose (Dordrecht) 30:5589–5606

Andritsou V, De Melo EM, Tsouko E, Ladakis D, Maragkoudaki S, Koutinas AA, Matharu AS (2018) Synthesis and characterization of bacterial cellulose from citrus-based sustainable resources. ACS Omega 3:10365–10373

Bae SO, Shoda M (2005) Production of bacterial cellulose by Acetobacter xylinum BPR2001 using molasses medium in a jar fermentor. Appl Microbiol Biotechnol 67:45–51

Battad-Bernardo E, McCrindle SL, Couperwhite I, Neilan BA (2004) Insertion of an E. coli lacZ gene in Acetobacter xylinus for the production of cellulose in whey. FEMS Microbiol Lett 231:253–260

Betlej I, Rybak K, Nowacka M, Antczak A, Borysiak S, Krochmal-Marczak B, Lipska K, Boruszewski P (2022) Structural properties of bacterial cellulose film obtained on a substrate containing sweet potato waste. Crystals 12:1191

Bilgi E, Bayir E, Sendemir-Urkmez A, Hames EE (2016) Optimization of bacterial cellulose production by *Gluconacetobacter xylinus* using carob and haricot bean. Int J Biol Macromol 90:2–10

Cakar F, Kat A, Ozer I, Demirbag DD, Sahin F, Aytekin AO (2014) Newly developed medium and strategy for bacterial cellulose production. Biochem Eng J 92:35–40

Carreira P, Mendes JAS, Trovatti E, Serafim LS, Freire CSR, Silvestre AJD, Neto CP (2011) Utilization of residues from agro-forest industries in the production of high value bacterial cellulose. Bioresour Technol 102:7354–7360

Carvalho RSFD, Mahnke LC, Palacio SB, Barbosa WT, Hodel KVS, Barbosa JDV, Melo FDAD, Chorilli M, Meneguin AB, Pinto FCM, Morais MAD Jr, Aguiar JLDA (2025) Bacterial cellulose hydrogel produced by *Gluconacetobacter hansenii* using sugarcane molasses as medium: physicochemical characterization for wound healing applications. Carbohydr Polym Technol Appl 9:100632

Cavalcanti YF, Amorim JD, Medeiros AD, Da Silva CJ Jr, Durval IJ, Costa AF, Sarubbo LA (2023) Microbial cellulose production with tomato (solanum Lycopersicum) residue for industrial applications. Chem Eng Trans 100:409–414

Cavka A, Guo X, Tang S-J, Winestrand S, Joensson LJ, Hong F (2013a) Production of bacterial cellulose and enzyme from waste fiber sludge. Biotechnol Biofuels 6:1–10

Cavka A, Guo X, Tang S-J, Winestrand S, Joensson LJ, Hong F (2013b) Production of bacterial cellulose and enzyme from waste fiber sludge. Biotechnol Biofuels 6:25

Cazon P, Puertas G, Vazquez M (2022) Production and characterization of active bacterial cellulose films obtained from the fermentation of wine bagasse and discarded potatoes by Komagateibacter xylinus. Polymers (Basel) 14:5194

Cerrutti P, Roldan P, Garcia RM, Galvagno MA, Vazquez A, Foresti ML (2016) Production of bacterial nanocellulose from wine industry residues: importance of fermentation time on pellicle characteristics. J Appl Polym Sci 133. https://doi.org/10.1002/app.43109

Chen G, Chen L, Wang W, Hong FF (2018) Evaluation of six ionic liquids and application in pretreatment of sweet sorghum bagasse for bacterial nanocellulose production. J Chem Technol Biotechnol 93:3452–3461

Chen G, Wu G, Chen L, Wang W, Hong FF, Joensson LJ (2019a) Comparison of productivity and quality of bacterial nanocellulose synthesized using culture media based on seven sugars from biomass. Microb Biotechnol 12:677–687

Chen J, Chen C, Liang G, Xu X, Hao Q, Sun D (2019b) In situ preparation of bacterial cellulose with antimicrobial properties from bioconversion of mulberry leaves. Carbohydr Polym 220:170–175

Chen L, Hong F, Yang X-X, Han S-F (2013) Biotransformation of wheat straw to bacterial cellulose and its mechanism. Bioresour Technol 135:464–468

Cheng Z, Yang R, Liu X, Liu X, Chen H (2017) Green synthesis of bacterial cellulose via acetic acid pre-hydrolysis liquor of agricultural corn stalk used as carbon source. Bioresour Technol 234:8–14

Chundawat SPS, Donohoe BS, Sousa LDC, Elder T, Agarwal UP, Lu F, Ralph J, Himmel ME, Balan V, Dale BE (2011) Multi-scale visualization and characterization of lignocellulosic plant cell wall deconstruction during thermochemical pretreatment. Energy Environ Sci 4:973–984

Ciecholewska-Jusko D, Broda M, Zywicka A, Styburski D, Sobolewski P, Goracy K, Migdal P, Junka A, Fijalkowski K (2021) Potato juice, a starch industry waste, as a cost-effective medium for the biosynthesis of bacterial cellulose. Int J Mol Sci 22:10807

Czaja WK, Young DJ, Kawecki M, Brown RM Jr (2007) The future prospects of microbial cellulose in biomedical applications. Biomacromolecules 8:1–12

Devanthi PVP, Kho K, Nurdiansyah R, Briot A, Taherzadeh MJ, Aslanzadeh S (2021) Do Kombucha symbiotic cultures of bacteria and yeast affect bacterial cellulose yield in molasses? J Fungi 7:705

Dhar P, Pratto B, Goncalves Cruz AJ, Bankar S (2019) Valorization of sugarcane straw to produce highly conductive bacterial cellulose/graphene nanocomposite films through in situ fermentation: kinetic analysis and property evaluation. J Clean Prod 238:117859

Dorame-Miranda RF, Gamez-Meza N, Medina-Juarez LA, Ezquerra-Brauer JM, Ovando-Martinez M, Lizardi-Mendoza J (2019) Bacterial cellulose production by *Gluconacetobacter entanii* using pecan nutshell as carbon source and its chemical functionalization. Carbohydr Polym 207:91–99

Dubey S, Singh J, Singh RP (2018) Biotransformation of sweet lime pulp waste into high-quality nanocellulose with an excellent productivity using *Komagataeibacter europaeus* SGP37 under static intermittent fed-batch cultivation. Bioresour Technol 247:73–80

El-Gendi H, Salama A, El-Fakharany EM, Saleh AK (2023) Optimization of bacterial cellulose production from prickly pear peels and its ex situ impregnation with fruit byproducts for antimicrobial and strawberry packaging applications. Carbohydr Polym 302:120383

El-Gendi H, Taha TH, Ray JB, Saleh AK (2022) Recent advances in bacterial cellulose: a low-cost effective production media, optimization strategies and applications. Cellulose (Dordrecht) 29:7495–7533

Erbas Kiziltas E, Kiziltas A, Gardner DJ (2015) Synthesis of bacterial cellulose using hot water extracted wood sugars. Carbohydr Polym 124:131–138

Fernandes IDAA, Pedro AC, Ribeiro VR, Bortolini DG, Ozaki MSC, Maciel GM, Haminiuk CWI (2020) Bacterial cellulose: from production optimization to new applications. Int J Biol Macromol 164:2598–2611

Gladysheva EK, Skiba EA, Zolotukhin VN, Sakovich GV (2018) Study of the conditions for the biosynthesis of bacterial cellulose by the producer *Medusomyces gisevii* Sa-12. Appl Biochem Microbiol 54:179–187

Goelzer FDE, Faria-Tischer PCS, Vitorino JC, Sierakowski M-R, Tischer CA (2009) Production and characterization of nanospheres of bacterial cellulose from Acetobacter xylinum from processed rice bark. Mater Sci Eng C 29:546–551

Gomes FP, Silva NHCS, Trovatti E, Serafim LS, Duarte MF, Silvestre AJD, Neto CP, Freire CSR (2013) Production of bacterial cellulose by *Gluconacetobacter sacchari* using dry olive mill residue. Biomass Bioenergy 55:205–211

Gorgieva S, Jancic U, Cepec E, Trcek J (2023) Production efficiency and properties of bacterial cellulose membranes in a novel grape pomace hydrolysate by *Komagataeibacter melomenusus* AV436T and *Komagataeibacter xylinus* LMG 1518. Int J Biol Macromol 244:125368

Guo X, Cavka A, Jonsson LJ, Hong F (2013) Comparison of methods for detoxification of spruce hydrolysate for bacterial cellulose production. Microb Cell Factories 12:93

Guo X, Chen L, Tang J, Joensson LJ, Hong FF (2016) Production of bacterial nanocellulose and enzyme from [AMIM]Cl-pretreated waste cotton fabrics: effects of dyes on enzymatic saccharification and nanocellulose production. J Chem Technol Biotechnol 91:1413–1421

Guzel M, Akpinar O (2019) Production and characterization of bacterial cellulose from citrus peels. Waste Biomass Valoriz 10:2165–2175

Guzel M, Akpinar O (2020) Preparation and characterization of bacterial cellulose produced from fruit and vegetable peels by *Komagataeibacter hansenii* GA2016. Int J Biol Macromol 162:1597–1604

Ha JH, Shehzad O, Khan S, Lee SY, Park JW, Khan T, Park JK (2008) Production of bacterial cellulose by a static cultivation using the waste from beer culture broth. Korean J Chem Eng 25:812–815

Hasanin MS, Abdelraof M, Hashem AH, El Saied H (2023) Sustainable bacterial cellulose production by Achromobacter using mango peel waste. Microb Cell Factories 22:24

Hong F, Guo X, Zhang S, Han S-F, Yang G, Joensson LJ (2012) Bacterial cellulose production from cotton-based waste textiles: enzymatic saccharification enhanced by ionic liquid pretreatment. Bioresour Technol 104:503–508

Hong F, Han S-F (2011) Biorefinery of bacterial cellulose from rice straw: enhanced enzymatic saccharification by ionic liquid pretreatment. Eng Sci 9:23–26

Hong F, Qiu K (2008) An alternative carbon source from konjac powder for enhancing production of bacterial cellulose in static cultures by a model strain *Acetobacter aceti* subsp. *xylinus* ATCC 23770. Carbohydr Polym 72:545–549

Hong F, Zhu YX, Yang G, Yang XX (2011) Wheat straw acid hydrolysate as a potential cost-effective feedstock for production of bacterial. J Chem Technol Biotechnol 86:675–680

Huang C, Guo H-J, Xiong L, Wang B, Shi S-L, Chen X-F, Lin X-Q, Wang C, Luo J, Chen X-D (2016) Using wastewater after lipid fermentation as substrate for bacterial cellulose production by *Gluconacetobacter xylinus*. Carbohydr Polym 136:198–202

Huang C, Yang XY, Xiong L, Guo HJ, Luo J, Wang B, Zhang HR, Lin XQ, Chen XD (2015) Evaluating the possibility of using acetone-butanol-ethanol (ABE) fermentation wastewater for bacterial cellulose production by *Gluconacetobacter xylinus*. Lett Appl Microbiol 60:491–496

Hussain Z, Sajjad W, Khan T, Wahid F (2019) Production of bacterial cellulose from industrial wastes: a review. Cellulose (Dordrecht) 26:2895–2911

Hyun JY, Mahanty B, Kim CG (2014) Utilization of Makgeolli sludge filtrate (MSF) as low-cost substrate for bacterial cellulose production by *Gluconacetobacter xylinus*. Appl Biochem Biotechnol 172:3748–3760

Jacek P, Dourado F, Gama M, Bielecki S (2019) Molecular aspects of bacterial nanocellulose biosynthesis. Microb Biotechnol 12:633–649

Jahan F, Kumar V, Saxena RK (2018) Distillery effluent as a potential medium for bacterial cellulose production: A biopolymer of great commercial importance. Bioresour Technol 250:922–926

Ji L, Xue W, Zhu L, Jiang J (2022) An alternative carbon source from cassava residue saccharification liquid for in-situ fabrication of polysaccharide macromolecule/bacterial cellulose composite hydrogel: a comparative study. Sustainability 14:14277

Joensson LJ, Martin C (2016) Pretreatment of lignocellulose: formation of inhibitory by-products and strategies for minimizing their effects. Bioresour Technol 199:103–112

Kalashnikova O, Pankova E, Sukhikh S, Babich O, Samusev I, Tcibulnikova A, Ivanova S, Kriger O (2024) Production of bacterial cellulose using a symbiotic consortium of bacteria and yeast on soybean molasses medium. LWT Food Sci Technol 205:116480

Kamal T, Ul-Islam M, Fatima A, Ullah MW, Manan S (2022) Cost-effective synthesis of bacterial cellulose and its applications in the food and environmental sectors. Gels 8:552

Kashcheyeva EI, Gismatulina YA, Budaeva VV (2019) Pretreatments of non-woody cellulosic feedstocks for bacterial cellulose synthesis. Polymers (Basel) 11:1645

Khan H, Saroha V, Raghuvanshi S, Bharti AK, Dutt D (2021) Valorization of fruit processing waste to produce high value-added bacterial nanocellulose by a novel strain Komagataeibacter xylinus IITR DKH20. Carbohydr Polym 260:117807

Khattak WA, Khan T, Ul-Islam M, Ullah MW, Khan S, Wahid F, Park JK (2015) Production, characterization and biological features of bacterial cellulose from scum obtained during preparation of sugarcane jaggery (gur). J Food Sci Technol (New Delhi) 52:8343–8349

Kim H, Son J, Lee J, Yoo HY, Lee T, Jang M, Oh J-M, Park C (2021) Improved production of bacterial cellulose through investigation of effects of inhibitory compounds from lignocellulosic hydrolysates. GCB Bioenergy 13:436–444

Kongruang S (2008) Bacterial cellulose production by *Acetobacter xylinum* strains from agricultural waste products. Appl Biochem Biotechnol 148:245–256

Kose R, Sunagawa N, Yoshida M, Tajima K (2013) One-step production of nanofibrillated bacterial cellulose (NFBC) from waste glycerol using Gluconacetobacter intermedius NEDO-01. Cellulose (Dordrecht) 20:2971–2979

Kuo C-H, Huang C-Y, Shieh C-J, Wang H-MD, Tseng C-Y (2019) Hydrolysis of orange peel with cellulase and pectinase to produce bacterial cellulose using *Gluconacetobacter xylinus*. Waste Biomass Valoriz 10:85–93

Kuo C-H, Lin P-J, Lee C-K (2010) Enzymatic saccharification of dissolution pretreated waste cellulosic fabrics for bacterial cellulose production by *Gluconacetobacter xylinus*. J Chem Technol Biotechnol 85:1346–1352

Kurosumi A, Sasaki C, Yamashita Y, Nakamura Y (2009) Utilization of various fruit juices as carbon source for production of bacterial cellulose by *Acetobacter xylinum* NBRC 13693. Carbohydr Polym 76:333–335

Lee AC, Salleh MM, Ibrahim MF, Bahrin EK, Jenol MA, Abd-Aziz S (2024) Pineapple peel as alternative substrate for bacterial nanocellulose production. Biomass Convers Biorefinery 14:5541–5549

Leonarski E, Cesca K, Zanella E, Stambuk BU, De Oliveira D, Poletto P (2021) Production of kombucha-like beverage and bacterial cellulose by acerola byproduct as raw material. LWT Food Sci Technol 135:110075

Li W, Zhang S, Zhang T, Shen Y, Han L, Peng Z, Xie Z, Zhong C, Jia S (2021) Bacterial cellulose production from ethylenediamine pretreated *Caragana korshinskii* Kom. Ind Crop Prod 164:113340

Li W, Shen Y, Liu H, Huang X, Xu B, Zhong C, Jia S (2023) Bioconversion of lignocellulosic biomass into bacterial nanocellulose: challenges and perspectives. Green Chem Eng 4:160–172

Lin D, Lopez-Sanchez P, Li R, Li Z (2014) Production of bacterial cellulose by *Gluconacetobacter hansenii* CGMCC 3917 using only waste beer yeast as nutrient source. Bioresour Technol 151:113–119

Lotfiman S, Awang Biak DR, Ti TB, Kamarudin S, Nikbin S (2018) Influence of date syrup as a carbon source on bacterial cellulose production by *Acetobacter xylinum* 0416. Adv Polym Technol 37:1085–1091

Lu T, Gao H, Liao B, Wu J, Zhang W, Huang J, Liu M, Huang J, Chang Z, Jin M, Yi Z, Jiang D (2020) Characterization and optimization of production of bacterial cellulose from strain CGMCC 17276 based on whole-genome analysis. Carbohydr Polym 232:115788

Luo M-T, Zhao C, Huang C, Chen X-F, Huang Q-L, Qi G-X, Tian L-L, Xiong L, Li H-L, Chen X-D (2017) Efficient using durian shell hydrolysate as low-cost substrate for bacterial cellulose production by *Gluconacetobacter xylinus*. Indian J Microbiol 57:393–399

Ma X, Yuan H, Wang H, Yu H (2021) Coproduction of bacterial cellulose and pear vinegar by fermentation of pear peel and pomace. Bioprocess Biosyst Eng 44:2231–2244

Machado RTA, Meneguin AB, Sabio RM, Franco DF, Antonio SG, Gutierrez J, Tercjak A, Berretta AA, Ribeiro SJL, Lazarini SC, Lustri WR, Barud HS (2018) *Komagataeibacter rhaeticus* grown in sugarcane molasses-supplemented culture medium as a strategy for enhancing bacterial cellulose production. Ind Crop Prod 122:637–646

Modenbach AA, Nokes SE (2014) Effects of sodium hydroxide pretreatment on structural components of biomass. Trans ASABE 57:1187–1198

Molina-Ramirez C, Enciso C, Torres-Taborda M, Zuluaga R, Ganan P, Rojas OJ, Castro C (2018) Effects of alternative energy sources on bacterial cellulose characteristics produced by *Komagataeibacter medellinensis*. Int J Biol Macromol 117:735–741

Moon S-H, Park J-M, Chun H-Y, Kim S-J (2006) Comparisons of physical properties of bacterial celluloses produced in different culture conditions using saccharified food wastes. Biotechnol Bioprocess Eng 11:26–31

Moradali MF, Rehm BH (2020) Bacterial biopolymers: from pathogenesis to advanced materials. Nat Rev Microbiol 18:195–210

Nakai T, Tonouchi N, Konishi T, Kojima Y, Tsuchida T, Yoshinaga F, Sakai F, Hayashi T (1999) Enhancement of cellulose production by expression of sucrose synthase in *Acetobacter xylinum*. Proc Natl Acad Sci USA 96:14–18

Ogrizek L, Lamovsek J, Cus F, Leskovsek M, Gorjanc M (2021) Properties of bacterial cellulose produced using white and red grape bagasse as a nutrient source. PRO 9:1088

Oz YE, Kalender M (2023) A novel static cultivation of bacterial cellulose production from sugar beet molasses: series static culture (SSC) system. Int J Biol Macromol 225:1306–1314

Pacheco G, Nogueira CR, Meneguin AB, Trovatti E, Silva MCC, Machado RTA, Ribeiro SJL, Da Silva Filho EC, Barud HDS (2017) Development and characterization of bacterial cellulose produced by cashew tree residues as alternative carbon source. Ind Crop Prod 107:13–19

Park SM, Yoon SJ, Son HJ, Lee CY, Kim HS (2010) Properties of bacterial cellulose cultured in different carbon sources. Polymer (Korea) 34:522–526

Podgorbunskikh EM, Bychkov AL, Lomovsky OI (2019) Determination of surface accessibility of the cellulose substrate according to enzyme sorption. Polymers (Basel) 11:1201

Premjet S, Premjet D, Ohtani Y (2007) The effect of ingredients of sugar cane molasses on bacterial cellulose production by *Acetobacter xylinum* ATCC 10245. Sen'i Gakkaishi 63:193–199

Rahman SSA, Vaishnavi T, Vidyasri GS, Sathya K, Priyanka P, Venkatachalam P, Karuppiah S (2021) Production of bacterial cellulose using *Gluconacetobacter kombuchae* immobilized on Luffa aegyptiaca support. Sci Rep 11:2912

Raiszadeh-Jahromi Y, Rezazadeh-Bari M, Almasi H, Amiri S (2020) Optimization of bacterial cellulose production by *Komagataeibacter xylinus* PTCC 1734 in a low-cost medium using optimal combined design. J Food Sci Technol (New Delhi) 57:2524–2533

Rani MU, Appaiah KAA (2013) Production of bacterial cellulose by *Gluconacetobacter hansenii* UAC09 using coffee cherry husk. J Food Sci Technol 50:755–762

Revin V, Liyaskina E, Nazarkina M, Bogatyreva A, Shchankin M (2018) Cost-effective production of bacterial cellulose using acidic food industry by-products. Braz J Microbiol 49:151–159

Revin VV, Dolganov AV, Liyaskina EV, Nazarova NB, Balandina AV, Devyataeva AA, Revin VD (2021) Characterizing bacterial cellulose produced by *Komagataeibacter sucrofermentans* H-110 on molasses medium and obtaining a biocomposite based on it for the adsorption of fluoride. Polymers (Basel) 13:1422

Saavedra-Sanabria OL, Duran D, Cabezas J, Hernandez I, Blanco-Tirado C, Combariza MY (2021) Cellulose biosynthesis using simple sugars available in residual cacao mucilage exudate. Carbohydr Polym 274:118645

Sadalage PS, Pawar KD (2021) Production of microcrystalline cellulose and bacterial nanocellulose through biological valorization of lignocellulosic biomass wastes. J Clean Prod 327:129462

Salari M, Sowti Khiabani M, Rezaei Mokarram R, Ghanbarzadeh B, Samadi Kafil H (2019) Preparation and characterization of cellulose nanocrystals from bacterial cellulose produced in sugar beet molasses and cheese whey media. Int J Biol Macromol 122:280–288

Saleh AK, El-Gendi H, Ray JB, Taha TH (2023) A low-cost effective media from starch kitchen waste for bacterial cellulose production and its application as simultaneous absorbance for methylene blue dye removal. Biomass Convers Biorefinery 13:12437–12449

Santoso SP, Lin S-P, Wang T-Y, Ting Y, Hsieh C-W, Yu R-C, Angkawijaya AE, Soetaredjo FE, Hsu H-Y, Cheng K-C (2021) Atmospheric cold plasma-assisted pineapple peel waste hydrolysate detoxification for the production of bacterial cellulose. Int J Biol Macromol 175:526–534

Sharifah FSM, Shahril M, Junaidi Z (2021) Oil palm frond juice and coconut water as alternative fermentation substrate for bacterial cellulose production. IOP Conf Ser Mater Sci Eng 1092:012055

Singh S, Simmons BA, Vogel KP (2009) Visualization of biomass solubilization and cellulose regeneration during ionic liquid pretreatment of switchgrass. Biotechnol Bioeng 104:68–75

Skiba EA, Budaeva VV, Ovchinnikova EV, Gladysheva EK, Kashcheyeva EI, Pavlov IN, Sakovich GV (2020) A technology for pilot production of bacterial cellulose from oat hulls. Chem Eng J (Amsterdam) 383:123128

Skiba EA, Shavyrkina NA, Skiba MA, Mironova GF, Budaeva VV (2023) Biosynthesis of bacterial nanocellulose from low-cost cellulosic feedstocks: effect of microbial producer. Int J Mol Sci 24:14401

Soemphol W, Hongsachart P, Tanamool V (2018) Production and characterization of bacterial cellulose produced from agricultural by-product by *Gluconacetobacter* strains. Mater Today Proc 5:11159–11168

Song Z, Yang G, Liu X, Yan Z, Yuan Y, Liao Y (2014) Comparison of seven chemical pretreatments of corn straw for improving methane yield by anaerobic digestion [Erratum to document cited in CA161:502821]. PLoS One 9

Souza EF, Furtado MR, Carvalho CWP, Freitas-Silva O, Gottschalk LMF (2020) Production and characterization of *Gluconacetobacter xylinus* bacterial cellulose using cashew apple juice and soybean molasses. Int J Biol Macromol 146:285–289

Sulaiman F (2023) Wastewater from the Arenga starch industry as a potential medium for bacterial cellulose and cellulose acetate production. Polymers (Basel) 15:870

Sun S, Sun S, Cao X, Sun R (2016) The role of pretreatment in improving the enzymatic hydrolysis of lignocellulosic materials. Bioresour Technol 199:49–58

Szczerbowski D, Pitarelo AP, Zandona Filho A, Ramos LP (2014) Sugarcane biomass for biorefineries: comparative composition of carbohydrate and non-carbohydrate components of bagasse and straw. Carbohydr Polym 114:95–101

Tonouchi N, Horinouchi S, Tsuchida T, Yoshinaga F (1998) Increased cellulose production from sucrose by Acetobacter after introducing the sucrose phosphorylase gene. Biosci Biotechnol Biochemist 62:1778–1780

Tsouko E, Kourmentza C, Ladakis D, Kopsahelis N, Mandala I, Papanikolaou S, Paloukis F, Alves V, Koutinas A (2015) Bacterial cellulose production from industrial waste and by-product streams. Int J Mol Sci 16:14832–14849

Tyagi N, Suresh S (2016) Production of cellulose from sugarcane molasses using *Gluconacetobacter intermedius* SNT-1: optimization & characterization. J Clean Prod 112:71–80

Ul-Islam M, Ullah MW, Khan S, Park JK (2020) Production of bacterial cellulose from alternative cheap and waste resources: a step for cost reduction with positive environmental aspects. Korean J Chem Eng 37:925–937

Uraki Y, Morito M, Kishimoto T, Sano Y (2002) Bacterial cellulose production using monosaccharides derived from hemicelluloses in water-soluble fraction of waste liquor from atmospheric acetic acid pulping. Holzforschung 56:341–347

Urbina L, Hernandez-Arriaga AM, Eceiza A, Gabilondo N, Corcuera MA, Prieto MA, Retegi A (2017) By-products of the cider production: an alternative source of nutrients to produce bacterial cellulose. Cellulose (Dordrecht) 24:2071–2082

Uzyol HK, Sacan MT (2017) Bacterial cellulose production by *Komagataeibacter hansenii* using algae-based glucose. Environ Sci Pollut Res 24:11154–11162

Vasconcellos VM, Farinas CS, Ximenes E, Slininger P, Ladisch M (2019) Adaptive laboratory evolution of nanocellulose-producing bacterium. Biotechnol Bioeng 116:1923–1933

Vazquez A, Foresti ML, Cerrutti P, Galvagno M (2013) Bacterial cellulose from simple and low cost production media by *Gluconacetobacter xylinus*. J Polym Environ 21:545–554

Wang Q, Nnanna PC, Shen F, Huang M, Tian D, Hu J, Zeng Y, Yang G, Deng S (2021a) Full utilization of sweet sorghum for bacterial cellulose production: A concept of material crop. Ind Crop Prod 162:113256

Wang S, Sun X, Yuan Q (2018) Strategies for enhancing microbial tolerance to inhibitors for biofuel production: a review. Bioresour Technol 258:302–309

Wang Y, Liu P, Zhang G, Yang Q, Lu J, Xia T, Peng L, Wang Y (2021b) Cascading of engineered bioenergy plants and fungi sustainable for low-cost bioethanol and high-value biomaterials under green-like biomass processing. Renew Sust Energ Rev 137:110586

Wu J-M, Liu R-H (2013) Cost-effective production of bacterial cellulose in static cultures using distillery wastewater. J Biosci Bioeng 115:284–290

Xu S, Xu S, Ge X, Tan L, Liu T (2022) Low-cost and highly efficient production of bacterial cellulose from sweet potato residues: optimization, characterization, and application. Int J Biol Macromol 196:172–179

Yang F, Cao Z, Li C, Chen L, Wu G, Zhou X, Hong FF (2023) A recombinant strain of *Komagataeibacter xylinus* ATCC 23770 for production of bacterial cellulose from mannose-rich resources. New Biotechnol 76:72–81

Yang XY, Huang C, Guo HJ, Xiong L, Li YY, Zhang HR, Chen XD (2013) Bioconversion of elephant grass (*Pennisetum purpureum*) acid hydrolysate to bacterial cellulose by *Gluconacetobacter xylinus*. J Appl Microbiol 115:995–1002

Ye J, Zheng S, Zhang Z, Yang F, Ma K, Feng Y, Zheng J, Mao D, Yang X (2019) Bacterial cellulose production by *Acetobacter xylinum* ATCC 23767 using tobacco waste extract as culture medium. Bioresour Technol 274:518–524

Zaki ZAA, Francis F, Zaini NAM (2023) Bacterial cellulose production from oil palm empty fruit bunch (OPEFB) hydrolysate using *Komagataeibacter xylinus* strain. Biomass Convers Biorefinery 13:14629–14640

Zhang J, Chen H, Jin M, Zheng F (2024) Biosynthesis of antibacterial bacterial cellulose in-situ using yellow wine wastewater through co-fermentation. Ind Crop Prod 222:119715

Zhang S, Winestrand S, Chen L, Li D, Jonsson LJ, Hong F (2014a) Tolerance of the nanocellulose-producing bacterium *Gluconacetobacter xylinus* to lignocellulose-derived acids and aldehydes. J Agric Food Chem 62:9792–9799

Zhang S, Winestrand S, Guo X, Chen L, Hong F, Jonsson LJ (2014b) Effects of aromatic compounds on the production of bacterial nanocellulose by *Gluconacetobacter xylinus*. Microb Cell Factories 13:62/1

Zhao H, Xia J, Wang J, Yan X, Wang C, Lei T, Xian M, Zhang H (2018) Production of bacterial cellulose using polysaccharide fermentation wastewater as inexpensive nutrient sources. Biotechnol Biotechnol Equip 32:350–356

Zou X, Wu G, Stagge S, Chen L, Jonsson LJ, Hong FF (2017) Comparison of tolerance of four bacterial nanocellulose-producing strains to lignocellulose-derived inhibitors. Microb Cell Factories 16:229/1

Chapter 5
Bacterial Cellulose as a Stabilizer of Multiphase Systems

Daniela Martins, Ana Cristina Rodrigues, Miguel Gama, and Fernando Dourado

Abstract Most cosmetic formulations are composed of various substances including water, oils and other moisturizing ingredients, active agents (anti-wrinkle, antioxidant, makeup, sunlight protection, cleansing), preservatives, perfumes, and texturing agents with specific functions. Such complex multiphase systems are conventionally stabilized by surfactant molecules, which may cause irritancy or other adverse reactions to the skin. Current trends in cosmetics, driven by the increasing demands of more informed and awareness among consumers and stricter regulations, include more sustainable, natural, and vegan formulations, non-irritant or skin sensitizing ingredients, and substitutes for ionic surfactants (Bianchet et al., Biotechnol Rep 27:e00502, 2020; Bom et al., J Clean Prod 225:270–290, 2019; Tournilhac and Lorant, Espec. Cosmet. Uses 1(12), 2003)). Oil-in-water emulsions are more frequent in the cosmetic field because water as the external phase provides a cooler, less greasy, and lighter feel when applied to the skin. Besides function, stability, and organoleptic properties of these products, shear-thinning behavior and particle suspension capabilities are among the desired properties of many cosmetic emulsions. The suspended materials can be functional (exfoliants, oils, encapsulates, and active agents) or simply aesthetic, and the shear-thinning response improves the spreadability and facilitates rubbing of the product on the skin. In this scope, natural solid particles—particularly BC—emerge as promising stabilizers and rheology modifiers for the production of green and safe cosmetic emulsions, a topic addressed in this chapter, including applications in other multiphasic systems (solid dispersions in liquids and foams).

Parts of this chapter are reproduced from the PhD Thesis available Open Access at (Martins 2021). https://hdl.handle.net/1822/76945

D. Martins · A. C. Rodrigues · M. Gama (✉) · F. Dourado
CEB – Centre of Biological Engineering, University of Minho, Campus de Gualtar, Braga, Portugal

LABBELS – Associate Laboratory in Biotechnology and Bioengineering and Microelectromechanical Systems, Braga, Guimarães, Portugal
e-mail: dsr.martins@ceb.uminho.pt; fmgama@deb.uminho.pt; fdourado@ceb.uminho.pt

F. Dourado et al. (eds.), *Advances in Bacterial Cellulose*,
https://doi.org/10.1007/978-3-032-07042-5_5

Keywords Bacterial nanocellulose · Multiphase systems · Interfacial stabilization · Pickering emulsions · Pickering foams

1 Introduction

1.1 Global Challenge: The Need for Sustainable Stabilizers in Multiphase Systems

Multiphase systems, characterized by the coexistence of two or more immiscible phases, are inherently unstable and pose a significant challenge across various industries, such as chemical processing, pharmaceuticals, food, and cosmetics. In these industries, ensuring proper distribution and interaction of phases is essential for product quality, process efficiency, and safety. There are several mechanisms responsible for the instability of these systems. These include gravitational separation (creaming or sedimentation) and Ostwald ripening (diffusion of droplets). The overall performance of the system as well as the effectiveness of various processes and products may be greatly impacted by such instability, which can result in phenomena like interfacial deformation, coalescence (the merging of droplets), and eventually phase separation. (Brennen 2005; McClements 2015; McClements and Jafari 2018; Murray et al. 2011; Taylor 1998). A lack of stabilization, for instance, can result in phase separation in cosmetic formulations like lotions and creams. This can change the texture and decrease the formulation's effectiveness, and if preservatives are not evenly distributed between phases, it may even lead to microbial growth. These failures may compromise consumer trust and present regulatory and financial risks (Binks and Horozov 2006; Michalek et al. 2019). Emulsions used in food applications, like those of salad dressings and milk-based drinks, are also very prone to destabilization. Phase separation in these products can result in undesirable changes in their appearance, taste, and shelf-life. For instance, in dairy-based emulsions, creaming or sedimentation can lead to the formation of a serum layer (fat-free liquid), negatively impacting consumer acceptance and product functionality (Dickinson 2009b). These instabilities can also affect the bioavailability of nutrients encapsulated within the dispersed phase, further complicating their use in functional foods (McClements 2018).

Traditionally, surfactants have been employed to stabilize multiphase systems by reducing the interfacial tension between two immiscible phases (McClements 2015). However, surfactants primarily act at the interface and do not contribute directly to the bulk properties of the system (Dickinson 2013). Also, surfactants can desorb from the interface under certain conditions, for instance, due to changes in temperature, pH, and ionic strength, leading to instability (Lopetinsky et al. 2006; Tadros 2013). Moreover, the global production and use of surfactants raise significant sustainability concerns, as many commercial surfactants are derived from petrochemicals, posing risks related to skin irritation, environmental persistence, and non-biodegradability (Bom et al. 2019). Their long-term sustainability is further threatened by consumer preferences for "clean-label" products and regulatory pressures (Badmus et al. 2021; Lochhead 2007; Lu et al. 2024; Michalek et al. 2019). This global trend of sustainability is pushing industries worldwide to look for more environmentally friendly options.

Due to these limitations, bio-based alternatives like proteins, biosurfactants, and polysaccharides (e.g., chitosan and alginate) are becoming increasingly popular. In particular, particle-based stabilizers (such as cellulose, silica, and clay) are gaining more interest given their functionalization potential, biocompatibility, and biodegradability. The latter method, known as Pickering stabilization, creates a physical barrier that inhibits coalescence and phase separation by using solid particles to adsorb at the interface between two immiscible phases. The effectiveness of particle stabilizers is due to their capacity to form a rigid, mechanically strong interfacial layer that can improve the long-term stability of emulsions, compared to surfactant-stabilized systems. Also, the particles can improve the stability of the biphasic system by structuring the continuous phase, increasing the viscosity, and preventing phase separation (Binks and Horozov 2006; Capron et al. 2017; Dickinson 2012; R. Jiang 2024; Tadros 2013; Viseras et al. 2021).

In multiphase systems, bacterial cellulose (BC) has clear advantages over synthetic stabilizers and plant cellulose. Unlike plant cellulose, which requires intensive processing to remove impurities, BC is synthesized by bacteria as a pure nanofibrillar network. Its unique structure provides exceptional mechanical strength and interfacial cohesion, forming dense barriers that resist coalescence more effectively than plant-based or synthetic counterparts. Also, BC is biodegradable in nature and can be produced from a variety of renewable waste feedstocks, obviating or reducing the environmental drawbacks of synthetic surfactants (Cruz et al. 2024; Kumar et al. 2024; Zhong 2020). Its hypoallergenic nature and biocompatibility make it an ideal candidate for sensitive applications like cosmetics. BC can also improve product functionality by stabilizing emulsions, providing texture and enabling the encapsulation of active ingredient. Its technological potential as a stabilizer and suspending agent for cosmetic creams and cleansing formulations has been mostly leveraged by multinational companies, the generated knowledge being found in the form of intellectual property (Caggioni et al. 2014; Heath et al. 2011; Swazey and Morrison 2018). Examples include patent US 2007/019777 A1 from CP Kelco, regarding BC containing formulations for use as rheology modifiers in products such as hair conditioners and hair styling products (Yang et al. 2007); patent WO 2011/019876 A2 from The Procter and Gamble Company, referring to personal cleansing formulations containing BC in combination with a cationic polymer as an external structuring system (Heath et al. 2011); patent US 6534071 B1 from L'Oréal, concerning oil-in-water emulsions for cosmetic uses containing commercial BC formulations (Tournilhac and Lorant 2003); and others, such as Colgate Palmolive (D'Ambrogio et al. 2011) and Unilever (Hu et al. 2012). These documents all point to a high level of industry interest in BC formulations for personal care and cosmetic products, which must be further investigated and reported systematically in the scientific literature. However, the widespread adoption of BC as a commercial stabilizer is contingent on overcoming economic and environmental hurdles. Currently, the production of BC is more expensive than that of plant cellulose, mainly due to the high costs associated with fermentation and purification processes. The use of low-cost feedstocks and the genetic strain engineering are among the promising advancements in biotechnology to reduce the BC production costs. Life cycle assessments (LCA) have shown that BC production has lower environmental impacts compared to other nanocelluloses in several mid-point impact categories (Aramwit 2016; Bianchet et al. 2020; Dourado et al. 2016; Martins et al. 2021; Oliveira et al. 2022; Silva et al. 2025; Zhong 2020). Ultimately, BC might be able to meet the co-occurrent demands of ecological responsibility and high-performance stabilization. Its unique biosynthesis, structural precision, and multifunctionality position BC as a transformative alternative in industries aiming to balance technical efficacy with sustainability imperatives.

This chapter aims to provide a comprehensive overview of the use of BC as a stabilizer in multiphase systems. It discusses the fundamental principles behind BC's stabilization mechanisms, reviews recent studies on its effectiveness, and explores the challenges and opportunities for its industrial application. With this, the chapter aims to contribute to the growing body of knowledge on sustainable materials and inspire further research and development in this area.

1.2 Multiphase Systems: Definition and Characteristics

Thermodynamically metastable mixtures of two or more immiscible phases are known as heterogeneous systems. These can be classified as solid-in-liquid dispersions (suspensions), gas-in-liquid (foams), or liquid-in-liquid (emulsions) in biphasic systems. We speak of oil-in-water (o/w) emulsions when oil is the dispersed/internal phase, and water is the continuous phase, or water-in-oil (w/o) emulsions, when the opposite occurs (Lam et al. 2014; Santos et al. 2016). Emulsions are unstable without the addition of an emulsifying agent or surfactant. The amphiphilic nature of these molecules drives their arrangement at the oil–water interface, with each moiety oriented towards either phase, depending on its relative affinity. By surrounding the dispersed phase droplets and decreasing the interfacial energy, these tensioactive agents stabilize the emulsion, preventing coalescence (merging) of the droplets and phase separation (Lam et al. 2014; Santos et al. 2016). Similarly to emulsions, foams are dispersions of gas bubbles in a liquid continuous phase. The tension at the air–water interface is higher due to the large density differences between the fluids, making this system even less stable than emulsions (Lam et al. 2014).

A stable system has a homogeneously (or colloidally) dispersed internal phase in equilibrium, maintaining this state over time and through environmental changes due to the action of stabilizing agents that prevent thermodynamic phenomena leading to phase separation ("breaking" of the system). The mechanisms of instability in biphasic systems are gravitational separation, coalescence, flocculation, and Ostwald ripening (Costa et al. 2019).

- Gravitational destabilization occurs as sedimentation when the internal phase has higher density than the continuous phase, or as creaming when reversed. Creaming occurs in o/w emulsions, as oil droplets ascend and form a top layer, while sedimentation occurs in w/o emulsions with water droplets depositing at the bottom.
- Ostwald ripening is the mass diffusion of the internal phase from smaller to larger droplets, due to pressure differences have higher Laplace pressure (internal pressure) from increased curvature, causing molecules to migrate from high-pressure regions (small droplets) to lower-pressure regions (larger droplets), leading to larger droplets growing at the expense of smaller ones (Ariyaprakai and Dungan 2010; Koroleva and Yurtov 2003). This phenomenon may be also explained, in a more general way, as arising from the higher specific energy of smaller particles, giving rise to mass transfer from those smaller particles (dispersed liquid or solid) to larger ones, reducing the interfacial energy of the system.
- Finally, flocculation, or coagulation/aggregation, of dispersed droplets occurs when Van der Waals attractive forces dominate over electrostatic repulsions, as described by the Derjaguin–Landau–Verwey–Overbeek (DLVO) theory. This theory differentiates between coagulation and flocculation based on the stability of the resulting aggregates. Coagulation creates stable structures in a primary energy minimum, requiring significant energy to disrupt, while flocculation involves weaker particle interactions in a secondary energy minimum, which can be easily reversed with minimal energy input (e.g., agitation) (Salou et al. 1998; Schowalter and Eidsath 2001).

The above phenomena can increase the risk of coalescence, resulting in a smaller surface area. Coalescence can also occur when no steric barrier exists between droplets and is favored when the concentration of the stabilizing agent is too low to fully saturate the interface between phases. To minimize energy imbalances (i.e., instability), molecules at the interface may redistribute themselves. Partially covered droplets may also undergo bridging flocculation by sharing the adsorbed stabilizing agent without coalescing. These interactions lead to the accumulation of the dispersed phase, eventually leading to phase separation. Smaller droplets with some surface charge and a monodisperse size distribution are less affected by these mechanisms, thus more stable (Chevalier and Bolzinger 2013; Costa et al. 2019; Dickinson 2009a, 2012).

2 Particle-Stabilized Systems: Pickering Emulsions and Foams

Recent years have shown a shift towards studying biological materials for stabilizing foams and emulsions in food, cosmetic, and drug delivery applications. Biosurfactants, for example, are less toxic, more biodegradable and can act in a wider range of conditions. Their biotechnological production is more environmentally friendly than chemically produced surfactants, and industrial by-products can be used in the culture media (Santos et al. 2016). On the other hand, particle-stabilized emulsions and foams—known as Pickering emulsions and foams—have also gained increasing interest. Examples of organic solid particles include biopolymers such as polysaccharides, aromatic macromolecules, and polypeptides). These particles irreversibly adsorb at the interface and form a film (monolayer or multilayer) around the dispersed phase droplets. The solid particle coating on the droplets surface acts as a mechanical barrier, preventing coagulation and coalescence phenomena (Binks 2002; Chevalier and Bolzinger 2013; Hu et al. 2015; Lam et al. 2014).

As a general rule, the size of the particles suitable for Pickering stabilization must be smaller than that of the dispersed phase droplets. First, because large or denser particles are more susceptible to the effect of gravity, leading to poor suspension stability and sedimentation; second, because the particles must adsorb to a curved surface, and large particles would protrude or stick out of the droplet surface (or the droplets would have to be much larger to accommodate the entirety of a stiff particle on its surface). Smaller particles can arrange more easily, packing side by side at the interface, covering it more efficiently without deforming it. Due to their small sizes, nanocelluloses are particularly advantageous for Pickering stabilization. Although they can have several μm in length, some studies have noted that they are flexible enough to adjust to the curved interface, slightly bending on top of the oil droplets (Capron et al. 2017). This behavior has been observed both in BC nanofibrils (BCNF) (Li et al. 2019) and BC nanocrystals (BCNC) (Kalashnikova et al. 2011, 2013), among others. Thus, flexibility arises as a relevant feature for Pickering stabilization.

Besides size, the morphology and affinity for each phase influence how solid particles organize at the interface of immiscible phases. Pickering stabilization occurs under the partial wetting conditions of solid colloidal particles, requiring some degree of amphiphilicity (Chevalier and Bolzinger 2013). The formation of o/w or w/o emulsions, foams, or aerosols (water-in-air dispersions) depends on the particle's wettability, measured by contact angle (Fig. 5.1). Hydrophilic particles are mostly wet by water and have a water contact angle below 90°; hydrophobic ones reside mostly in the oil or air phase, originating a contact angle above 90°. The particle layer around the disperse phase droplets curves in a way that the larger area of the particle surface remains on the external side, facing the phase with higher affinity, originating an air-in-water (a/w) or o/w system when particles are hydrophilic (<90°) and water-in-air (w/a) or w/o system for hydrophobic ones (>90°). However, extremely hydrophilic or hydrophobic particles may produce very large drops, more prone to coalescence and loss of stability, or disperse completely in one phase, without emulsification (Binks 2002; Chevalier and Bolzinger 2013; Lam et al. 2014). Cellulose has some degree of amphiphilic character, due to the different density of hydroxyl groups in the different faces of the crystal, as discussed ahead in more detail.

Foams are more challenging to stabilize than emulsions due to structural factors. First, gas–liquid interfaces have higher surface tension than liquid–liquid interfaces, making it harder for particles to adsorb and form stable barriers (Zhao et al. 2019).

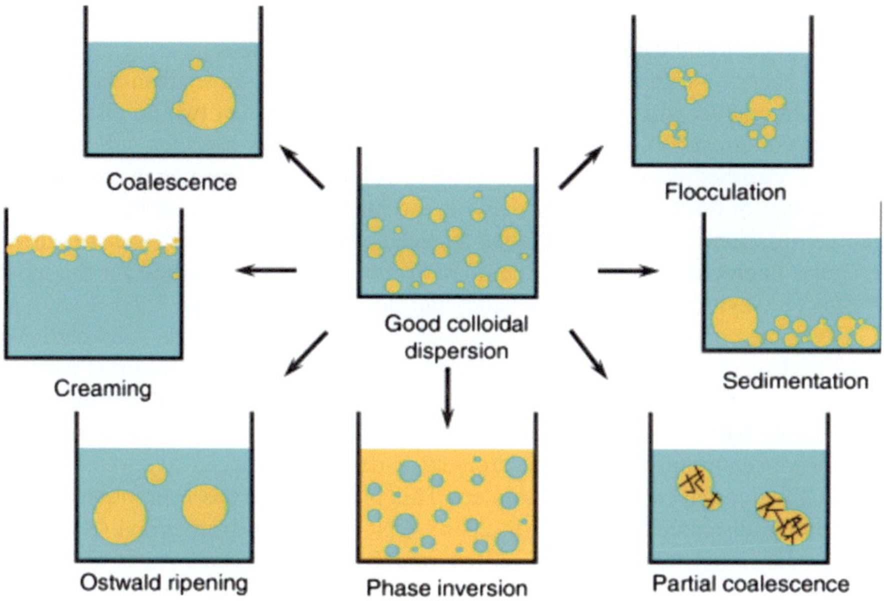

Fig. 5.1 Mechanisms of destabilization in colloidal systems. Reproduction from https://doi.org/10.1016/j.cis.2015.09.003, authorized by Elsevier

Additionally, gas bubbles are more prone to Ostwald ripening (gas diffusion between bubbles) due to the higher solubility and diffusivity of gases compared to oil molecules, complicating the control of the bubbles' size and increasing the risk of their collapse (Fameau et al. 2023). Second, foam films are thinner and more fragile than emulsion interfaces, making them more susceptible to rupture under gravitational drainage or mechanical stress (Zhao et al. 2019). Furthermore, the low viscosity of gas phases provides minimal viscous drag on the particles, facilitating the displacement of the particles from the interface (Sakthivel and Babu Salin 2021). Given these challenges, particles that successfully stabilize emulsions may fail to stabilize foams unless they possess specific characteristics, including appropriate contact angle (close to 90°), suitable size (typically submicron), and the ability to form dense interfacial networks that can withstand high capillary pressures (Binks 2002; Fameau et al. 2023; Hunter et al. 2008; Kruglyakov et al. 2011; Lam et al. 2014; Zhu et al. 2015). The stricter geometric constraints imposed by bubble curvature and the lower viscosity of gas phases further explain why particles must meet more demanding criteria for effective foam stabilization.

Theoretically, the free energy of spontaneous desorption—the energy required to remove a spherical particle from an interface—is given by Eq. (5.1):

$$\Delta G_d = \pi r^2 \gamma_{1,2} \left(1 - |\cos\theta|\right)^2 \tag{5.1}$$

where r is the particle radius, $\gamma_{1,2}$ is the interfacial tension between the two phases, and θ, the contact angle of the particle relatively to the aqueous phase (Binks 2002;

Dickinson 2012; Lam et al. 2014). This equation only applies to particles small enough to neglect the effect of gravity (under 1 μm in diameter).

This energy—with a maximum at 90° angle—is much larger than the thermal energy. In practice, this makes the adsorption at an interface of two immiscible liquids appear irreversible. According to the equation, larger particles will be more strongly adsorbed since the desorption energy is higher, but even nanoparticles have free energy of desorption higher than the thermal energy. For particles with other geometries, such as cellulose (found as needle-like nanocrystals, rod-shaped or irregular microparticles, fibers), the relationship needs to be adapted for the morphology (Aveyard et al. 2003; Chevalier and Bolzinger 2013) (Fig. 5.2).

Given the quasi-irreversible nature of adsorption to the interface, solid particles stabilized systems show a high resistance to coalescence, even at minimal particle concentrations—only enough to cover the interfacial area, even if partially. The concentration of solid particles also influences the system's interfacial area; increasing the concentration increases the droplets' surface area that can be covered, originating smaller and more stable droplets. This phenomenon has been observed for milled cellulose particles (Lu et al. 2018), CNCs obtained from wood (Bai et al. 2019) and cotton (Cherhal et al. 2016), and BCNC (Kalashnikova et al. 2011).

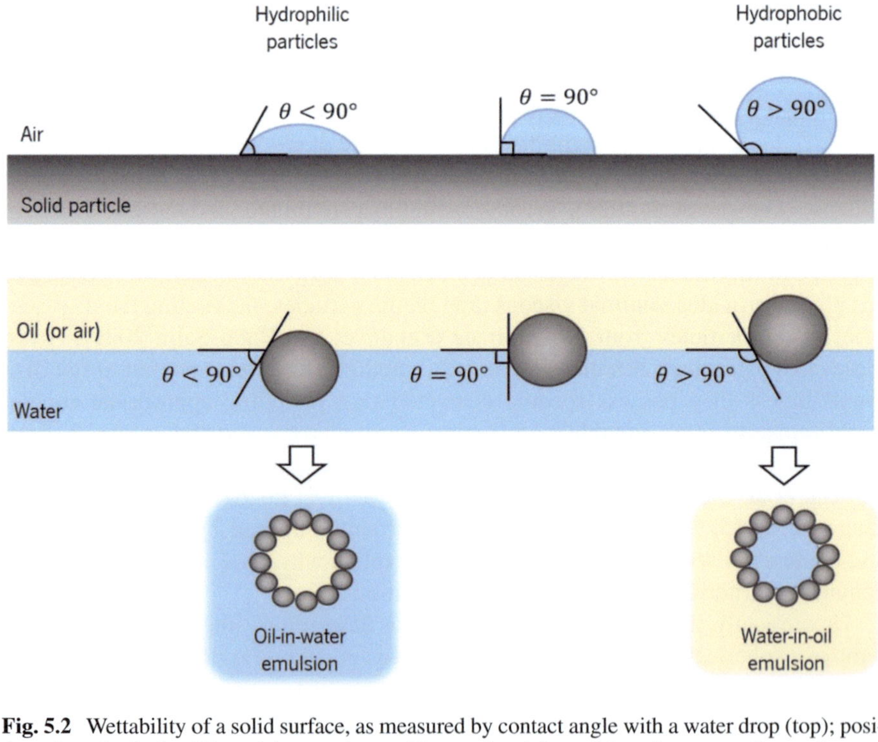

Fig. 5.2 Wettability of a solid surface, as measured by contact angle with a water drop (top); position of a spherical solid particle at a planar oil/water (or air/water) interface, according to its contact angle, and corresponding position at a curved interface resulting in oil-in-water or water-in-oil emulsions (bottom)

Further increases in the concentration, beyond the limit of the interfacial area, lead to the dispersion and accumulation of the solid particles in the continuous phase, further increasing the stabilization through steric effects, as discussed ahead (Binks 2002; Chevalier and Bolzinger 2013). Using rod-shaped cotton cellulose nanocrystals, Cherhal et al. (2016) observed that the droplet size of o/w emulsions is contingent upon the amount of oil and concentration of solid particles needed to cover the interfacial area; once adsorption occurs, increasing the water content has no effect on the system's stability. Besides concentration, the particle size determines the droplet size of the Pickering emulsion: smaller particles yield smaller droplets, which ultimately enhance their stability against gravitational forces (Binks and Lumsdon 2001; Pan et al. 2023; Tavacoli et al. 2012). Moreover, Niu et al. (2018) prepared Pickering o/w emulsions with CNCs produced from increasing hydrolysis durations. They observed that with longer hydrolysis time, the morphology of the resulting emulsion droplets became much more uniform, with reduced flocculation, attributing this to the higher number of smaller particles capable of covering a larger surface area.

Along with stabilization by adsorption to the interface of two fluids, solid particles possess an alternative stabilization mechanism: the structuring of the continuous phase. When the interfacial area of a biphasic system is completely covered, excess particles remain in the continuous phase. These particles can form a 3D network leading to an increase in the viscosity of the continuous phase. This results in a higher support to the disperse phase droplets and a reduction in coalescence and creaming. Several works have discussed the networking behavior of cellulose particles in aqueous media, a mechanism dependent on the particles' aspect ratio and concentration (Kalashnikova et al. 2013; Lu et al. 2018; Martins et al. 2020a; Souza et al. 2020; Winuprasith and Suphantharika 2015; Zhao et al. 2011). Hence, thickening of the emulsions and foams decelerates the processes that lead to phase separation and improves long-term storage. This thickening mechanism is especially valuable in cosmetic applications, ensuring that lotions, creams, hair conditioners, and foaming facial cleansers remain effective and aesthetically appealing throughout storage and use (Binks 2002; Chevalier and Bolzinger 2013; Lam et al. 2014). Similarly, in food applications, the structuring of the continuous phase by solid particles can significantly enhance the stability and texture of various products, for instance, in mayonnaise, dressings and sauces, and in foamed food products like mousses and whipped toppings (Abdul Hadi et al. 2024; Chen et al. 2020; Kelmer Müller and Costa 2024).

2.1 Relevant Properties of Cellulose in Interface Stabilization

Recent research has been focusing on the use of cellulose nanoparticles as stabilizers of biphasic systems, particularly emulsions. Nanocelluloses possess particular characteristics that are well-suited for stabilizing o/w interfaces. Other features must also be controlled to ensure the best outcome, as detailed below.

One critical property of cellulose particles is their structural anisotropy, leading to an amphiphilic character. As referred before, unlike strictly hydrophilic materials, which cannot adsorb at o/w or a/w interface, cellulose crystals are amphiphilic. This amphiphilicity arises because the smaller edge planes of the crystallites are richer in less polar domains (-CH), due to molecular alignment, creating a hydrophobic region. Conversely, the larger planes are rich in polar groups (-OH), imparting a hydrophilic character. This means that only a small area of the molecule has affinity to oil (or air), while the larger area is preferentially wetted by water (CA < 90°). This makes cellulose nanoparticles ideal for systems with an aqueous continuous phase, and a hydrophobic disperse phase such as oil or air (Capron et al. 2017; Costa et al. 2019; Fujisawa et al. 2017).

A second key property is the anisotropic shape of cellulose nanoparticles, specifically their high aspect ratio. As anisotropic particles with high aspect ratio, cellulose nanofibers, and nanocrystals tend to one-dimensionality since the width and height can be negligible in comparison to their length (Capron et al. 2017). This characteristic allows these particles to organize in different orientations and interact with each other at the interface, creating a steric barrier even at low concentrations. When the interface is mostly covered, the excess particles interact in the continuous phase forming self-assembled networks or gels.

Beyond their intrinsic properties, the source of cellulose and extraction/purification processes marginally influences their stabilizing efficacy. Native celluloses share a common crystal allomorph, cellulose I, with varying proportions of Iα and Iβ structures. For instance, BC is richer in Iα (60% and above), while wood and plant celluloses are richer in Iβ (Kondo et al. 2016; Moon et al. 2011; Saxena and Brown 2005). On the other hand, the size, aspect ratio, and surface charge of the particles, which depend on the source, crystallinity, and extraction/purification processes, can influence their organization in the interface, their interaction in the continuous phase, and therefore the stabilizing efficacy.

3 Studies on the Use of BC in Multiphase Systems

The potential of BC as a Pickering stabilizer and emulsifier has been widely acknowledged. BC has been investigated as an interfacial stabilizer in various forms, including as mechanically ground particles or chemically hydrolyzed nanofibrils and nanocrystals (Jia et al. 2016; Kalashnikova et al. 2011; Martins et al. 2020b). Grinding mechanically disintegrates BC into smaller fragments, breaking its fibrous network and reducing BC to bundles of different sizes, thereby improving dispersibility (Li et al. 2019; Li et al. 2023b; Martins et al. 2020a). Additionally, its surface can be modified with alkyl chains or surfactants to adjust its hydrophobicity (Bae et al. 2022). This versatility allows for its application in virtually all types of interfaces, either as the main stabilizing agent or as a co-adjuvant.

Martins and co-workers (Martins et al. 2020a) studied the effect of different wet-grinding methods on BC membranes, varying in applied shear force from hand

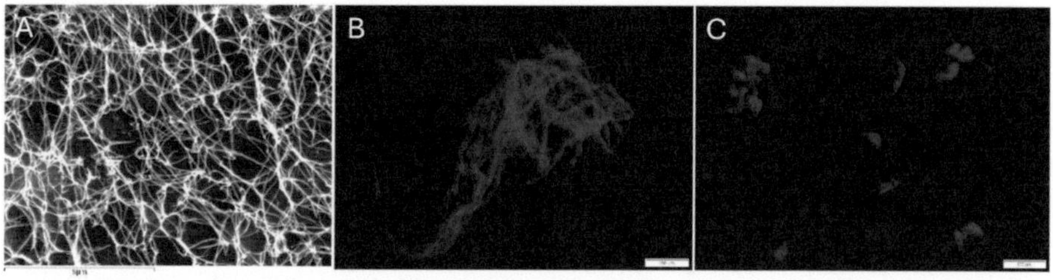

Fig. 5.3 Scanning electron micrograph of BC membrane (**a**); fluorescence micrographs of fiber bundles (stained with Calcofluor white), obtained from BC membranes processed using a hand blender (**b**) or 6 cycles of HPH (**c**). Scale bars correspond to 8 μm for A, 100 μm for B and C

blender to several cycles of high-pressure homogenization (HPH). After mechanical shear, the entangled nanofibrillar network was broken down into flocs, composed mainly of fiber bundles of variable sizes and shapes (Fig. 5.3). These bundles appeared as loose fiber branches emerging from a deeply entangled core; increasing grinding shear led to further fragmentation and decrease of the fiber bundles, accompanied by a decrease in functionality.

In food applications, a BC pulp prepared using high shear mixing and ultrasound, has been employed as a stabilizer for olive o/w (10:90) emulsions. Its performance was compared to that of carboxymethyl cellulose and hydroxypropyl methylcellulose (Paximada et al. 2016b). BC emulsions exhibited higher viscosity and stability than the soluble celluloses. Also, they were less affected by variations in pH, temperature, and salinity, formed larger droplets, yet remained more stable—with extremely low creaming and no signs of droplet coalescence. Li et al. (2019) used high-pressure homogenization (HPH) to obtain BC nanofibrils with different sizes. The HPH process further refined the ground BC, progressively reducing the size of the bundles with increasing homogenization cycles from 10 to 80. Specifically, the structure and morphology of the ground BC changed from flocculating, entangling, and bundle forming to a disordered and loose network by applying 40 cycles of HPH and was fully disrupted after 80 cycles. This treatment also reduced the apparent viscosity of the pseudoplastic BC pulp. O/w emulsions (10% dispersed phase) stabilized by 0.5% of BC nanofibers (BCNF) prepared with 40 HPH cycles became gel like, whereas with 0.1% the emulsions showed instability. The elongated shape and relative flexibility of the BCNF endowed them the ability to bend and overlap at the droplet surface, leading to an effective coverage (higher than 100%). BCNFs prepared with 20 or 40 cycles of HPH showed better emulsifying capacity, compared to those which underwent 10 or 80 cycles. Moreover, the nanofibers prepared by 40 cycles of HPH maintained the ability to form inter-droplet networks, further stabilizing the droplets by steric hindrance, and this system could resist changes in pH and ionic strength. Higher BCNF content led to the reduction of the average droplet size, as more particles were available to adsorb at the interface (Li et al. 2019).

A common modification of BC fibers involves adding negative surface charges, usually via acid hydrolysis or TEMPO-mediated oxidation, to improve their colloidal stability and functionality, particularly in o/w emulsions. As with physical disruption methods, hydrolysis or oxidation of BC affects the fibers' size and morphology, as well as their tendency for aggregation or dispersion. Hydrochloric acid hydrolysis of a BC pulp resulted in a random dispersion of nanofibers, with diameters within 30–80 nm and 100 nm to several μm in length, indicative of hydrogen bonds disruption and peeling of the crystalline cellulose fibrils (Zhai et al. 2018). The obtained nanofibers were able to significantly lower the IFT of peanut oil/w interface and produce food-grade nanoemulsions with low BC concentration. However, a pH-dependency was observed, as emulsions were only stable at pH ≥7. Within this pH range, the zeta potential of the emulsion droplets was negative, and its modulus increased with the concentration of hydrolyzed BCNF; at lower pH values, however, the net charge of the BC-covered droplets was much lower in modulus, and the emulsion instability in this pH range was attributed to the weak inter-droplet electrostatic repulsion (Zhai et al. 2018). In fact, at low pH, the negative charges imparted on BC by acid hydrolysis are neutralized due to charge screening by H^+. Thus, repulsive forces between interfacial fiber layers are weakened, and the entrapped emulsion droplets become more susceptible to flocculation and coalescence. A similar behavior has been observed by Yan et al. (2017) with BCNC prepared by sulfuric acid hydrolysis followed by oxidation of the pyranose ring with hydrogen peroxide, conferring carboxyl groups to BCNC (Yan et al. 2017). In this study, the properties and performance of the prepared BCNC were compared to unmodified BC in the emulsification of olive oil. Pristine BC exhibited an ultrafine network architecture, whereas BCNC appeared short and rod-like due to the removal of amorphous regions and cleavage of the crystalline nanofibers. Both cellulosic particles showed negative zeta potential values, but the charge modulus was higher for BCNC, derived from the surface carboxyl groups, which made the suspension more stable. At the same concentration, emulsions stabilized by BCNC had smaller droplets and narrower size distribution than those prepared with BC, indicating that BCNC possessed better emulsifying performance. Nevertheless, BCNC were more sensitive to changes in pH and ionic strength than BC, affecting the emulsion stability under certain conditions. The zeta potential of the BCNC became more negative with increasing pH, induced by the progressive increase of charged carboxyl groups, which generated a sufficient electrostatic repulsion force to disintegrate BCNC aggregates. This enhanced their emulsifying ability (increasing the emulsion serum index, ESI) with increasing pH, reaching an optimum at around pH 8; further increase was detrimental to the ESI, as excess charge density hindered the adsorption of BCNC at the o/w interface. On the other hand, increasing ionic strength (by addition of NaCl) led to an electrostatic shielding effect of the BCNC surface charges and consequent decrease of the zeta potential modulus towards zero, increasing particle aggregation. Emulsion stability was improved at low salt concentrations (below 80 mM), as some degree of charge screening promoted the adsorption of BCNCs at the oil/water interface. Higher salt concentration induced fast particle aggregation, causing insufficient surface coverage and a steep decrease in the ESI.

TEMPO-mediated oxidation is another common modification of BC fibers to improve their emulsifying properties, imparting a negative charge by the creation of carboxylic groups at the fibers surface. The works of Jia et al. (2016, 2019) and Wu et al. (2021) have focused on preparing BCNF with different carboxylate contents, by controlling the oxidation degree. The impact of the charged particles on emulsion formation and stability was studied. After oxidation, the dense porous BC network changed from showing coarse fibril bundles to an ultrafine structure of individualized and smaller nanofibrils, assigned to the partial dissolution of amorphous regions among fibrils or interfibrillar electrostatic repulsion provided by carboxylate groups. Jia and co-workers found that TEMPO-oxidized BC (TOBC) with increasingly higher carboxylate contents was more hydrophilic, which lowered the detachment energy of the particles from the interface, resulting in unstable paraffin o/w (1:1) emulsions. TOBC with lower oxidation degree (less charged) was shown to be better Pickering emulsion stabilizers than unmodified BC or highly charged TOBC. However, following a similar protocol, Wu's group obtained different results for dodecane-in-water (1:9) emulsions. Although those authors acknowledge that highly negative charges (leading to strong inter-fibril electrostatic repulsion) would hinder TOBC adsorption at the o/w interface, their findings showed that BC exhibited the highest surface activity and emulsifying ability. This was followed by TOBC with a higher oxidation degree, which also demonstrated improved interfacial behavior. The excellent stability of BC Pickering emulsions was attributed to the large fiber bundles creating a three-dimensional network and increasing viscosity, while the emulsions stabilized with highly charged TOBC benefitted from narrower emulsion sizes with increased inter-droplet distances due to electrostatic repulsion. It is worth to mention that these works differ on the o/w ratio and the type of oil used, which could help explain the seemingly contradictory results. Furthermore, the emulsification capacity of TOBC could also be regulated by modulating the dissociation level of carboxylic groups, varying pH and ionic strength conditions. For the highly charged TOBCs, emulsions prepared at pH 3.0 exhibited the highest viscosity, due to weakened electrostatic repulsions by protonation of the carboxylic groups favoring inter-fibril hydrogen bonding. This led to the effective emulsion stability despite the larger emulsion droplet sizes; at pH 11.0, emulsion viscosity and ESI were the lowest, as a result from the destruction of the hydrogen bonds, and the more pronounced creaming effect was attributed to decreased electrostatic repulsive force between droplets. With increasing ionic strength, the charge screening effect by the presence of NaCl facilitated the adsorption of TOBC at the interface and led to the formation of fibril clusters and a denser particle layer, accompanied by a tighter fibril network on the continuous phase. Emulsion droplet size increased with NaCl concentration, but the emulsion viscosity and elasticity also increased, leading to an overall positive effect to emulsion stability which translated to high ESI values (Wu et al. 2021).

The work of Pinto et al. (2024) focused on the use of BC nanocrystals and nanofibrils (prepared by acid hydrolysis and TEMPO-oxidation, respectively) for the stabilization of food-grade low o/w emulsions (1% oil). Examples of such systems include beverages and soft drinks, as well as low-fat dressings and sauces. The

BCNF were individualized and well distributed in appearance, with an average diameter of 61 nm and length reaching several micrometers; BCNC had an average length of 425 nm and an average width of 40 nm. Both BCNC and BCNF exhibited a high negative surface charge and were able to stabilize low-oil Pickering emulsions, showing a dependence on ionic strength. The authors observed that these particles were preferentially located at the surface of the emulsion droplets in the presence of salt, creating a robust barrier against coalescence and Ostwald ripening. While the presence of salt was determinant for the long-term stability of emulsions with lower concentrations of BC particles, higher concentrations were also capable of maintaining emulsion stability without salt, most probably due to a superior coating of the interface and enhanced structuring of the continuous phase.

Another strategy to enhance the functionality of BC in the stabilization of emulsions involves the addition of a second component. For food-grade emulsions, BC has been combined with proteins and soluble polysaccharides. For instance, Paximada et al. (2016a) prepared olive oil-in-water (10:90) emulsions using whey protein isolate (WPI, within 2–5% in the emulsion) as the primary stabilizer, and BC was subsequently added. The addition of BC (up to 1% m/m BC in the emulsion) improved the physical properties of the emulsions by acting as a thickener in the continuous phase. Specifically, BC enhanced the stability and viscosity of the emulsions, preventing phase separation and reducing creaming, an effect dependent on the concentration. BC can adsorb onto the WPI-coated droplets, forming a secondary layer that helps to reinforce the structure of the emulsion. At higher concentrations, BC also creates a network within the aqueous phase, further stabilizing the oil droplets by limiting their movement and preventing coalescence. A comparative rheological analysis of emulsions prepared with xanthan gum and locust bean gum (LBG) demonstrated that BC serves as an effective alternative thickener, even at lower concentrations. According to the study, BC at 0.1% wt was sufficient to obtain the same yield stress as XG at 0.7% wt and LBG at 1% wt (Paximada et al. 2016a). Liu et al. (2021) prepared composite colloidal particles of BCNC, obtained through acid hydrolysis, and soybean protein isolate (SPI) using an antisolvent precipitation approach and freeze-drying, yielding stable dispersions at varying BCNC-to-SPI ratios. An increase in the BCNC-to-SPI ratio enhanced the hydrophobicity of the composite particles, leading to a significant rise in the water contact angle measured on films of freeze-dried particles. A maximum contact angle of 84.5° at a BCNC-to-SPI ratio of 7:25 was measured. These particles effectively stabilized high internal phase emulsions containing 75% sunflower oil. The BCNC-to-SPI ratio of 7:25 exhibited the highest emulsifying capacity among all tested formulations, producing emulsions with small droplet sizes and a uniform microstructure. BCNCs played a crucial role in enhancing the emulsifying performance of SPI by forming a gel-like network at the oil–water interface, reinforcing the structural integrity of the emulsions. Additionally, BCNC functioned as rheological modifiers, improving the viscoelastic properties of SPI solutions and contributing to the long-term stability (2 months, at room temperature) of the emulsions by preventing phase separation and maintaining droplet size distribution (Liu et al. 2021).

A distinct effect of BC was described in the work of Hou et al. (2021) as they investigated the role of BC fibers (up to 0.3% m/m/ in the emulsion) in stabilizing cyclodextrin-based soybean oil-in-water emulsions (10% m/m soybean and 0.8% m/m cyclodextrin) by preventing droplet flocculation. BC was introduced into the pre-prepared emulsions during a second homogenization step, allowing it to interact with the cyclodextrin inclusion complexes (ICs) at the droplets' surface. At low concentrations, BC enhanced the packing density of ICs at the surface of oil droplets. In this scenario, BC acted as a non-adsorbing polymer, remaining dispersed in the continuous phase. Because BC fibers take up space in the continuous phase, they create a "crowding effect," pushing the ICs closer together at the droplet interface. This phenomenon, known as excluded volume repulsion, forces the ICs to pack more densely, forming a stronger and more uniform protective layer around the droplets. This effect improves the stability of the emulsion. At higher concentrations, BC formed an entangled network within the continuous phase, restricting droplet mobility and further suppressing flocculation. These dual bulk and interfacial contributions of BC significantly improved emulsion stability, with effective suppression of phase separation even at concentrations as low as 0.01 wt% (Hou et al. 2021).

Different from the excluded volume repulsion effect, the depletion stabilization mechanism has been described for several Pickering emulsions, including plant celluloses (Bai et al. 2018). The addition of a non-adsorbing particle or polymer into the continuous phase creates depletion zones around droplets or colloids (a region where the particles concentration is lower than the bulk). When two droplets approach closely, their depletion zones overlap. This reduces the total free volume available to the particles in the system, leading to a higher effective concentration in the surrounding bulk phase. This higher concentration creates an osmotic pressure on the droplet surfaces facing the particle-depleted gap. Since there are no particles within the gap to exert counter-pressure, this net outward pressure generates an attractive force, leading to flocculation due to depletion attraction, until a threshold value above which the flocculation rate starts to decrease; then, for higher concentrations, stability improves by the so-called depletion stabilization mechanism—the depletion of polymer molecules from the space between particles creates a repulsive force that pushes the particles apart (Shahbazi et al. 2024). In emulsions prone to creaming, three concentration-dependent depletion regimes can be observed: in the first type, at very low concentrations of the non-adsorbing polymer, it does not significantly affect the emulsion behavior nor its stability, and the emulsions undergo creaming; in the second regime, the concentration of the non-adsorbing polymer is sufficient to induce significant phase behavior changes; droplet flocculation is visible and proportional to polymer concentration. This flocculation can lead to the formation of weak, interconnected networks that resist gravitational settling, thereby gradually reducing creaming; in the third regime, emulsions become fully stable against creaming and sedimentation due to dominant depletion stabilization. Further increase in the concentration of polymer results only in rheological changes, leading to the formation of emulsion gels (Bai et al. 2018) (Fig. 5.4).

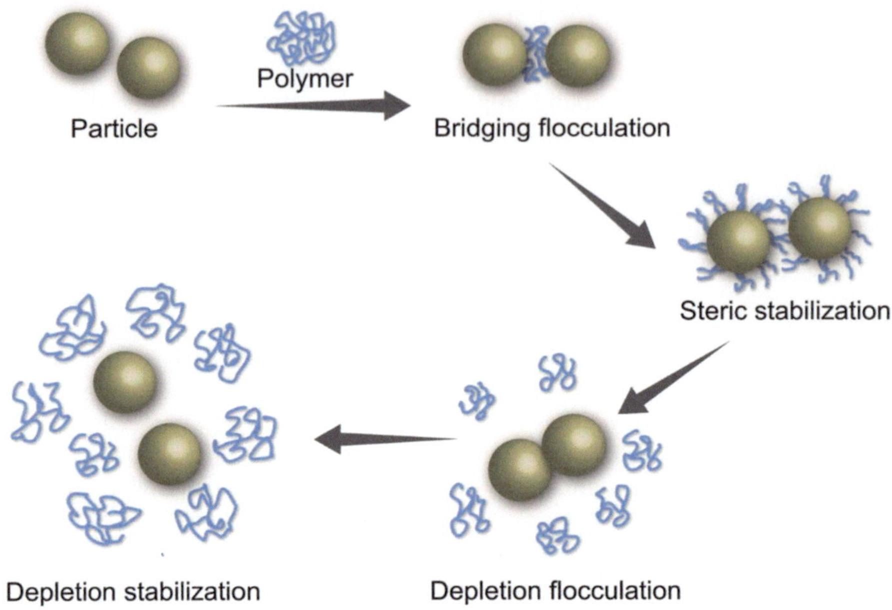

Fig. 5.4 The effects of polymer on colloid stability mechanism. Different stabilizing or flocculating effects are highlighted, according to the polymer's concentration and ability to adsorb on the colloid. Reproduction from https://doi.org/10.1016/j.colsurfa.2024.133316, authorized by Elsevier

The work of Zhang and colleagues (Zhang et al. 2022) highlighted the depletion stabilization effect attained with BC and carboxymethyl chitosan (CCS) complexes, prepared via electrostatic interaction. Briefly, BC's nanofiber structure (0.0–0.2% m/m in the final emulsion) interacted with CCS (0.5% m/m in the final emulsion), an anionic polysaccharide, leading to pH-dependent assembly. At lower pH, stronger electrostatic attraction created tighter complexes, while at higher pH, repulsion prevented aggregation, promoting depletion stabilization. These complexes contributed to emulsion stability by either adsorbing at the o/w interface or remaining dispersed in the continuous phase, influencing droplet interactions. The soybean oil concentration in the emulsion was 50 wt%, as it was mixed with the BC/CCS complex dispersions at a 1:1 volume ratio during formulation The adsorption behavior of BC/CCS complexes at the soybean oil–water interface was analyzed using quartz crystal microbalance with dissipation monitoring (QCM-D), revealing that adsorption occurred at pH 7.0 but was negligible at pH 9.6. Despite this, stable emulsions were obtained at both pH values, suggesting distinct stabilization mechanisms. At pH 7.0, stabilization was primarily achieved through steric hindrance, as BC/CCS complexes adsorbed at the interface and formed a protective layer around the oil droplets. In contrast, at pH 9.6, where adsorption was minimal, stabilization was driven by depletion effects, with BC/CCS complexes dispersed in the continuous phase restricting droplet movement and preventing coalescence. The magnitude of this depletion effect was directly correlated with the concentration of BC; emulsions lacked stability without BC, but increasing its concentration enhanced the stability by suppressing droplet flocculation. While CCS alone was sufficient to stabilize emulsions at pH 7.0, the presence of BC at higher concentrations contributed to the formation of a crosslinked network in the continuous phase, further improving emulsion stability and reducing creaming.

Other authors have added polysaccharides to the fermentation media. This in situ (during fermentation) modification method has been used, for example, to produce BC composites with sodium alginate of different molecular weights (Jiang et al. 2020). The sodium alginate endowed BC with a hydrogen bonding system resulting from the carboxylic acid groups, altering its amphiphilicity. The BC composites with high molecular weight alginate displayed narrower fibers and less intramolecular hydrogen bonds, producing emulsions with smaller and more uniform oil droplets.

Using hydrophobically modified BCNF, grafted with a C18 alkyl chain, Seo et al. (2021) produced N-decane o/w emulsions (20% oil) containing avobenzone, a lipophilic chemical UV filter. The C18BCNF showed increased surface activity and was able to create a thin membrane at the oil–water interface, yielding highly stable Pickering emulsions, resistant to freeze-thawing cycles. The C18BCNF-armored Pickering emulsions could entrap metal ions, limiting its influx into the internal phase, thus preventing the denaturation of avobenzone. This way, the UV-blocking performance of the emulsions was improved in 20% after 1 h of UV irradiation, compared to conventional Tween 80-stabilized emulsions.

Besides o/w emulsions, with great applicability in several areas as shown in this chapter, some works have studied the utilization of BC in different emulsion systems. High internal phase emulsions (HIPE), for example, differ from common o/w emulsions due to the high oil volume fraction (more than 74%), which leads to close packing of the oil droplets. The stabilization of HIPE systems generally requires a high concentration of surfactants and is prone to phase inversion (Capron and Cathala 2013). Li and co-workers (Li et al. 2021) showed that HPH-treated BCNF can be used as a green alternative to stabilize such systems. HIPE containing 75% dodecane were produced with only 0.3% BCNF, showing excellent physical stability and no creaming behavior.

Other authors have succeeded in stabilizing w/o emulsions using BC, either by hydrophobization of the nanofibrils by acetylation (Sousa et al. 2017) or their combination with cationic surfactant-coated particles (Cho et al. 2021). Furthermore, Lei et al. (2022) have demonstrated that TEMPO-oxidized BC nanofibrils are capable of stabilizing water-in-water (w/w) emulsions, composed of immiscible dextran and maltodextrin solutions. This uncommon system was proposed for the encapsulation of water-soluble bioactive ingredients in food-grade aqueous products. The formation of a viscoelastic network structure was proposed as the main stabilization mechanism, preventing the system's destabilization for a period of at least 7 days.

Still regarding food emulsions, recent studies have also been oriented towards BC's effect as a lipid and nutrient protector, being capable of improving the stability and preventing the oxidation of sensitive oils (Wang et al. 2024), liposoluble vitamins (Bae et al. 2022), antioxidants, and other bioactive molecules (Li et al. 2023b;

Shen et al. 2023). This is assigned to the mechanical barrier created by the BC around the dispersed phase droplets, slowing down lipid digestion and improving the bioaccessibility of compounds of interest (Shen et al. 2023; Wu et al. 2025). Furthermore, BC-based Pickering emulsions have been applied in the development of food packaging solutions (Chen et al. 2025; Li et al. 2020; Li et al. 2023a; Xia et al. 2023), and even on 3D-printed meat substitutes (Wang et al. 2023), demonstrating a wide applicability and functionality in the food industry.

Cellulose has also captured interest as a foam stabilizer. Although some reports concern dry aerogels (Cervin et al. 2013; Zhang et al. 2020), cellulose particles have also been applied to aqueous foams (Capron et al. 2017; Cervin et al. 2015; Hu et al. 2016). As mentioned before, the stabilization of foams is more demanding than emulsions. Despite the amphiphilic character of cellulose molecules, sufficient for stabilizing emulsions without substantial modification, the high hydrophilicity drives their accumulation in the plateau borders of foams; for their adsorption at the air/water interface, hydrophobization or combination with other surface-active molecules is necessary (Cervin et al. 2015; Hu et al. 2016; Lam et al. 2014). Foaming also requires the fast adsorption of the stabilizer to the a/w interface, which is sometimes impaired by the size of the solid particles. Despite the drawbacks of native cellulose particles as foaming agents, they have proved to be great foam stabilizers. More recently, BC has been used in the stabilization of protein food foams, without modification (Martins et al. 2024). The addition of 0.1% BC increased the foam capacity of egg white protein (EWP) and soy protein isolate (SPI) and improved the foam stability by arranging in the plateau borders and nodes of the foams, clogging the water channels and preventing drainage, also contributing for the structuring of the continuous phase. The foam stabilization was further improved by increasing BC concentration. Interestingly, BC improved the foam capacity of each protein at its most favorable pH. Larger BC bundles yielded higher foam overrun, so the authors hypothesized that larger flakes were acting as anchoring sites for the air bubbles during foam formation. Other works have also demonstrated the stabilization and strengthening effect of BC modified with surfactants (Rühs et al. 2018) highlighting the importance of the filamentous nanoparticle films formed at the interface, reducing the contact between bubbles (and their coalescence), preventing dehydration of the aqueous film, and ultimately improving stability (Huang et al. 2024).

4 Market-Oriented Developments

From the above review, not much emphasis was placed on the suitability of such materials for real industrial application. In an effort to enhance the appeal of BC to industry stakeholders and facilitate its adoption in large-scale applications, Gama and co-workers (Martins et al. 2020a, b) focused on the development of a powdered BC formulation, easily dispersible and able to retain the functional properties after drying and rehydration. Upon drying, native cellulose particles tend to irreversibly aggregate due to the formation of inter-chain hydrogen bonds, a process known as hornification. A substantial amount of energy is required to rehydrate and redisperse the cellulose into a colloidal suspension. To prevent this, the authors mixed never-dried BC with carboxymethyl cellulose (CMC), a soluble and negatively charged polymer. CMC adsorbs onto BC, imparting a surface charge that improves the fibers' dispersion due to electrostatic repulsive forces and prevents hornification during drying. The dried (and milled) BC:CMC formulation (at a ratio of 1:1) could be completely redispersed in water within 5 min, under low-energy magnetic stirring, recovering the original three-dimensional fiber structure, preserving its rheological and functional properties. Powdered BC:CMC possessed crucial characteristics for an effective Pickering stabilization such as adequate surface charge, increased hydrophobicity compared to BC, the ability to reduce the oil–water interface energy (contrarily to CMC alone), and thickening and structuring capacity. Its emulsifying and stabilizing properties were benchmarked against commercially available plant-derived celluloses, showing a superior performance even at lower concentration. This enhanced functionality was attributed to the high aspect ratio and flexible nature of the BC fibers and their ability to structure the continuous phase. In a more in-depth comparison with xanthan gum—a staple of hydrocolloids and food industry additives—BC:CMC proved to have a more consistent behavior under varying environmental conditions. While both CMC and xanthan gum are polyelectrolytes and have their behavior affected by pH and ionic strength, after rehydration of the BC:CMC formulation, the BC fibers are responsible for the material's rheological, emulsifying, and structuring properties. This means that even at very low pH values or high salt concentrations, a short high-shear treatment of the fibers in aqueous media is sufficient to activate their assembly into a three-dimensional network. Thus, contrarily to xanthan gum, the viscosity of BC:CMC was mostly unaffected by pH (in the range from 2.8 to 10.4), ionic strength (from 0 to 350 mM NaCl), or temperature (between 6 and 80 °C). It was also capable of stabilizing sunflower oil-in-water emulsions with salt and an acidic pH, emulating food emulsions with the more demanding conditions. By increasing the BC to CMC ratio (up to 1:0.5), the rheological and emulsifying properties of the mixture were further improved, without losing the ability to redisperse in water under low-energy mixing. Contrarily to xanthan gum, these materials had a measurable yield stress—the minimum stress required for the material to initiate flow, and indicator of the system's integrity and stability. The BC:CMC formulation with higher fiber content also presented high viscosity, the strongest elastic gel behavior, and the best emulsifying performance (producing smaller oil droplets, more consistent in size, and emulsions resistant to centrifugation). The robust stability of BC:CMC's rheological properties in different environments could represent an advantage in comparison to xanthan gum and other common hydrocolloids for food emulsions and other multiphase systems (Martins et al. 2023).

Given the promising results here described, it is not surprising that a few BC-based products are being marketed already. It is the case of cellulon or arabalon, liquid formulations manufactured by CPKelco, used for detergent formulations. The great potential is drawing the interest of investors, and several start-ups are active in the field. The coming years will probably see new products and more players advancing the translation of BC to the cosmetic market, whereas in the food market regulatory constraints in Europe and the USA will refrain such applications for longer time.

5 Conclusions

This chapter presented a comprehensive review of BC as an effective stabilizer of multiphase systems, with applications in cosmetic and food and emphasized its unique properties and its dual mechanism of stabilization of these systems. It discussed its effectiveness in different forms, including mechanically ground particles and chemically modified nanofibrils and nanocrystals. It further acknowledged the key challenges related to its production costs and environmental impacts, proposing that advancements such as the use of low-cost feedstocks and genetic strain engineering could be used to overcome these challenges.

In parallel, this chapter also looked at practical considerations for the industrialization of BC formulations, particularly the development of an easily dispersible dried BC formulations, through co-processing with carboxymethyl cellulose. Such major advancement may set the stage for BC to expand beyond current niche applications and become a mainstream sustainable ingredient.

In sum, despite existing challenges in achieving a cost-effective mass production and overcoming current regulatory constraints, the well-established advantages of BC strongly support its premium positioning by meeting the growing demands for ecologically responsible and high-performance stabilizers across various industries.

References

Abdul Hadi N, Ashaari A, Matos M, Wan Rasdi N (2024) Exploring particle-based stabilisation of Pickering emulsions in food, aquaculture, and industrial applications. Int J Food Sci Technol 59:6834–6855. https://doi.org/10.1111/ijfs.17478

Aramwit P (2016) Introduction to biomaterials for wound healing. In: Ågren MS (ed) Wound healing biomaterials. Volume 2: functional biomaterials. Woodhead Publishing. https://doi.org/10.1016/B978-1-78242-456-7.00001-5

Ariyaprakai S, Dungan SR (2010) Influence of surfactant structure on the contribution of micelles to Ostwald ripening in oil-in-water emulsions. J Colloid Interface Sci 343(1):102–108. https://doi.org/10.1016/j.jcis.2009.11.034

Aveyard R, Binks BP, Clint JH (2003) Emulsions stabilised solely by colloidal particles. Adv Colloid Interf Sci 100–102:503–546. https://doi.org/10.1016/S0001-8686(02)00069-6

Badmus SO, Amusa HK, Oyehan TA, Saleh TA (2021) Environmental risks and toxicity of surfactants: overview of analysis, assessment, and remediation techniques. Environ Sci Pollut Res 28(44):62085–62104. https://doi.org/10.1007/s11356-021-16483-w

Bae J, Seo HM, Shin K, Choi J, Lee DR, Jiang Z, Shen D, Kim JW (2022) Hydrophobically modified cellulose nanofibers-enveloped solid lipid microparticles for improved antioxidant cargo retention. Macromol Rapid Commun 43(7):1–8. https://doi.org/10.1002/marc.202100917

Bai L, Huan S, Xiang W, Rojas OJ (2018) Pickering emulsions by combining cellulose nanofibrils and nanocrystals: phase behavior and depletion stabilization. Green Chem 20(7):1571–1582. https://doi.org/10.1039/c8gc00134k

Bai L, Lv S, Xiang W, Huan S, McClements DJ, Rojas OJ (2019) Oil-in-water Pickering emulsions via microfluidization with cellulose nanocrystals: 2. In vitro lipid digestion. Food Hydrocoll 96(March):709–716. https://doi.org/10.1016/j.foodhyd.2019.04.039

Bianchet RT, Vieira Cubas AL, Machado MM, Siegel Moecke EH (2020) Applicability of bacterial cellulose in cosmetics – bibliometric review. Biotechnol Rep 27:e00502. https://doi.org/10.1016/j.btre.2020.e00502

Binks BP, Lumsdon SO (2001) Pickering emulsions stabilized by monodisperse latex particles: effects of particle size. Langmuir 17(15):4540–4547. https://doi.org/10.1021/la0103822

Binks BP (2002) Particles as surfactants - similarities and differences. Curr Opin Colloid Interface Sci 7:21–41. https://doi.org/10.2118/141256-PA

Binks BP, Horozov TS (2006) Colloidal particles at liquid interfaces. Cambridge University Press. https://doi.org/10.1017/CBO9780511536670

Bom S, Jorge J, Ribeiro HM, Marto J (2019) A step forward on sustainability in the cosmetics industry: a review. J Clean Prod 225:270–290. https://doi.org/10.1016/j.jclepro.2019.03.255

Brennen CE (2005) System instabilities. In: Fundamentals of multiphase flow. Cambridge University Press. https://doi.org/10.1017/CBO9780511807169.016

Caggioni M, Ortiz R, Barnabas FA, Nunes RV, Flood JA, Corominas F (2014) US 8716213 B2 - Liquid detergent composition comprising an external structuring system comprising a bacterial cellulose network. 2(12)

Capron I, Cathala B (2013) Surfactant-free high internal phase emulsions stabilized by cellulose nanocrystals. Biomacromolecules 14(2):291–296. https://doi.org/10.1021/bm301871k

Capron I, Rojas OJ, Bordes R (2017) Behavior of nanocelluloses at interfaces. Curr Opin Colloid Interface Sci 29:83–95. https://doi.org/10.1016/j.cocis.2017.04.001

Cervin NT, Andersson L, Ng JBS, Olin P, Bergström L, Waìšgberg L (2013) Lightweight and strong cellulose materials made from aqueous foams stabilized by nanofibrillated cellulose. Biomacromolecules 14(2):503–511. https://doi.org/10.1021/bm301755u

Cervin NT, Johansson E, Benjamins JW, Wågberg L (2015) Mechanisms behind the stabilizing action of cellulose nanofibrils in wet-stable cellulose foams. Biomacromolecules 16(3):822–831. https://doi.org/10.1021/bm5017173

Chen L, Ao F, Ge X, Shen W (2020) Food-grade Pickering emulsions: preparation, stabilization and applications. Molecules 25(14). https://doi.org/10.3390/molecules25143202

Chen Q, Yang F, Hou Y, Li Z, Yuan D, Liu C, Hu F, Zhao R, Wang H, Liu W, Zhang F, Wang Q, Yu D (2025) Bacterial cellulose-based Pickering emulsions reinforced with silver and silica nanoparticles for advanced antibacterial and hydrophobic food packaging solutions. Carbohydr Polym 355(November 2024):123357. https://doi.org/10.1016/j.carbpol.2025.123357

Cherhal F, Cousin F, Capron I (2016) Structural description of the interface of Pickering emulsions stabilized by cellulose nanocrystals. Biomacromolecules 17(2):496–502. https://doi.org/10.1021/acs.biomac.5b01413

Chevalier Y, Bolzinger MA (2013) Emulsions stabilized with solid nanoparticles: Pickering emulsions. Colloids Surf A Physicochem Eng Asp 439:23–34. https://doi.org/10.1016/j.colsurfa.2013.02.054

Cho YS, Lee SH, Seo HM, Shin K, Kang MH, Lee M, Park J, Kim JW (2021) Structuring Pickering emulsion interfaces with bilayered coacervates of cellulose nanofibers and hectorite nanoplatelets. Langmuir 37(13):3828–3835. https://doi.org/10.1021/acs.langmuir.0c03082

Costa C, Medronho B, Filipe A, Mira I, Lindman B, Edlund H, Norgren M (2019) Emulsion formation and stabilization by biomolecules: the leading role of cellulose. Polymers 11(10):1–18. https://doi.org/10.3390/polym11101570

Cruz MA, Flor-Unda O, Avila A, Garcia MD, Cerda-Mejía L (2024) Advances in bacterial cellulose production: a scoping review. Coatings 14(11). https://doi.org/10.3390/coatings14111401

D'Ambrogio R, Peru DA, Gambogi JE, Kinscherf KM, Patel D, Tavares R (2011) WO 2011/056951 A1 - Microfibrous cellulose having a particle size distribution for structured surfactant compositions

Dickinson E (2009a) Hydrocolloids and emulsion stability. In: Philips GO, Williams PA (eds) Handbook of hydrocolloids, 2nd edn. Woodhead Publishing. https://doi.org/10.1533/9781845695873.23

Dickinson E (2009b) Hydrocolloids as emulsifiers and emulsion stabilizers. Food Hydrocoll 23(6):1473–1482. https://doi.org/10.1016/j.foodhyd.2008.08.005

Dickinson E (2012) Use of nanoparticles and microparticles in the formation and stabilization of food emulsions. Trends Food Sci Technol 24(1):4–12. https://doi.org/10.1016/j.tifs.2011.09.006

Dickinson E (2013) Stabilising emulsion-based colloidal structures with mixed food ingredients. J Sci Food Agric 93(4):710–721. https://doi.org/10.1002/jsfa.6013

Dourado F, Fontão A, Leal M, Rodrigues AC, Gama M (2016) Chapter 12 - Process modeling and techno-economic evaluation of an industrial bacterial nanocellulose fermentation process. In: Bacterial nanocellulose. Elsevier, pp 199–214

Fameau A-L, Guzmán E, Ritacco HA, Saint-Jalmes A (2023) Editorial: liquid foams and emulsions stabilized by bio-based particles. Front Soft Matter 3(April):1–2. https://doi.org/10.3389/frsfm.2023.1193657

Fujisawa S, Togawa E, Kuroda K (2017) Nanocellulose-stabilized Pickering emulsions and their applications. Sci Technol Adv Mater 18(1):959–971. https://doi.org/10.1080/14686996.2017.1401423

Heath BP, Coffindaffer TW, Kyte KE, Smith ED, McConaughy SD (2011) WO 2011/019876 A2 - Personal cleansing compositions comprising a bacterial cellulose network and cationic polymer. https://patents.google.com/patent/US8097574B2/en

Hou J, Chu C, Xu HN, Zhang L (2021) Bulk and interfacial contributions to stabilization of cyclodextrin-based emulsions mediated by bacterial cellulose. Langmuir 37(5):1961–1969. https://doi.org/10.1021/acs.langmuir.0c03478

Hu YT, Palla-Venkata CS, Vethamuthu MS (2012) Liquid surfactant compositions structured with fibrous polymer and further comprising citrus fibers having no flow instability or shear banding

Hu Z, Ballinger S, Pelton R, Cranston ED (2015) Surfactant-enhanced cellulose nanocrystal Pickering emulsions. J Colloid Interface Sci 439:139–148. https://doi.org/10.1016/j.jcis.2014.10.034

Hu Z, Xu R, Cranston ED, Pelton RH (2016) Stable aqueous foams from cellulose nanocrystals and methyl cellulose. Biomacromolecules 17(12):4095–4099. https://doi.org/10.1021/acs.biomac.6b01641

Huang M, Zhang C, Hou F, Yang H, Ding N (2024) Stabilization and strengthening effects of filamentous nanocellulose in the foam forming of quartz paper. Int J Biol Macromol 263(P2):130251. https://doi.org/10.1016/j.ijbiomac.2024.130251

Hunter TN, Pugh RJ, Franks GV, Jameson GJ (2008) The role of particles in stabilising foams and emulsions. Adv Colloid Interf Sci 137(2):57–81. https://doi.org/10.1016/j.cis.2007.07.007

Jia Y, Zhai X, Fu W, Liu Y, Li F, Zhong C (2016) Surfactant-free emulsions stabilized by tempo-oxidized bacterial cellulose. Carbohydr Polym 151:907–915. https://doi.org/10.1016/j.carbpol.2016.05.099

Jia Y, Zheng M, Xu Q, Zhong C (2019) Rheological behaviors of Pickering emulsions stabilized by TEMPO-oxidized bacterial cellulose. Carbohydr Polym 215(March):263–271. https://doi.org/10.1016/j.carbpol.2019.03.073

Jiang R (2024) Applications of nanoscale silica (silicon dioxide) in the cosmetics industry. In: Proceedings of the 4th international conference on materials chemistry and environmental engineering. https://doi.org/10.54254/2755-2721/60/20240890

Jiang Y, Yu G, Zhou Y, Liu Y, Feng Y, Li J (2020) Effects of sodium alginate on microstructural and properties of bacterial cellulose nanocrystal stabilized emulsions. Colloids Surf A Physicochem Eng Asp 607(May):125474. https://doi.org/10.1016/j.colsurfa.2020.125474

Kalashnikova I, Bizot H, Bertoncini P, Cathala B, Capron I (2013) Cellulosic nanorods of various aspect ratios for oil in water Pickering emulsions. Soft Matter 9(3):952–959. https://doi.org/10.1039/c2sm26472b

Kalashnikova I, Bizot H, Cathala B, Capron I (2011) New Pickering emulsions stabilized by bacterial cellulose nanocrystals. Langmuir 27(12):7471–7479. https://doi.org/10.1021/la200971f

Kelmer Müller F, Costa FF (2024) Innovations and stability challenges in food emulsions. Sustain Food Technol 3:96–122. https://doi.org/10.1039/d4fb00201f

Kondo T, Rytczak P, Bielecki S (2016) Bacterial nanocellulose characterization. In: Gama M, Dourado F, Bielecki S (eds) Bacterial nanocellulose: from biotechnology to bio-economy. Elsevier B. V. https://doi.org/10.1016/B978-0-444-63458-0.00004-4

Koroleva MY, Yurtov EV (2003) Effect of ionic strength of dispersed phase on Ostwald ripening in water-in-oil emulsions. Colloid J 65(1):40–43. https://doi.org/10.1023/A:1022362807131

Kruglyakov PM, Elaneva SI, Vilkova NG (2011) About mechanism of foam stabilization by solid particles. Adv Colloid Interf Sci 165(2):108–116. https://doi.org/10.1016/j.cis.2011.02.003

Kumar M, Kumar V, Saran S (2024) Current research and advances in the molecular aspect of bacterial cellulose synthesis with special emphasis on the production of bacterial cellulose from waste and its food applications. Cellulose 31(6):3323–3351. https://doi.org/10.1007/s10570-024-05842-8

Lam S, Velikov KP, Velev OD (2014) Pickering stabilization of foams and emulsions with particles of biological origin. Curr Opin Colloid Interface Sci 19(5):490–500. https://doi.org/10.1016/j.cocis.2014.07.003

Lei C, Xie Y, Wu Y, Li Y, Li B, Pei Y, Liu S (2022) Properties and stability of water-in-water emulsions stabilized by microfibrillated bacterial cellulose. Food Hydrocoll 130(November 2021):107698. https://doi.org/10.1016/j.foodhyd.2022.107698

Li Q, Ma Q, Wu Y, Li Y, Li B, Luo X, Liu S (2020) Oleogel films through the Pickering effect of bacterial cellulose nanofibrils featuring interfacial network stabilization. J Agric Food Chem 68(34):9150–9157. https://doi.org/10.1021/acs.jafc.0c03214

Li Q, Wang Y, Wu Y, He K, Li Y, Luo X, Li B, Wang C, Liu S (2019) Flexible cellulose nanofibrils as novel Pickering stabilizers: the emulsifying property and packing behavior. Food Hydrocoll 88(September 2018):180–189. https://doi.org/10.1016/j.foodhyd.2018.09.039

Li Q, Wu Y, Shabbir M, Pei Y, Liang H, Li J, Chen Y, Li Y, Li B, Luo X, Liu S (2021) Coalescence behavior of eco-friendly Pickering-MIPES and HIPEs stabilized by using bacterial cellulose nanofibrils. Food Chem 349(July 2020). https://doi.org/10.1016/j.foodchem.2021.129163

Li S, Wang X, Luo Y, Chen Z, Yue T, Cai R, Muratkhan M, Zhao Z, Wang Z (2023a) A green versatile packaging based on alginate and anthocyanin via incorporating bacterial cellulose nanocrystal-stabilized camellia oil Pickering emulsions. Int J Biol Macromol 249(July):126134. https://doi.org/10.1016/j.ijbiomac.2023.126134

Li Z, Hu W, Dong J, Azi F, Xu X, Tu C, Tang S, Dong M (2023b) The use of bacterial cellulose from kombucha to produce curcumin loaded Pickering emulsion with improved stability and antioxidant properties. Food Sci Human Wellness 12(2):669–679. https://doi.org/10.1016/j.fshw.2022.07.069

Liu Z, Lin D, Shen R, Yang X (2021) Bacterial cellulose nanofibers improved the emulsifying capacity of soy protein isolate as a stabilizer for Pickering high internal-phase emulsions. Food Hydrocoll 112(August 2020):106279. https://doi.org/10.1016/j.foodhyd.2020.106279

Lochhead RY (2007) The role of polymers in cosmetics: recent trends. In: Morgan SE, Havelka KO, Lochhead RY (eds) Cosmetic nanotechnology: polymers and colloids. Am Chem Soc. https://doi.org/10.1021/bk-2007-0961.ch001

Lopetinsky RJG, Masliyah JH, Xu Z (2006) Solids-stabilized emulsions: a review. In: Binks BP, Horozov TS (eds) Colloidal particles at liquid interfaces. Cambridge University Press. https://doi.org/10.1017/CBO9780511536670.007

Lu M, Zhang G, Holmberg K (2024) Toxicity and environmental aspects of surfactants. Tenside Surfactant Deterg 61. https://doi.org/10.1515/tsd-2024-2624

Lu X, Zhang H, Li Y, Huang Q (2018) Fabrication of milled cellulose particles-stabilized Pickering emulsions. Food Hydrocoll 77:427–435. https://doi.org/10.1016/j.foodhyd.2017.10.019

Martins D (2021) Bacterial cellulose: one material, multiple products. PhD thesis, University of Minho. Available Open Access at: https://hdl.handle.net/1822/76945

Martins D, Carvalho Ferreira D, Gama M, Dourado F (2020a) Dry bacterial cellulose and carboxymethyl cellulose formulations with interfacial-active performance: processing conditions and redispersion. Cellulose 27(11):6505–6520. https://doi.org/10.1007/s10570-020-03211-9

Martins D, Dourado F, Gama M (2023) Effect of ionic strength, pH and temperature on the behaviour of re-dispersed BC:CMC - a comparative study with xanthan gum. Food Hydrocoll 135(June 2022):108163. https://doi.org/10.1016/j.foodhyd.2022.108163

Martins D, Estevinho B, Rocha F, Dourado F, Gama M (2020b) A dry and fully dispersible bacterial cellulose formulation as a stabilizer for oil-in-water emulsions. Carbohydr Polym 230(July 2019):115657. https://doi.org/10.1016/j.carbpol.2019.115657

Martins D, Khodamoradi N, Silva-Carvalho R, Gama M, Moradi M, Dourado F (2024) Effect of bacterial cellulose on the foaming properties of egg white and soy proteins. Carbohydr Polym Technol Appl 7(May):100517. https://doi.org/10.1016/j.carpta.2024.100517

Martins D, Rocha C, Dourado F, Gama M (2021) Bacterial cellulose-carboxymethyl cellulose (BC:CMC) dry formulation as stabilizer and texturizing agent for surfactant-free cosmetic formulations. Colloids Surf A Physicochem Eng Asp 617(September 2020):126380. https://doi.org/10.1016/j.colsurfa.2021.126380

McClements DJ (2015) Interfacial properties and their characterization. In: Food emulsions - principles, practices and techniques, 3rd edn. Taylor and Francis Group

McClements DJ (2018) Encapsulation, protection, and delivery of bioactive proteins and peptides using nanoparticle and microparticle systems: a review. Adv Colloid Interf Sci 253:1–22. https://doi.org/10.1016/j.cis.2018.02.002

McClements DJ, Jafari SM (2018) Improving emulsion formation, stability and performance using mixed emulsifiers: a review. Adv Colloid Interf Sci 251:55–79. https://doi.org/10.1016/j.cis.2017.12.001

Michalek IM, John SM, Caetano dos Santos FL (2019) Microbiological contamination of cosmetic products – observations from Europe, 2005–2018. J Eur Acad Dermatol Venereol 33(11):2151–2157. https://doi.org/10.1111/jdv.15728

Moon RJ, Martini A, Nairn J, Simonsen J, Youngblood J (2011) Cellulose nanomaterials review: structure, properties and nanocomposites. Chem Soc Rev 40(7). https://doi.org/10.1039/c0cs00108b

Murray BS, Durga K, Yusoff A, Stoyanov SD (2011) Stabilization of foams and emulsions by mixtures of surface active food-grade particles and proteins. Food Hydrocoll 25(4):627–638. https://doi.org/10.1016/j.foodhyd.2010.07.025

Niu F, Han B, Fan J, Kou M, Zhang B, Feng ZJ, Pan W, Zhou W (2018) Characterization of structure and stability of emulsions stabilized with cellulose macro/nano particles. Carbohydr Polym 199(April):314–319. https://doi.org/10.1016/j.carbpol.2018.07.025

Oliveira TJ, Segato TCM, Machado GP, Grotto D, Jozala AF (2022) Evolution of bacterial cellulose in cosmetic applications: an updated systematic review. Molecules 27(23). https://doi.org/10.3390/molecules27238341

Pan J, Chen J, Wang X, Wang Y, Fan JB (2023) Pickering emulsion: from controllable fabrication to biomedical application. Interdiscip Med 1(3). https://doi.org/10.1002/INMD.20230014

Paximada P, Koutinas AA, Scholten E, Mandala IG (2016a) Effect of bacterial cellulose addition on physical properties of WPI emulsions. Comparison with common thickeners. Food Hydrocoll 54:245–254. https://doi.org/10.1016/j.foodhyd.2015.10.014

Paximada P, Tsouko E, Kopsahelis N, Koutinas AA, Mandala I (2016b) Bacterial cellulose as stabilizer of o/w emulsions. Food Hydrocoll 53:225–232. https://doi.org/10.1016/j.foodhyd.2014.12.003

Pinto NOF, Bourbon AI, Martins D, Pereira A, Cerqueira MA, Pastrana L, Gama M, Azeredo HMC, Rosa MF, Gonçalves C (2024) Bacterial cellulose nanocrystals or nanofibrils as

Pickering stabilizers in low-oil emulsions: a comparative study. Food Hydrocoll 157(July). https://doi.org/10.1016/j.foodhyd.2024.110427

Rühs PA, Storz F, López Gómez YA, Haug M, Fischer P (2018) 3D bacterial cellulose biofilms formed by foam templating. Npj Biofilms Microbiomes 4(1):1–6. https://doi.org/10.1038/s41522-018-0064-3

Sakthivel S, Babu Salin R (2021) Imidazolium based ionic liquid stabilized foams for conformance control: bulk and porous scale investigation. RSC Adv 11(47):29711–29727. https://doi.org/10.1039/d1ra04966f

Salou M, Siffert B, Jada A (1998) Study of the stability of bitumen emulsions by application of DLVO theory. Colloids Surf A Physicochem Eng Asp 142(1):9–16. https://doi.org/10.1016/S0927-7757(98)00406-3

Santos DKF, Rufino RD, Luna JM, Santos VA, Sarubbo LA (2016) Biosurfactants: multifunctional biomolecules of the 21st century. Int J Mol Sci 17(3):1–31. https://doi.org/10.3390/ijms17030401

Saxena IM, Brown RM (2005) Cellulose biosynthesis: current views and evolving concepts. Ann Bot 96(1):9–21. https://doi.org/10.1093/aob/mci155

Schowalter WR, Eidsath AB (2001) Brownian flocculation of polymer colloids in the presence of a secondary minimum. Proc Natl Acad Sci USA 98(7):3644–3651. https://doi.org/10.1073/pnas.061028498

Seo HM, Seo M, Shin K, Choi S, Kim JW (2021) Bacterial cellulose nanofibrils-armored Pickering emulsions with limited influx of metal ions. Carbohydr Polym 258(January):117730. https://doi.org/10.1016/j.carbpol.2021.117730

Shahbazi M, Jäger H, Huc-Mathis D, Asghartabar Kashi P, Ettelaie R, Sarkar A, Chen J (2024) Depletion flocculation of high internal phase Pickering emulsion inks: a colloidal engineering approach to develop 3D printed porous scaffolds with tunable bioactive delivery. ACS Appl Mater Interfaces 16. https://doi.org/10.1021/acsami.4c11035

Shen R, Yang X, Lin D (2023) PH sensitive double-layered emulsions stabilized by bacterial cellulose nanofibers/soy protein isolate/chitosan complex enhanced the bioaccessibility of curcumin: in vitro study. Food Chem 402(September 2022):134262. https://doi.org/10.1016/j.foodchem.2022.134262

Silva FAGS, Branco S, Dourado F, Neto B, Gama M (2025) Life cycle assessment of bacterial cellulose and comparison to other cellulosic sources. J Clean Prod 493(March 2024):144876. https://doi.org/10.1016/j.jclepro.2025.144876

Sousa AF, Ferreira S, Lopez A, Borges I, Pinto RJB, Silvestre AJD, Freire CSR (2017) Thermosetting AESO-bacterial cellulose nanocomposite foams with tailored mechanical properties obtained by Pickering emulsion templating. Polymer 118:127–134. https://doi.org/10.1016/j.polymer.2017.04.073

Souza AG, Ferreira RR, Paula LC, Setz LFG, Rosa DS (2020) The effect of essential oil chemical structures on Pickering emulsion stabilized with cellulose nanofibrils. J Mol Liq 320:114458. https://doi.org/10.1016/j.molliq.2020.114458

Swazey JM, Morrison NA (2018) US10030214B2 - Personal care products comprising microfibrous cellulose and methods of making the same

Tadros TF (2013) Emulsion formation, stability, and rheology. In: Emulsion formation and stability. Wiley-VCH Verlag

Tavacoli JW, Katgert G, Kim EG, Cates ME, Clegg PS (2012) Size limit for particle-stabilized emulsion droplets under gravity. Phys Rev Lett 108(26):1–5. https://doi.org/10.1103/PhysRevLett.108.268306

Taylor P (1998) Ostwald ripening in emulsions. Adv Colloid Interf Sci 75(2):107–163. https://doi.org/10.1016/S0001-8686(98)00035-9

Tournilhac F, Lorant R (2003) US 6534071 B1 - Composition in the form of an oil in-water emulsion containing cellulose fibrils, and its uses, especially cosmetic uses. 1(12)

Viseras C, Sánchez-Espejo R, Palumbo R, Liccardi N, García-Villén F, Borrego-Sánchez A, Massaro M, Riela S, López-Galindo A (2021) Clays in cosmetics and personal-care products. Clay Clay Miner 69(5):561–575. https://doi.org/10.1007/s42860-021-00154-5

Wang M, Liu Q, Zeng X, Chen R, Wang C, Li M, Wu X, Weng W, Zhang Y (2024) Bacterial cellulose nanofibrils for the physical and oxidative stability of fish oil-loaded Pickering emulsions. Colloids Surf A Physicochem Eng Asp 694(April):134154. https://doi.org/10.1016/j.colsurfa.2024.134154

Wang Y, Liu Q, Yang Y, Qiu C, Jiao A, Jin Z (2023) Fabrication of a double-network high internal phase emulsion gel stabilized by bacterial cellulose nanofibrils: enhancement of heat stability and 3D printing. Food Hydrocoll 143(May):108872. https://doi.org/10.1016/j.foodhyd.2023.108872

Winuprasith T, Suphantharika M (2015) Properties and stability of oil-in-water emulsions stabilized by microfibrillated cellulose from mangosteen rind. Food Hydrocoll 43:690–699. https://doi.org/10.1016/j.foodhyd.2014.07.027

Wu P, Song Z, Weng H, Wei X, Yu J, Nie C, Niu Y, Kang L, Wang L, Wang JG (2025) The properties of Pickering emulsions stabilized by bacterial cellulose nanofibrils and its retarding effect on lipid digestion. Int J Biol Macromol 305(P2):141335. https://doi.org/10.1016/j.ijbiomac.2025.141335

Wu Y, Zhang X, Qiu D, Pei Y, Li Y, Li B, Liu S (2021) Effect of surface charge density of bacterial cellulose nanofibrils on the rheology property of O/W Pickering emulsions. Food Hydrocoll 120(June):106944. https://doi.org/10.1016/j.foodhyd.2021.106944

Xia J, Sun X, Jia P, Li L, Xu K, Cao Y, Lü X, Wang L (2023) Multifunctional sustainable films of bacterial cellulose nanocrystal-based, three-phase Pickering nanoemulsions: a promising active food packaging for cheese. Chem Eng J 466(April). https://doi.org/10.1016/j.cej.2023.143295

Yan H, Chen X, Song H, Li J, Feng Y, Shi Z, Wang X, Lin Q (2017) Synthesis of bacterial cellulose and bacterial cellulose nanocrystals for their applications in the stabilization of olive oil Pickering emulsion. Food Hydrocoll 72:127–135. https://doi.org/10.1016/j.foodhyd.2017.05.044

Yang Z-F, Morrison NA, Talashek TA, Brinkmann DF, DiMasi D, Chen YL (2007) US 2007/019777 A1 - Bacterial cellulose-containing formulations. 1(19)

Zhai X, Lin D, Liu D, Yang X (2018) Emulsions stabilized by nanofibers from bacterial cellulose: new potential food-grade Pickering emulsions. Food Res Int 103(October 2017):12–20. https://doi.org/10.1016/j.foodres.2017.10.030

Zhang X, Zhou J, Chen J, Li B, Li Y, Liu S (2020) Edible foam based on Pickering effect of bacterial cellulose nanofibrils and soy protein isolates featuring interfacial network stabilization. Food Hydrocoll 100(July 2019):105440. https://doi.org/10.1016/j.foodhyd.2019.105440

Zhang Y, Yang S, Tang H, Wan S, Qin W, Zeng Q, Huang J, Yu G, Feng Y, Li J (2022) Depletion stabilization of emulsions based on bacterial cellulose/carboxymethyl chitosan complexes. Carbohydr Polym 297(March):119904. https://doi.org/10.1016/j.carbpol.2022.119904

Zhao GH, Kapur N, Carlin B, Selinger E, Guthrie JT (2011) Characterisation of the interactive properties of microcrystalline cellulose-carboxymethyl cellulose hydrogels. Int J Pharm 415(1–2):95–101. https://doi.org/10.1016/j.ijpharm.2011.05.054

Zhao W, Hou Q, Wang X (2019) The influence of gas diffusion mechanisms on foam stability for foam forming of paper products. Bioresources 14(4):9893–9903. https://doi.org/10.15376/biores.14.4.9893-9903

Zhong C (2020) Industrial-scale production and applications of bacterial cellulose. Front Bioeng Biotechnol 8(December):1–19. https://doi.org/10.3389/fbioe.2020.605374

Zhu Y, Pei X, Jiang J, Cui Z, Binks BP (2015) Responsive aqueous foams stabilized by silica nanoparticles hydrophobized in situ with a conventional surfactant. Langmuir 31(47):12937–12943. https://doi.org/10.1021/acs.langmuir.5b03681

Chapter 6
3D Bioprinting of Bacterial Cellulose for Biomedical Applications

Baneeprajnya Nayak, Fernando Dourado, Mohan Das, and Rintu Banerjee

Abstract 3D bioprinting is transforming tissue engineering and regenerative medicine by allowing the production of intricate, cell-filled constructs with precise spatial control. The growing need for biocompatible and structurally resilient scaffolds has sparked the interest in integrating bacterial cellulose (BC) as a component of bioinks for 3D bioprinting, given its unique combination of mechanical robustness, biocompatibility, high water retention capacity, and nanofibrillar architecture.

This chapter explores the use of BC as a component of bioinks in hydrogel-based formulations, with the potential to encapsulate cells and biomolecules for bioprinting into biofunctional constructs. The key topics here addressed include the critical design features of bioinks and the influence of the printing parameters on the design's fidelity and cellular performance. Several bioprinting technologies are reviewed for their compatibility with BC formulations. The chapter also highlights biomedical applications, with an emphasis on bone, cartilage, soft tissue engineering, and antimicrobial scaffold development. It also presents the main findings from *in vitro* and *in vivo* studies, demonstrating the promising outcomes in cartilage regeneration and bone repair.

While the number of studies on BC bioprinting remains limited, existing evidence strongly support its potential for clinical use. The challenges of BC's non-degradability, high viscosity and the lack of standardized methods are acknowledged, alongside emerging strategies to address them. The chapter concludes by outlining future directions such as scalable production and long-term *in vivo* validation.

B. Nayak · M. Das · R. Banerjee (✉)
Agricultural and Food Engineering Department, Indian Institute of Technology, Kharagpur, India
e-mail: rb@iitkgp.ac.in

F. Dourado
CEB – Centre of Biological Engineering, University of Minho, Campus de Gualtar, Braga, Portugal

LABBELS – Associate Laboratory in Biotechnology and Bioengineering and Microelectromechanical Systems, Braga, Guimarães, Portugal
e-mail: fdourado@ceb.uminho.pt

F. Dourado et al. (eds.), *Advances in Bacterial Cellulose*,
https://doi.org/10.1007/978-3-032-07042-5_6

Keywords Bacterial cellulose · 3D bioprinting · Bioink · Biomedical applications · Tissue engineering

1 Introduction

Tissue engineering and regenerative medicine have made significant progress in the past few decades in recreating functional constructs that mimic native tissues. Conventional methods typically involve prefabricated scaffolds that are then seeded with cells somewhat randomly and enhanced with growth factors. However, this approach has had a limited capacity to reproduce the intricate 3D architecture of living tissues. These limitations include for instance an inadequate vascularization, poor integration with host tissue, and lack of structural complexity (Satpathy et al. 2019; Bose et al. 2013; Agarwal et al. 2020). In parallel, the use of time-consuming fabrication methods, scalability constraints, and high production costs further weight their limited feasibility in clinical practice (Singh and Williams 2008).

Additive manufacturing (AM) has emerged as an advanced fabrication technique that aimed to overcome those limitations. Briefly, AM allows to build complex biological structures from 3D models, layer by layer, in a precise and controlled manner (Melchels et al. 2012). Yet, AM can only fabricate non-living constructs. An extension of this technique is 3D bioprinting, a process by which active cells, growth factors, and biomaterials can be 3D-printed to create more functional, tissue-mimicking structures (Singh and Williams 2008; Murphy and Atala 2014; Xiongfa et al. 2018). Compared to conventional biomaterials, 3D bioinks, that are used in bioprinting, provide several advantages such as flexibility, improved mechanical properties, tunable biodegradability, and patient-specific customization. These advantages allow for more advanced applications in tissue transplantation and regenerative medicine (Ganpisetti and Lalatsa 2021). Bioprinting has been used in tissue engineering, regenerative medicine, and drug testing to produce functional tissues and organs (Santoni et al. 2021; Agarwal et al. 2020). In recent years, it has been extended to new areas including as biosensing and environmental remediation (Tibbetts 2021).

The bioinks are typically solutions of biomaterials in hydrogel form that may contain live cells for the growth of targeted tissues. Common biopolymers used in bioprinting include alginate, collagen, and gelatine, due to their biocompatibility and ability to mimic the extracellular matrix. Synthetic polymers, such as polyethylene glycol (PEG) and polycaprolactone (PCL), are used for their tunable mechanical properties and structural integrity (Santoni et al. 2021; Agarwal et al. 2020; Shiva et al. 2023). The successful printing and functionality of the constructs requires bioinks to meet specific criteria such as biocompatibility, structural integrity, and proper rheological properties. (Santoni et al. 2021; Tarassoli et al. 2021). Bioink formulations are designed to close-mimic the mechanical properties of the native tissues. While some tissues require more elasticity, other requires more structural integrity (Banigo et al. 2025). They are also selected based on their compatibility with the targeted tissue or organ to ensure their successful integration and functionality. Currently, several types of bioinks are already actively marketed and are part of a growing industry. Bioinks are sold for applications in tissue engineering, organ printing, drug testing, and personalized medicine. Specialized companies that focus on bioprinting and regenerative medicine include CELLINK (https://www.cellink.com/), CollPlant Biotechnologies (https://collplant.com/), and Allevi (https://www.allevi3d.com/) as prominent players in the bioink market.

While both natural and synthetic polymers have allowed significant advances in the development of bioink formulations, they also present certain limitations for functional tissue engineering. Natural polymers, although biocompatible and bioactive, often suffer from inadequate mechanical strength, batch-to-batch variability, and unpredictable degradation kinetics (Hospodiuk et al. 2017; Gungor-Ozkerim et al. 2018; Malda et al. 2013; Mahendiran et al. 2021). Although synthetic alternatives provide higher tunability and structural stability, they often lack biological recognition signals and can trigger inflammatory responses from residual crosslinking agents or degradation by-products (Zhu et al. 2016; Yue et al. 2015; Chimene et al. 2016). In this context, BC may present unique advantages as a biomaterial that combines the benefits of both material classes while bypassing their limitations. BC possesses an exceptional combination of intrinsic properties, such as superior mechanical strength, high purity, high water retention capacity, excellent biocompatibility with mammalian tissues, and extensive functionalization potential through its surface hydroxyl groups. These properties, position BC as an ideal candidate for the next generation of bioinks, particularly for applications demanding both structural integrity and biomimetic functionality, such as load-bearing tissue regeneration or complex organotypic modeling (Betlej et al. 2021; Choi et al. 2022; Vartanian 2023).

This chapter explores the application of 3D bioprinting techniques from BC-based bioinks. It addresses optimization strategies and critical printing parameters that influence process fidelity, and current approaches to scale-up and commercial implementation. Finally, it examines the biomedical potential of 3D-printed BC scaffolds, highlighting their relevance in tissue engineering and regenerative medicine and discusses future directions and challenges.

2 Bacterial Cellulose as a Component of Bioinks for Bioprinting

2.1 Relevant Properties of Bioinks for Bioprinting

The development of bioinks for bioprinting depends on the use of natural and/or synthetic biomaterials to modulate the rheological, mechanical, physicochemical, and biological properties of the material for tissue engineering purposes. These biomaterials are also combined along with growth factors and nutrients to promote cell growth, differentiation, and survival. Hybrid bioinks have emerged as a promising approach that combines the biofunctionality of natural and generally biodegradable polymers with the mechanical and physicochemical tunability of synthetic matrices, resulting in composite systems that better mimic the complexity of the extracellular matrix (ECM) of native tissues (Ng et al. 2020; Teixeira et al. 2022; Morgan et al. 2020; Shiva et al. 2023; Tarassoli et al. 2021; Ganpisetti and Lalatsa 2021; Gouripriya et al. 2024; Cesur and Laçin 2022). Most bioinks are developed in the form of hydrogels, and this physical state is critical for their success in bioprinting applications, as they aim to closely mimic the hydration and mechanical characteristics of the native extracellular environment (Teixeira et al. 2022; Zhou et al. 2022). The high water content of hydrogels not only facilitates an efficient diffusion of nutrients and gases but also supports the encapsulation and viability of living cells throughout the printing and post-printing processes (Gungor-Ozkerim et al. 2018).

One of the defining rheological properties that makes hydrogels suitable for bioprinting is their shear-thinning behavior (Fig. 6.1). This property ensures that the bioink can be smoothly extruded through a nozzle and solidified upon deposition to preserve the structural fidelity of the printed construct (Lee et al. 2019). The mechanical strength, that can be tailored by adjusting their composition, concentration, and crosslinking density, is also a critical property of bioinks, particularly for high load-bearing scenarios such as bone and cartilage (Teixeira et al. 2022; Zhou et al. 2022; Morgan et al. 2020). Another key property is biodegradability, to ensure

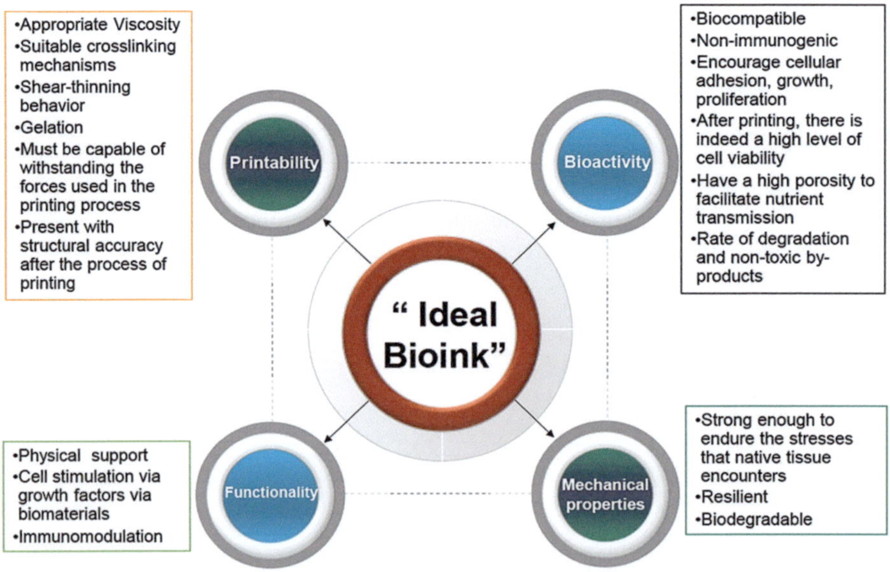

Fig. 6.1 Desired properties of bioinks. Reproduced from Ghosh and Yi (2022), licensed under CC BY 4.0 (https://creativecommons.org/licenses/by/4.0/)

that the bioink scaffold provides temporary support during the healing process, while ultimately being resorbed by the body, thus minimizing long-term foreign body reactions (Lee et al. 2019). An ideal bioink should degrade at a rate that matches that of tissue regeneration, allowing for the gradual replacement of the scaffold material with newly formed tissue (Malda et al. 2013).

The bioactivity of a bioink concerns its ability to support specific cellular functions and behaviors, such as adhesion, proliferation, and differentiation (Rastogi and Kandasubramanian 2019). The incorporation of bioactive molecules such as growth factors or ECM-derived peptides into the bioink formulation can significantly improve its bioactivity (Ng et al. 2020).

A well-designed bioink should allow to create a porous structure that facilitates nutrient and oxygen diffusion, while allowing for waste removal from cellular metabolism. A higher porosity has been shown to enhance cell infiltration and tissue vascularization, which are essential for the successful regeneration of the growing tissue (Zhou et al. 2022). The pore size and interconnectivity can be manipulated through various printing parameters, including the extrusion speed and nozzle diameter, as well as the intrinsic properties of the bioink (Teixeira et al. 2022).

Lastly, hydrogels exhibit versatile crosslinking mechanisms, such as ionic, thermal, enzymatic, or photo-induced processes, which can be used to stabilize the printed constructs. This tunability allows to modulate the mechanical properties and control the degradation rates in alignment with specific tissue regeneration requirements (Teixeira et al. 2022; Morgan et al. 2020).

2.2 *Potential and Limitations of Bacterial Cellulose in Bioink Development*

BC has emerged as a promising component in bioink formulations for 3D bioprinting. Its well-known exceptional properties make it structurally robust and functionally adaptable for a variety of tissue engineering applications (Lu et al. 2024; Swingler et al. 2021; Grunberg et al. 2024; Utoiu et al. 2024). In bioprinting, BC is mainly integrated as a nanofibrillar additive to enhance the rheological properties, mechanical strength, and print fidelity of the bioinks. After mechanical disintegration, BC suspensions exhibit shear-thinning behavior, a critical property for extrusion-based bioprinting (Grunberg et al. 2024; Sharma et al. 2025). The nanostructured architecture of BC provides unique advantages for bioprinting. Its 3D network offers enhanced mechanical stability, crucial for maintaining the structural integrity of the printed scaffold (Grunberg et al. 2024; Utoiu et al. 2024). BC-based bioinks exhibit exceptional mechanical robustness, offering superior stiffness and toughness compared to conventional hydrogel bioinks such as alginate or agarose (Grunberg et al. 2024).

Huang et al. (2019) demonstrated that BC-reinforced silk-based hydrogels maintain high shape fidelity and pore architecture under stress, which are key factors for the structural success of printed constructs (Table 6.1). The addition of BC to polyvinyl alcohol (PVA) matrices led to enhanced compressive strength and biocompatibility, supporting its application in bone tissue engineering (Aki et al. 2020). TEMPO (2,2,6,6-tetramethylpiperidine-1-oxyl)-mediated oxidation led to BC's improved dispersibility and the formation of stable hydrogels with shear-thinning behavior (Wei et al. 2020; Wang et al. 2021; Sharma et al. 2025). Also, surface-carboxylation of BC with maleic acid treatment resulted in homogeneous BC dispersions with enhanced rheological properties and printability when combined with gelatine (Wang et al. 2021).

The capacity of BC to interact with other functional nanomaterials, such as hydroxyapatite or copper nanostructures, has allowed the development of scaffolds with osteoinductive or antimicrobial properties (Turlybekuly et al. 2020; Gutierrez et al. 2019). For instance, Apelgren et al. (2019) reported the successful human cartilage formation within 3D bioprinted constructs using a BC-based bioink, demonstrating its potential as a support matrix for chondrogenic applications. Wang et al. (2021) further validated the osteogenic potential of BC-gelatine bioinks, which promoted *in vivo* bone regeneration and vascularization.

Despite the above promising attributes, several limitations preclude a wider adoption of BC in bioprinting. Although BC hydrogels are printable, their high viscosity may pose challenges for cell encapsulation (Shariati et al. 2015). To compensate for the higher viscosity, higher shear forces can be used during extrusion. However, this may damage cells, leading to their reduced viability (Hölzl et al. 2016). For example, high concentrations of nanofibrillated cellulose from plants (>2 wt%) improve printability, but drastically lower cell survival rates (<50%) due to mechanical stress (Markstedt et al. 2015). Blending BC with softer hydrogels such as alginate or gelatine (Table 6.1) may mitigate this issue, but often at the cost of reduced mechanical strength (Rumon et al. 2025). Lin et al. (2023) also highlighted the need for combining BC with other cellulose derivatives or thermosensitive polymers to improve its extrusion capabilities and crosslinking behavior.

Another constraint of BC is its limited cell interactivity. While non-cytotoxic, BC lacks specific bioactive motifs that promote cell adhesion and proliferation, requiring surface functionalization with bioactive molecules or co-formulation with ECM-derived materials to enhance its biological performance (McCarthy et al. 2019; Utoiu et al. 2024). Also, native BC is non-degradable *in vivo*. While this persistence can be beneficial for long-term structural support (e.g., bone or cartilage tissue engineering), it may restrict its use in applications where scaffold resorption and tissue remodeling are required within specific timeframes (e.g., in vascular or neural tissues) (Lu et al. 2024). TEMPO-mediated oxidation of BC may alter its mechanical integrity (Lai et al. 2013; Cui et al. 2014; Mazega et al. 2025) which can be modulated by varying the charge density of the introduced carboxyl groups (Sharma et al. 2025).

Processing challenges also exist. The isolation and purification of BC at a scale compatible with reproducible bioink production can be resource-intensive, and the variability introduced by different microbial strains, culture conditions, and post-processing techniques may arguably affect batch-to-batch consistency (Lu et al.

2024). Finally, while some innovative strategies such as support bath printing and solid matrix-assisted techniques have improved the structural control of BC-containing constructs (Shin et al. 2019), standardized protocols for printing with BC-based bioinks are still under development.

3 Bioprinting of Bacterial Cellulose-Based Bioinks

As mentioned, bioprinting allows for the precise deposition of cells within a viscous biomaterial in a predetermined 3D arrangement, through the use of computer-aided printing technologies. In contrast to conventional 3D printing, bioprinting requires increasingly advanced techniques to meet the dual demands of biological functionality (such as cell adhesion, proliferation, and viability) and mechanical stability of the printed constructs (Jovic et al. 2019; Liu et al. 2023). Bioprinting technology has evolved to incorporate various advanced techniques such as inkjet bioprinting, laser-assisted bioprinting, SMAP, and extrusion-based bioprinting (Figs. 6.2, 6.3, and 6.4), among others, differing in their mechanisms of material deposition, resolution, and suitability for specific biological and mechanical requirements. Each technique allows the creation of specific high-resolution constructs to achieve proper structural integrity and biofunctionalization (Liu et al. 2023; Lee and Yeong 2016).

In 3D bioprinting, optimization of the print parameters, such as extrusion speed, nozzle diameter, and printing temperature are crucial for the effective and precise printing process. The extrusion rate and nozzle speed are interdependent, as they determine the amount of material to be extruded per unit length and time. Both the printing and extrusion speeds must be synchronized to ensure the uniform deposition of the bioink (Feng et al. 2019). The nozzle diameter also influences the fineness and precision of the printed construct. A smaller nozzle diameter allows for higher precision, but it increases the printing time and feed pressure. This increase in pressure can lead to machine overload and wear, as well as non-uniform bioink deposition. Thus, when selecting the nozzle diameter, it is essential to consider the bioink's viscosity and the maximum pressure that the printer can withstand to prevent bioink blockage (Feng et al. 2019). The nozzle height—the distance between the nozzle tip and the printing bed—also affects the printing quality. However, studies have shown that this parameter is less critical than the nozzle's diameter for printing accuracy. At a constant nozzle height and diameter, the balance between extrusion speed and nozzle movement rate is the key factor for printing quality (Yap et al. 2020). Ideally, the diameter of the printed material should match that of the nozzle diameter, and the nozzle height should be minimized to ensure the proper adhesion of the bioink to the beneath layer. This balance ensures that the layer thickness, nozzle height, and nozzle diameter are consistent (El-Sayegh et al. 2020).

Table 6.1 shows a comparative analysis of the bioprinting techniques for BC in biomedical applications. Despite the limited number of studies available, extrusion-based bioprinting has been the most explored technique.

Table 6.1 Comparative analysis of bioprinting techniques for bacterial cellulose in biomedical applications

Bioprinting method	Bioink components	Study design	Clinical relevance	Cell type	Bioprinting parameters	Printing resolution
Extrusion-based	BC disassembled by Aqueous CounterCollision, calcium chloride, and sodium hydrogen phosphate, forming hydroxyapatite within the BC matrix	Development and evaluation of BC-based bioink for cartilage tissue engineering	Cartilage repair and regenerative medicine	Human nasal chondrocytes	Nozzle size: 0.41 mm, Print speed: 3.0–5.2 mm/s, Pressure: 1.0–2.0 bar; Grid constructs ($5 \times 5 \times 1$ mm^3)	High fidelity with precise structural integrity
	BC, gelatin methacryloyl (GelMA), hyaluronic acid methacrylate (HAMA)	Development and characterization of GelMA/HAMA/BC composite hydrogels for nasal cartilage tissue engineering	Nasal cartilage repair and regenerative medicine	Mouse chondrocyte cell line (ATDC5)	Nozzle size: 0.41 mm, Print speed: 3.0–5.2 mm/s, Pressure: 1.0–2.0 bar; Fiber spacing of 700 μm	Line width of 350 ± 23 μm; center distance of parallel lines was 732 ± 40 μm.
	BC, alginate	Development of a low-cost 3D bioprinting platform with in situ crosslinking capabilities	Scalable and affordable bioprinting applications	None	0.85 mm nozzle, 0.005 mm^3/s extrusion rate, 1 bar extrusion pressure	High resolution, up to 72 wt% BC in dried structures
	BC, alginate, copper nanoparticles	Ionic crosslinking with calcium ions followed by ion exchange with copper ions (method A) and direct ionic crosslinking with copper ions (method B)	Antimicrobial wound dressing and infection prevention		0.25 mm nozzle, 0.005 mm^3/s extrusion rate, 1 bar extrusion pressure	Mesh-like inner pattern with 1.5 mm thread spacing
	BC, Polyvinyl Alcohol (PVA), Hexagonal Boron Nitride (hBN)	Fabrication of scaffolds with controlled porosity using 3D printing technology	Bone tissue engineering applications		0.4 mm nozzle, 0.3 mL/h flow rate for 12 wt% PVA, 0.5 mL/h for 12 wt% PVA/0.25 wt% hBN, 1 mL/h for 12 wt% PVA/0.25 wt% hBN/0.5 wt% BC	Nozzle diameter of 0.4 mm; pore size ranging from 265.68 ± 15.39 μm to 291.71 ± 19.94 μm

	BC, silk fibroin, gelatin, and glycerol	Development and characterization of BC-enhanced silk fibroin/gelatin bioinks for soft tissue engineering	Soft tissue engineering and regenerative medicine		Nozzle diameter of 0.41 mm, dosing pressure of 1.0–2.0 bar, and dispensing speed of 3.0–5.2 mm/s	Line width of approximately 400 μm, close to the nozzle diameter
	BC, gelatin, and TEMPO-mediated oxidized bacterial cellulose (TO-BC) or maleic acid-treated BC (MA-BC)	Comparison of neat BC, TO-BC, MA-BC in 3D scaffolds	Bone tissue engineering		Air pressure extrusion with a 500 μm nozzle; print angles: 0° and 90°	Scaffold thickness set at 500 μm
	TO-BC, alginate, and laponite nanoclay	Development and characterization of TOBC/SA/nanoclay composite hydrogels for enhanced stability and protein delivery	Tissue engineering and drug delivery systems		Nozzle diameter of 410 μm; printing speed of 4 mm/s	Strand width of approximately 588 μm
Inkjet printing	BC, hydroxyapatite (HA), calcium chloride, and sodium hydrogen phosphate	Ink formulation by mixing BC in calcium chloride and sodium hydrogen phosphate solutions, freeze-drying	Regenerative bone tissue applications and improving osteogenic regeneration		Nozzle size: 0.41 mm, Print speed: 3.0–5.2 mm/s, Pressure: 1.0–2.0 bar; Air-compressed piston-driven extrusion, printing performed on a liquid nitrogen-cooled substrate to freeze the scaffold	Dimensions of the particles ranged between 2.98 μm and 3.36 μm
Solid Matrix-Assisted Printing (SMAP)	PTFE microparticles, cellulose nanofibers (CNF), *G. xylinus*	Fabrication of spheres, tubes, coils, rings	Potential applications in vascular tissue engineering and regenerative medicine	*G. xylinus*	Speed and pressure adjusted to control feature size	500–2500 μm

continued

Table 6.1 (continued)

Bioprinting method	Post-printing treatment	Key advantage	Properties	*In vivo* assays	*In vitro* assays	Reference
Extrusion-based	Crosslinking with calcium chloride	High water retention, excellent printability, and mechanical stability	High mechanical strength	Successful integration, structural stability, and chondrocyte proliferation in nude mice	Enhanced chondrocyte viability and proliferation in vitro	Apelgren et al. (2019)
	UV crosslinking after each layer for 15 s	Enhanced mechanical properties and improved printing fidelity	High compressive modulus, stable internal structure	Not evaluated	Enhanced cell viability, proliferation, and expression of cartilage-specific proteins (type II collagen and aggrecan) with ATDC5 cells	Sang et al. (2023)
	Crosslinking with calcium chloride	High BC content, high mechanical properties (qualitative data)	Self-supporting and flexible scaffold (qualitative data)	Not evaluated	Not evaluated	Grunberg et al. (2024)
	Sodium borohydride reduction	Antimicrobial properties, improved printability	Enhanced mechanical stability, antimicrobial activity, and structural integrity	Not evaluated	Not evaluated	Gutierrez et al. (2019)
	Drying at room temperature for 24 h	Enhanced cell viability and adhesion, suitable pore size for vascularization and nutrient transport	High tensile strength, good elongation at break	Not evaluated	Increased human osteoblast cell viability and adhesion	Aki et al. (2020)

	Freeze-drying followed by ethanol immersion and genipin crosslinking	Improved mechanical properties, shape fidelity, and hierarchical porous structure	High tensile strength, structural stability, and interconnected pores (10–20 μm for cell infiltration and 300–600 μm for nutrient supply)	Tissue integration and minimal inflammation in mice	Enhanced cell viability and proliferation with L929 fibroblasts	Huang et al. (2019)
	Crosslinking with N-hydroxysuccinimide (NHS) and 1-(3-dimethylaminopropyl)-3-ethylcarbodiimide hydrochloride (EDC)	Enhanced osteogenic differentiation and mineralized nodule formation	Compression modulus of BC: 426 kPa; TO-BC 426 kPa; MA-BC 707 kPa	MA-BC-based gelatin scaffolds significantly promoted bone regeneration in rat calvaria models, resulting in higher bone mineral density and increased trabecular thickness compared to TO-BC scaffolds (8-week-old Sprague-Dawley rats)	Enhanced osteoblast viability (MC3T3-E1 cells), ALP activity, and mineralized nodule formation	Wang et al. (2021)
	Crosslinking with calcium chloride	Enhanced structural stability and sustained protein release behavior	High shape fidelity, mechanical strength, and stability in physiological conditions	Not evaluated	Enhanced cell viability and proliferation with L929 fibroblasts	Wei et al. (2020)
Inkjet printing	Freeze-drying	Homogeneous distribution of HA particles within the BC matrix, along with enhanced structural integrity	Homogeneous microstructure, and stable particle integration	Not evaluated	Not evaluated	Turlybekuly et al. (2020)
Solid Matrix-Assisted Printing (SMAP)	Incubation for biosynthesis of BC at the surface of the printed structures	Controlled 3D BC hydrogel fabrication with oxygen permeability	High mechanical strength, structural stability, and interconnected network of BC fibers	Not evaluated	Not evaluated	Shin et al. (2019)

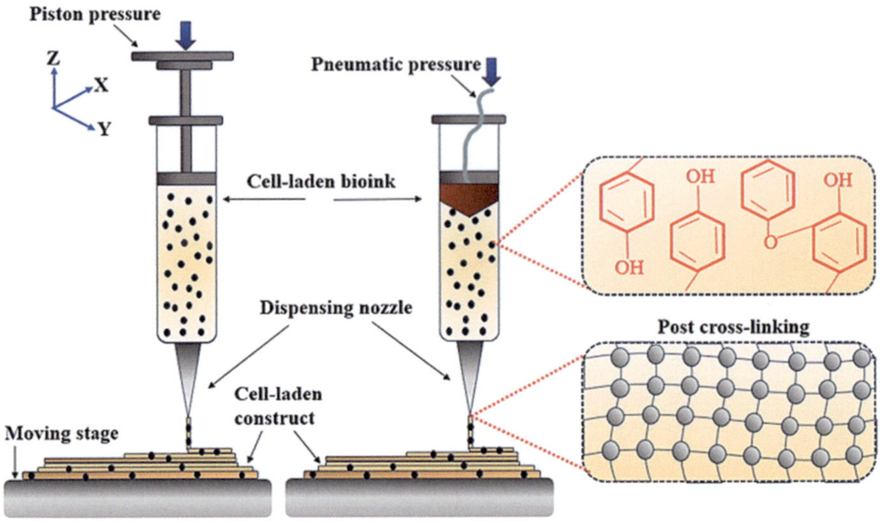

Fig. 6.2 Extrusion-based bioprinting. Adapted from Amoli et al. (2022), licensed under CC BY 4.0

3.1 *Extrusion-Based Bioprinting*

The extrusion-based bioprinting (EBB) is a method optimized for viscous bioinks. In this method, bioinks are loaded into a syringe and deposited under controlled temperature and pressure conditions to preserve the biomolecular integrity and cell viability (Fig. 6.2). The versatility of EBB lies in its ability to adapt to a wide range of bioinks, including hydrogels, cell-laden matrices, and composite materials, making it suitable for several biomedical applications. The extrusion process can be driven by several mechanisms such as pneumatic pressure, piston-based systems, or screw-driven mechanisms, each affecting the precision and the consistency of the extruded material. Generally, the achievable resolution ranges from 15 to 400 μm. This range is largely determined by the nozzle diameter and printing parameters, such as layer thickness, printing path height, and path spacing. The bioink's viscosity, shear-thinning behavior, gelation kinetics, and fluid dynamics must also be carefully optimized to achieve a precise deposition. One technical difficulty of this method concerns with printing constructs with geometric complexities, especially at microscopic scales, such as structures with complex channels. Another limitation of the EBB is nozzle clogging, which becomes more problematic as the print core diameter decreases (Lee and Yeong 2016; Liu et al. 2023; Gu et al. 2020).

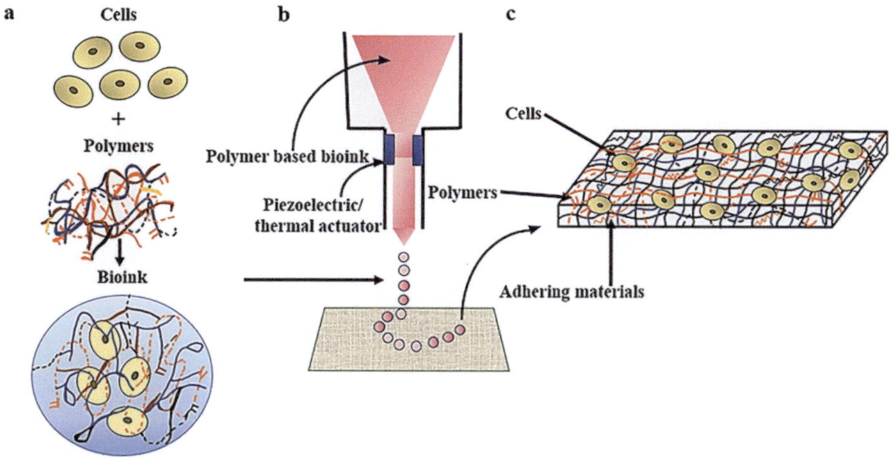

Fig. 6.3 Inkjet-based bioprinting (**a**) Bioink constituting of cells and polymers, (**b**) Bioprinting of scaffold, (**c**) Scaffold containing network of cells and polymers. Adapted from Gu et al. 2020, CC BY 4.0

3.2 Inkjet-Based Bioprinting

Inkjet-based bioprinting (IBB) belongs to the broader category of droplet-based bioprinting. This method can create intricate 3D structures by ejecting picoliter-sized droplets of bioink on a surface, through a nozzle, using thermal, piezoelectric, or electromagnetic mechanisms (Fig. 6.3). IBB is well-suited for applications requiring high resolution and controlled spatial distribution of cells and biomaterials (Liu et al. 2023). The achievable resolution with inkjet bioprinting typically ranges around 100 μm, the smallest building unit being a droplet (Lee and Yeong 2016).

Critical aspects for this high-precision method are the low viscosity and low concentration of the bioink to ensure a smooth droplet formation and to prevent nozzle clogging. Such low viscosity, however, presents a challenge related to the uniform distribution of cells or microorganisms within the bioink. The size of the cells is also an important factor. Given the small size of the orifice of inkjet printheads (typically between 50 and 300 μm), cells must be appropriately sized to avoid clogging during ejection. Also, the size of the cells must be harmonized with the rheological properties of the bioink to ensure smooth extrusion while preserving cell viability. This compatibility will also influence cell survival, organization, and function within the printed construct. For instance, smaller cells, better matching with the bioink's viscosity and flow characteristics, are better suited to form intricate structures, as they can be dispersed more evenly. Contrarily, larger cells may tend to cluster, preventing a uniform construct architecture, proper nutrient diffusion, and affecting other physiological functions (Ng and Shkolnikov 2024; Xu et al. 2019; Lee and Yeong 2016; Liu et al. 2023).

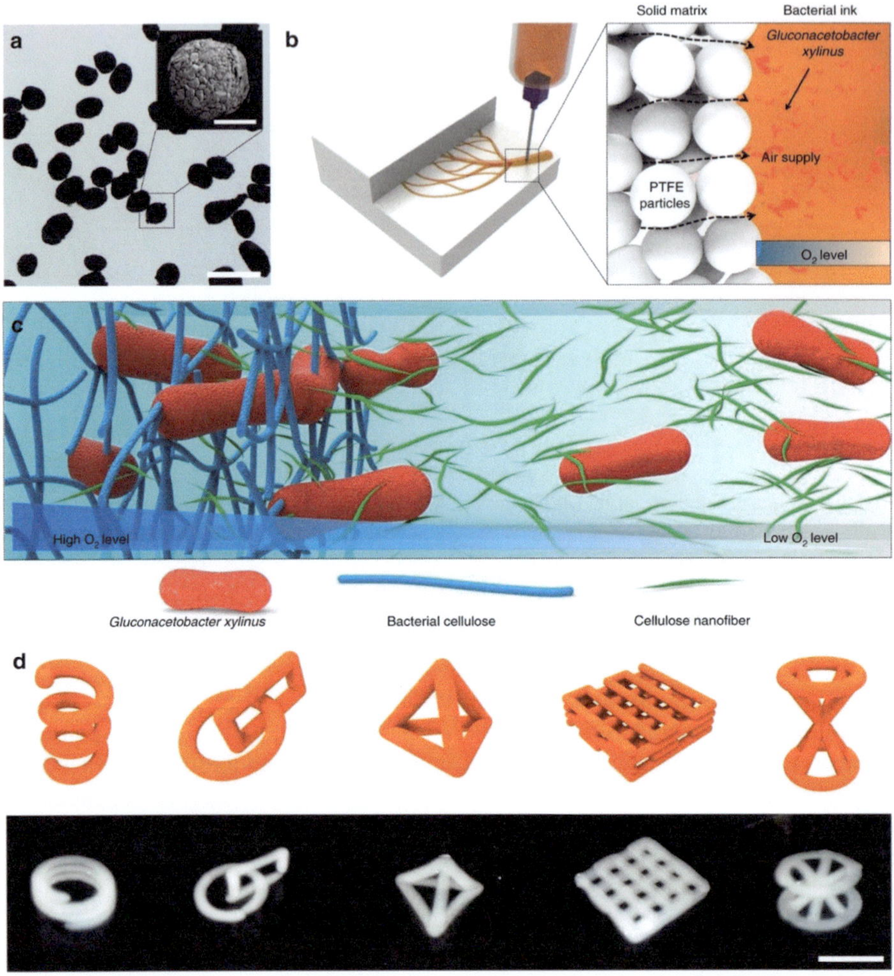

Fig. 6.4 Solid matrix-assisted 3D printing of bacteria containing ink. (**a**) Optical and electron microscopic images of PTFE microparticles used as a solid matrix. (**b**) Bacteria containing ink were printed inside of solid matrix. Oxygen was supplied through the solid particles to the surface of the CNF ink containing *G. xylinus*, allowing the bacteria to metabolize. (**c**) BC was produced at the surface of the printed CNF hydrogel ink in high oxygen conditions. (**d**) Coil, connected ring, tetrahedron, stacked lattice, and sandglass structure of CNF/BC after incubation as printed according to the CAD designs. Scale bars are 500 μm in the optical microscopic image and 100 μm in the FE-SEM image of (**a**), and 1 cm in the optical microscopic images of (**d**). Figure reproduced from Shin et al. 2019, CC BY 4.0

Given the low viscosity of the bioinks, upon deposition the constructs typically lack proper integrity. Post-processing crosslinking strategies are used to physically or chemically strengthen and stabilize the 3D structure of the printed composites. A common method is photo-crosslinking, which uses UV or visible light to quickly cure methacrylate polymers like GelMA (gelatin methacryloyl) and PVAMA (poly(vinyl alcohol) methacrylate) (Lai et al. 2024). Visible light setups, such as the Ruthenium (Ru) and sodium persulfate (SPS)-based systems, are also useful as they are more biocompatible and can solidify decellularized bioinks in seconds (Zheng et al. 2021). Thermal crosslinking is another option, where for instance, heat-treated GelMA improves the printability and solidifies after printing (Suntornnond et al. 2022). While ionic crosslinking is usually used in extrusion printing, it can also work in inkjet setups. This involves immersing the printed structures in calcium chloride or using aerosol sprays, especially for alginate-based inks (Rahman et al. 2024). Finally, enzymatic crosslinking can also be used, for instance in horseradish peroxidase (HRP)-driven dityrosine bonding, that uses UV and riboflavin to activate the enzymes and trigger the gelation process of tyramine-modified alginate (Perin et al. 2022).

3.3 Solid Matrix-Assisted Printing (SMAP)

The Solid Matrix-Assisted Printing (SMAP) involves the deposition of bioinks within a hydrophobic solid matrix, such as polytetrafluoroethylene (PTFE) microparticles. This matrix provides a supportive environment for maintaining the integrity during and after the printing process (Fig. 6.4). The SMAP method is particularly useful for applications requiring precise control over feature size and topology, as the solid matrix prevents distortions caused by gravity or due to substrate contact (Shin et al. 2019; Ee and Yau Li 2021).

SMAP has been used to fabricate free-form structures of hydrogels containing aerobic microorganisms, and in the fabrication of complex geometries, such as tubular scaffolds, for vascular tissue engineering (Shin et al. 2019; Ee and Yau Li 2021; Colly et al. 2023).

4 Biomedical Applications of 3D Bioprinted Bacterial Cellulose

Recent progress in AM has made it possible to create biofunctional materials using BC-based bioinks for a range of medical applications. The addition of BC to 3D-printed hydrogels has been shown to improve the mechanical strength, printing accuracy, and compatibility with living tissue—especially in areas like bone and cartilage repair, skin regeneration, and wound healing (Table 6.1).

4.1 Bone Tissue Engineering

BC-based bioinks have been shown to hold strong potential for bone repair (Boyetey et al. 2023). In a proof of concept, Turlybekuly et al. (2020) used jet printing to create scaffolds combining BC with in situ synthesized hydroxyapatite (HA). The result was a mineralized structure where the formed HA particles were evenly distributed throughout the BC network. These authors have demonstrated that by increasing the BC content in the BC/HA bioink, the mean size of the formed HA microparticles increased (from 2.98 μm to 3.36 μm).

In another approach, Aki et al. (2020) developed a composite bioink made of polyvinyl alcohol, hexagonal boron nitride (hBN), and varying content of BC to build scaffolds. Characterization of the mechanical properties of the scaffolds obtained by extrusion-based 3D printing showed that with 0.5 wt% BC into the PVA/hBN matrix, the construct had the highest elongation at break (93%) and a moderate tensile strength (0.10 MPa), indicating improved flexibility without severely compromising strength. Also, the in vitro study showed that the scaffolds containing 0.5 wt% BC supported the best the growth of human osteoblast cells after 72 h.

Wang et al. (2021) took a similar route, using extrusion-based bioprinting to make BC-infused scaffolds for a rat model with skull defects. The bioink combined gelatine with BC either TEMPO-oxidized or modified with maleic acid (MA), resulting in an homogeneous hydrogel. Bioprinting was done with an extrusion-based 3D printer, where the gel was pushed through a 500 μm nozzle allowing to create porous scaffolds, one layer at a time. After printing, the hydrogel was cross-linked with NHS and EDC. In vitro assays with MC3T3-E1 pre-osteoblast cells showed that both TEMPO-oxidized and maleic acid (MA)-treated BC supported cell growth. However, the MA-BC led to a greater formation of mineralized nodules and a higher expression of osteogenic markers such as ALP, Runx2, and COL1. Also, MA-BC scaffolds implanted into skull defects in rats promoted the highest bone healing, with the highest bone density and thicker trabecular structures.

4.2 Cartilage Tissue Engineering

BC has also been shown to play a key role in the preparation of detailed scaffolds for cartilage repair, by enabling the formation of strong and hydrated structures. Apelgren et al. (2019) used BC-based scaffolds combined with human nasal chondrocytes, successfully generating cartilage tissue in nude mice. In this work, the bioink was composed exclusively of BC, homogenized by aqueous countercollision to produce a suspension of BC fibrils suitable for bioprinting. The bioink was then combined with human nasal chondrocytes. Bioprinting was done using an extrusion-based system. The constructs were implanted subcutaneously into nude mice and then explanted at 30 and 60 days to evaluate *in vivo* cartilage formation. Results have shown that BC scaffold had sufficient mechanical stability and structural integrity to allow the proliferation of human nasal chondrocytes and to enable the development of neocartilage with increasing cell density and extracellular matrix over time.

In another study, Sang et al. (2023) developed a photocurable hydrogel bioink composed of GelMA, HAMA, and BC for nasal cartilage repair. After extrusion-based printing, the structure was crosslinked with light, resulting in a mechanically strong scaffold. After a 7-day in vitro study, the scaffold showed good viability for ATDC5 mouse chondrocyte cells. Also, the sustained expression of type II collagen (Col2A1), aggrecan and Sox9 by the chondrocytes, suggests that the scaffold supported cartilage-like tissue development.

4.3 *Soft Tissue Engineering and Hierarchical Scaffolds*

BC-based bioinks have shown high promise in soft tissue repair, especially for wound healing and skin regeneration. Their ability to form porous, layered structures makes them ideal for these applications. Huang et al. (2019) used extrusion-based bioprinting to create silk-based hydrogel scaffolds reinforced with BC nanofibers. The BC nanofibers, incorporated at different concentrations, aimed at reinforcing the mechanical properties and improving the printing fidelity. Bioprinting resulted in the formation of porous scaffolds. L929 mouse fibroblasts were seeded onto the scaffolds to evaluate cell adhesion, spreading, and viability. Fibroblasts were observed to adhered well, spread uniformly, and remained highly viable. Also, the constructs were implanted subcutaneously in ICR (Institute of Cancer Research) mice to assess biocompatibility and tissue integration. No signs of significant inflammation were observed; moreover, the scaffolds supported tissue infiltration and integration over a 4-week period.

In a novel proof of concept, Shin et al. (2019) developed a bioink composed of a hydrogel containing a culture medium, *Gluconacetobacter xylinus* cells and carboxymethylated cellulose nanofibers (CMFs). Given their negative charge, the CMFs were highly water dispersible, improving the rheological properties of the ink. This viscoelastic ink was printed using SMAP, with loosely packed PTFE particles used as the support matrix. These microparticles formed a porous, loosely packed structure with interconnected porosity, allowing oxygen to diffuse through the gaps until it reached the printed hydrogel. After bioprinting, the construct was left for cultivation for 7 days. Then, it was rinsed with water to remove the bacteria and immersed in a Ca^{2+} solution to crosslink the carboxymethyl groups of the CMFs. This helped to reinforce the BC/CMFs matrix and to improve its structural stability. The outer layer of the obtained scaffold was mainly composed of a more compact BC pellicle, whereas its internal structure was composed of a more porous matrix containing interwoven BC and CMFs. These authors also fabricated a collagen-functionalized BC/CMF tubular scaffold by SMAP. Here, post-scaffold production, collagen was covalently immobilized onto the luminal surface. As assessed by Live/Dead staining, in vitro seeding of the tubular scaffold with NIH 3 T3 fibroblasts showed strong cell adhesion, high viability, and uniform distribution along the collagen-functionalized luminal surface. These results demonstrated the scaffold's suitability to support mammalian cell growth in tissue engineering applications.

4.4 Antimicrobial Scaffolds and Drug Delivery

BC has also been explored as a bioink component for the fabrication of antimicrobial scaffolds and targeted drug delivery systems. Gutierrez et al. (2019) explored two pre-printing treatments to produce alginate/BC scaffolds containing in situ formed copper nanoparticles. In one method (method A), the alginate/BC hydrogel was first ionically crosslinked with calcium ions, then subjected to ion exchange with copper ions by submerging the scaffold in a copper(II) chloride solution. Sodium borohydride was then added to reduce in situ the copper into nanoparticles. In another method (Method B), the hydrogel was ionically crosslinked directly using a copper(II) chloride solution, allowing Cu^{2+} ions to form the primary crosslinking network. Likewise, sodium borohydride was added to chemically reduce the embedded copper ions, resulting in the formation in situ of copper nanoparticles within the hydrogel matrix. In either method, the bioinks were subjected to extrusion-based 3D printing. Constructs produced by method A had better shape fidelity and structural robustness. The bioprinted scaffolds showed antimicrobial efficacy against *E. coli* and *S. aureus*.

In a different approach, Wei et al. (2020) prepared a bioink by adding TEMPO-oxidized BC (TOBC) to a mixture of sodium alginate (SA) and laponite nanoclay (XIs). The TOBC provided nanofibrous reinforcement, the SA offered biocompatibility and crosslinking capability and XIs also enhanced mechanical stability and allowed for a sustained protein release. Bovine serum albumin (BSA) was used as a model protein to evaluate the hydrogel's sustained release capability. The prepared ink was extruded using extrusion-based 3D printing. The obtained scaffold showed excellent printability, high shape fidelity, and good structural stability in aqueous environments. The hydrogel supported a sustained and prolonged release of BSA, with over 70% cumulative release achieved gradually over 14 days. An in vitro assay was done with L929 fibroblasts. From Live/Dead staining and phalloidin/DAPI cytoskeletal imaging, the scaffold showed excellent cell adhesion, viability, and spreading on its surface. CCK-8 assays indicated no cytotoxic effects and sustained cell proliferation over 5 days.

5 Conclusion

The integration of BC into 3D bioprinting represents a significant stride toward addressing the limitations of conventional biomaterials in tissue engineering and regenerative medicine. BC's unique combination of mechanical robustness, biocompatibility, and functionalization potential makes it an ideal candidate for bioink formulations, particularly in applications demanding structural integrity and biomimetic functionality. Extrusion-based bioprinting has been the most extensively explored technique for BC, yielding promising results in bone and cartilage regeneration, while emerging methods like solid matrix-assisted printing offer new avenues for fabricating complex architectures. However, challenges such as optimizing rheological properties, ensuring long-term cell viability, and *in vivo* biodegradability and achieving scalable production remain critical hurdles. Future research must focus on standardizing BC-based bioinks, refining crosslinking strategies, and conducting comprehensive *in vivo* studies to validate clinical efficacy. As the field advances, BC's versatility and sustainability position it as a cornerstone material for next-generation bioprinting solutions, with the potential to revolutionize personalized medicine and tissue repair.

Acknowledgment The authors gratefully acknowledge the financial support from the Science and Engineering Research Board (SERB), Department of Science and Technology (DST), Government of India, for funding this work.

References

Agarwal S, Saha S, Balla VK, Pal A, Barui A, Bodhak S (2020) Current developments in 3D bioprinting for tissue and organ regeneration–a review. Front Mech Eng 6. https://doi.org/10.3389/fmech.2020.589171

Aki D, Ulag S, Unal S, Sengor M, Ekren N, Lin C-C, Yılmazer H, Ustundag CB, Kalaskar DM, Gunduz O (2020) 3D printing of PVA/hexagonal boron nitride/bacterial cellulose composite scaffolds for bone tissue engineering. Mater Des 196. https://doi.org/10.1016/j.matdes.2020.109094

Amoli MS, EzEldeen M, Jacobs R, Bloemen V (2022) Materials for dentoalveolar bioprinting: current state of the art. Biomedicine 10(1):71. https://doi.org/10.3390/biomedicines10010071

Apelgren P, Karabulut E, Amoroso M, Mantas A, Martinez Avila H, Kolby L, Kondo T, Toriz G, Gatenholm P (2019) In vivo human cartilage formation in three-dimensional bioprinted constructs with a novel bacterial Nanocellulose bioink. ACS Biomater Sci Eng 5(5):2482–2490. https://doi.org/10.1021/acsbiomaterials.9b00157

Banigo AT, Nauta L, Zoetebier B, Karperien M (2025) Hydrogel-based BioINKs for coaxial and triaxial bioprinting: a review of material properties, printing techniques, and applications. Polymers 17(7):917. https://doi.org/10.3390/polym17070917

Betlej I, Zakaria S, Krajewski KJ, Boruszewski P (2021) Bacterial cellulose properties and its potential application. Sains Malaysiana 50(2):493–505. https://doi.org/10.17576/jsm-2021-5002-20

Bose S, Vahabzadeh S, Bandyopadhyay A (2013) Bone tissue engineering using 3D printing. Mater Today 16(12):496–504. https://doi.org/10.1016/j.mattod.2013.11.017

Boyetey M-JB, Torgbo S, Sukyai P (2023) Bio-scaffold for bone tissue engineering with focus on bacterial cellulose, biological materials for hydroxyapatite synthesis and growth factors. Eur Polym J 194:112168. https://doi.org/10.1016/j.eurpolymj.2023.112168

Cesur NP, Laçin NT (2022) Construction of vascular graft by 3D printing using BC formulation as bioink. Cellul Chem Technol 56(1–2):99–113. https://doi.org/10.35812/cellulosechemtechnol.2022.56.09

Chimene D, Lennox KK, Kaunas RR, Gaharwar AK (2016) Advanced BioInks for 3D printing: a materials science perspective. Ann Biomed Eng 44(6):2090–2102. https://doi.org/10.1007/s10439-016-1638-y

Choi SM, Rao KM, Zo SM, Shin EJ, Han SS (2022) BC and its applications. Polymers 14(6):1080. https://doi.org/10.3390/polym14061080

Colly A, Marquette C, Frances JM et al (2023) Classification of emerging freeform three-dimensional printing techniques. MRS Bull 48:69–92. https://doi.org/10.1557/s43577-022-00348-9

Cui Q, Zheng Y, Lin Q, Song W, Qiao K, Liu S (2014) Selective oxidation of bacterial cellulose by NO_2–HNO_3. RSC Adv 4:44516. https://pubs.rsc.org/en/content/articlelanding/2014/ra/c3ra44516j

Ee LY, Yau Li SF (2021) Recent advances in 3D printing of nanocellulose: structure, preparation, and application prospects. Nanoscale Adv 3(5):1167–1208. https://doi.org/10.1039/D0NA00408A

El-Sayegh S, Romdhane L, Manjikian S (2020) A critical review of 3D printing in construction: benefits, challenges, and risks. Arch Civil Mech Eng 20:1–25. https://doi.org/10.1007/s43452-020-00038-w

Feng C, Zhang M, Bhandari B (2019) Materials properties of printable edible inks and printing parameters optimization during 3D printing: a review. Crit Rev Food Sci Nutr 59(19):3074–3081. https://doi.org/10.1080/10408398.2018.1481823

Ganpisetti R, Lalatsa A (2021) Cellulose Bio–INK on 3D printing applications. J Young Pharm 13(1):1–6. https://doi.org/10.5530/jyp.2021.13.1

Ghosh S, Yi H-G (2022) A review on bioinks and their application in plant bioprinting. Int J Bioprinting 8(4):612. https://doi.org/10.18063/ijb.v8i4.612

Gouripriya DA, Adhikari J, Debnath P, Ghosh S, Ghosh P, Thomas S, Ghandilyan E, Gorbatov P, Kuchukyan E, Gasparyan S, Saha P (2024) 3D printing of BC for potential wound healing applications: current trends and prospects. Int J Biol Macromol 279(Pt 2):135213. https://doi.org/10.1016/j.ijbiomac.2024.135213

Grunberg N, McMeeking A, Dieckmann E, Cheeseman C (2024) Development of printable bacterial nanocellulose bioinks for bioprinting applications. Biotechnol Sustain Mater 1(1). https://doi.org/10.1186/s44316-024-00015-w

Gu Z, Fu J, Lin H, He Y (2020) Development of 3D bioprinting: from printing methods to biomedical applications. Asian J Pharm Sci 15(5):529–557. https://doi.org/10.1016/j.ajps.2019.11.003

Gungor-Ozkerim PS, Inci I, Zhang YS, Khademhosseini A, Dokmeci MR (2018) Bioinks for 3D bioprinting: an overview. Biomater Sci 6(5):915–946. https://doi.org/10.1039/c7bm00765e

Gutierrez E, Burdiles PA, Quero F, Palma P, Olate-Moya F, Palza H (2019) 3D printing of antimicrobial alginate/bacterial-cellulose composite hydrogels by incorporating copper nanostructures. ACS Biomater Sci Eng 5(11):6290–6299. https://doi.org/10.1021/acsbiomaterials.9b01048

Hölzl K, Lin S, Tytgat L, Van Vlierberghe S, Gu L, Ovsianikov A (2016) Bioink properties before, during and after 3D bioprinting. Biofabrication 8(3):032002. https://doi.org/10.1088/1758-5090/8/3/032002

Hospodiuk M, Dey M, Sosnoski D, Ozbolat IT (2017) The bioink: a comprehensive review on bioprintable materials. Biotechnol Adv 35(2):217–239. https://doi.org/10.1016/j.biotechadv.2016.12.006

Huang L, Du X, Fan S, Yang G, Shao H, Li D, Cao C, Zhu Y, Zhu M, Zhang Y (2019) Bacterial cellulose nanofibers promote stress and fidelity of 3D-printed silk based hydrogel scaffold with hierarchical pores. Carbohydr Polym 221:146–156. https://doi.org/10.1016/j.carbpol.2019.05.080

Jovic TH, Kungwengwe G, Mills AC, & Whitaker IS (2019) Plant-Derived Biomaterials: A review of 3D bioprinting and biomedical applications. Frontiers in Mechanical Engineering, 5. https://doi.org/10.3389/fmech.2019.00019

Lai C, Zhang S, Sheng L, Liao S, Xi T, Zhang Z (2013) TEMPO-mediated oxidation of bacterial cellulose in a bromide-free system. Colloid Polym Sci 291:2985–2992. https://doi.org/10.1007/s00396-013-3033-7

Lai Y, Xiao X, Huang Z, Duan H, Yang L, Yang Y, Li C, Feng L (2024) Photocrosslinkable biomaterials for 3D bioprinting: mechanisms, recent advances, and future prospects. Int J Mol Sci 25(23):12567. https://doi.org/10.3390/ijms252312567

Lee JM, Ng WL, Yeong WY (2019) Resolution and shape in bioprinting: strategizing towards complex tissue and organ printing. Biofabrication 11(4):045003. https://doi.org/10.1088/1758-5090/ab23cf

Lee JM, Yeong WY (2016) Design and printing strategies in 3D bioprinting of cell-hydrogels: a review. Adv Healthc Mater 5(22):2856–2865. https://doi.org/10.1002/adhm.201600435

Lin L, Jiang S, Yang J, Qiu J, Jiao X, Yue X, Ke X, Yang G, Zhang L (2023) Application of 3D-bioprinted nanocellulose and cellulose derivative-based bio-inks in bone and cartilage tissue engineering. Int J Bioprint 9(1):637. https://doi.org/10.18063/ijb.v9i1.637

Liu Y, Xia X, Liu Z, & Dong M (2023) The next frontier of 3D bioprinting: bioactive materials functionalized by bacteria. Small, 19(10). https://doi.org/10.1002/smll.202205949

Lu Y, Mehling M, Huan S, Bai L, Rojas OJ (2024) Biofabrication with microbial cellulose: from bioadaptive designs to living materials. Chem Soc Rev 53(14):7363–7391. https://doi.org/10.1039/d3cs00641g

Mahendiran B, Muthusamy S, Sampath S, Jaisankar SN, Popat KC, Selvakumar R, Krishnakumar GS (2021) Recent trends in natural polysaccharide based bioinks for multiscale 3D printing in tissue regeneration: a review. Int J Biol Macromol 183:564–588. https://doi.org/10.1016/j.ijbiomac.2021.04.179

Malda J, Visser J, Melchels FP, Jüngst T, Hennink WE, Dhert WJA, Groll J, Hutmacher DW (2013) 25th Anniversary article: engineering hydrogels for biofabrication. Adv Mater 25(36):5011–5028. https://doi.org/10.1002/adma.201302042

Markstedt K et al (2015) 3D bioprinting human chondrocytes with nanocellulose–alginate bioink for cartilage tissue engineering applications. Biomacromolecules 16(5):1489–1496. https://doi.org/10.1021/acs.biomac.5b00188

Mazega A, Lehrhofer AF, Aguado RJ, Potthast A, Marquez R, Rosenau T, Delgado-Aguilar M (2025) Key insights into TEMPO-mediated oxidation of cellulose: influence of the starting material. Cellulose 32:452–467. https://doi.org/10.1007/s10570-025-06477-z

McCarthy RR, Ullah MW, Booth P, Pei E, Yang G (2019) The use of bacterial polysaccharides in bioprinting. Biotechnol Adv 37(8):107448. https://doi.org/10.1016/j.biotechadv.2019.107448

Melchels FPW, Domingos MAN, Klein TJ, Malda J, Bartolo PJ, Hutmacher DW (2012) Additive manufacturing of tissues and organs. Prog Polym Sci 37(8):1079–1104. https://doi.org/10.1016/j.progpolymsci.2011.11.007

Morgan FLC, Moroni L, Baker MB (2020) Dynamic bioinks to advance bioprinting. Adv Healthc Mater 9(15):1901798. https://doi.org/10.1002/adhm.201901798

Murphy SV, Atala A (2014) 3D bioprinting of tissues and organs. Nat Biotechnol 32(8):773–785. https://doi.org/10.1038/nbt.2958

Ng WL, Chan A, Ong YS, Chua CK (2020) Deep learning for fabrication and maturation of 3D bioprinted tissues and organs. Virtual Phys Prototyp 15(3):340–358. https://doi.org/10.1080/17452759.2020.1771741

Ng WL, Shkolnikov V (2024) Optimizing cell deposition for inkjet-based bioprinting. Int J Bioprinting 10(2). https://doi.org/10.36922/ijb.2135

Perin F, Mota C, Mancini I, Motta A, Maniglio D (2022) Photo-enzymatic dityrosine crosslinking for bioprinting. Polymer 252:124941. https://doi.org/10.1016/j.polymer.2022.124941

Rahman TT, Wood N, Rahman AM, Pei Z, Qin H (2024) Applying in situ ionic crosslinking in bioprinting using algae cells. J Manuf Sci Eng 146(3):034501. https://doi.org/10.1115/1.4064221

Rastogi P, Kandasubramanian B (2019) Review of alginate-based hydrogel bioprinting for application in tissue engineering. Biofabrication 11(4):042001. https://doi.org/10.1088/1758-5090/ab331e

Rumon MMH, Rahman MS, Akib AA, Sohag MS, Rakib MRA, Khan AR, Yesmin F, Shakil MS, Khan MMR (2025) Progress in hydrogel toughening: addressing structural and crosslinking challenges for biomedical applications. Discov Mater 5(1). https://doi.org/10.1007/s43939-025-00178-x

Sang S, Ma Z, Cao Y, Shen Z, Duan J, Zhang Y, Wang L, An Y, Mao X, Zhang Q (2023) BC enhanced photocurable hydrogel based on 3D bioprinting for nasal cartilage repair. Int J Polym Mater Polym Biomater 72(9):702–713. https://doi.org/10.1080/00914037.2022.2097395

Santoni S, Gugliandolo SG, Sponchioni M, Moscatelli D, Colosimo BM (2021) 3D bioprinting: current status and trends—a guide to the literature and industrial practice. Bio-Des Manuf 5(1):14–42. https://doi.org/10.1007/s42242-021-00165-0

Satpathy A, Pal A, Sengupta S, Das A, Hasan MM, Ratha I et al (2019) Bioactive nanohydroxyapatite doped electrospun PVA-chitosan composite nanofibers for bone tissue engineering applications. J Indian Inst Sci 99:289–302. https://doi.org/10.1007/s41745-019-00118-8

Shariati SRP, Moeinzadeh S, Jabbari E (2015) Hydrogels for cell encapsulation and bioprintings. In: Stem cell biology and regenerative medicine. Springer, Cham, pp 89–108. https://doi.org/10.1007/978-3-319-21386-6_4

Sharma C et al (2025) Cellulose-based 3D printing bio-inks for biomedical applications: a review. Int J Biol Macromol 305(Pt 2):141174. https://doi.org/10.1016/j.ijbiomac.2025.141174

Shin S, Kwak H, Shin D, Hyun J (2019) Solid matrix-assisted printing for three-dimensional structuring of a viscoelastic medium surface. Nat Commun 10(1):4650. https://doi.org/10.1038/s41467-019-12585-9

Shiva S, Asuwin Prabu RG, Bajaj G, John AE, Chandran S, Kumar VV, Ramakrishna S (2023) A review on the recent applications of synthetic biopolymers in 3D printing for biomedical applications. J Mater Sci Mater Med 34(62). https://doi.org/10.1007/s10856-023-06765-9

Singh P, Williams DJ (2008) Cell therapies: realizing the potential of this new dimension to medical therapeutics. J Tissue Eng Regen Med 2(6):307–319. https://doi.org/10.1002/term.108

Suntornnond R, Ng WL, Huang X, Yeow CHE, Yeong WY (2022) Improving printability of hydrogel-based bio-inks for thermal inkjet bioprinting applications via saponification and heat treatment processes. J Mater Chem B 10(31):5944–5956. https://doi.org/10.1039/D2TB00442A

Swingler S, Gupta A, Gibson H, Kowalczuk M, Heaselgrave W, Radecka I (2021) Recent advances and applications of bacterial cellulose in biomedicine. Polymers 13:412. https://doi.org/10.3390/polym13030412

Tarassoli SP, Jessop ZM, Jovic T, Hawkins K, Whitaker IS (2021) Candidate BioINKs for extrusion 3D bioprinting—a systematic review of the literature. Front Bioeng Biotechnol 9. https://doi.org/10.3389/fbioe.2021.616753

Teixeira MC, Lameirinhas NS, Carvalho JP, Silva JM, Reis RL (2022) A guide to polysaccharide-based hydrogel bioinks for 3D bioprinting applications. Int J Mol Sci 23(12):6564. https://doi.org/10.3390/ijms23126564

Tibbetts JH (2021) The future of bioprinting. Bioscience 71(6):564–570. https://doi.org/10.1093/biosci/biab046

Turlybekuly A, Sagidugumar A, Otarov Y, Magazov N, Pogrebnjak A, Savitskaya I, Akatan K, Kistaubayeva A, Talipova A (2020) In: Pogrebnjak A, Pogorielov M, Viter R (eds) Bacterial cellulose/hydroxyapatite printed scaffolds for bone engineering. Nanomaterials in biomedical application and biosensors (NAP-2019). Springer, Singapore. https://doi.org/10.1007/978-981-15-3996-1_1

Utoiu E, Manoiu VS, Oprita EI, Craciunescu O (2024) Bacterial cellulose: a sustainable source for hydrogels and 3D-printed scaffolds for tissue engineering. Gels 10(6). https://doi.org/10.3390/gels10060387

Vartanian A (2023) Tissue not an issue. Nat Rev Mater 8(3):146. https://doi.org/10.1038/s41578-023-00548-y

Wang X, Tang S, Chai S, Wang P, Qin J, Pei W, Bian H, Jiang Q, Huang C (2021) Preparing printable bacterial cellulose based gelatin gel to promote in vivo bone regeneration. Carbohydr Polym 270:118342. https://doi.org/10.1016/j.carbpol.2021.118342

Wei J, Wang B, Li Z, Wu Z, Zhang M, Sheng N, Liang Q, Wang H, Chen S (2020) A 3D-printable TEMPO-oxidized bacterial cellulose/alginate hydrogel with enhanced stability via nanoclay incorporation. Carbohydr Polym 238:116207. https://doi.org/10.1016/j.carbpol.2020.116207

Xiongfa J, Hao Z, Liming Z, Jun X (2018) Recent advances in 3D bioprinting for the regeneration of functional cartilage. Regen Med 13(1):73–87. https://doi.org/10.2217/rme-2017-0106

Xu H et al (2019) Effects of printing conditions on cell distribution within microspheres during inkjet-based bioprinting. AIP Adv 9(9). https://doi.org/10.1063/1.5116371

Yap YL, Sing SL, Yeong WY (2020) A review of 3D printing processes and materials for soft robotics. Rapid Prototyp J 26(8):1345–1361. https://doi.org/10.1108/RPJ-11-2019-0302

Yue K, Santiago GT, Alvarez MM, Tamayol A, Annabi N, Khademhosseini A (2015) Synthesis, properties, and biomedical applications of gelatin methacryloyl (GelMA) hydrogels. Biomaterials 73:254–271. https://doi.org/10.1016/j.biomaterials.2015.08.045

Zheng Z, Eglin D, Alini M, Richards GR, Qin L, Lai Y (2021) Visible light-induced 3D bioprinting technologies and corresponding bioink materials for tissue engineering: A review. Engineering 7(7):966–978. https://doi.org/10.1016/j.eng.2020.05.021

Zhou K, Yu H, Xu L, Qian X, Wang D, Wang J (2022) Hydrogels for 3D embedded bioprinting: a focused review on bioinks and support baths. J Mater Chem B 10(11):1897–1907. https://doi.org/10.1039/D1TB02246B

Zhu W, Ma X, Gou M, Mei D, Zhang K, Chen S (2016) 3D printing of functional biomaterials for tissue engineering. Curr Opin Biotechnol 40:103–112. https://doi.org/10.1016/j.copbio.2016.03.014

Chapter 7
Bacterial Cellulose in Electronic, Optoelectronic, Photonics, and Energy Applications

Jing Li, Zhuotong Wu, and Shiyan Chen

Abstract Bacterial cellulose (BC), an eco-friendly nanomaterial produced by microbial fermentation, has emerged as a versatile platform for advanced applications in electronics, optoelectronics, photonics, and energy technologies. Its unique three-dimensional nanofibrillar network, combined with remarkable mechanical strength, high crystallinity, and tailorable surface chemistry, enables exceptional performance in multifunctional devices. In electronics, BC serves as an excellent substrate for flexible and transparent conductive films, stretchable sensors, and wearable electronics, due to its superior mechanical flexibility and compatibility with various conductive materials. For optoelectronic applications, BC-based composites exhibit tunable optical properties and efficient charge transport characteristics, making them promising candidates for organic photovoltaics and light-emitting devices. In photonics, BC's nanostructured matrix facilitates precise control over light-matter interactions, enabling applications in structural coloration, optical sensors, and biodegradable photonic crystals. The energy sector benefits from BC's hierarchical porosity and ion-conducting properties, where it demonstrates outstanding performance as separators in batteries (e.g., Li-ion, Zn-ion), electrodes in supercapacitors, and proton-exchange membranes in fuel cells. Particularly noteworthy is BC's role in emerging energy harvesting systems, including triboelectric nanogenerators and osmotic power generators, where its piezoelectric properties and ion selectivity enable efficient energy conversion. With ongoing advances in functionalization strategies and scalable production, BC is poised to play a pivotal

J. Li
Fujian Key Laboratory of Novel Functional Textile Fibers and Materials, Minjiang University, Fuzhou, People's Republic of China
e-mail: lijing3151@mju.edu.cn

Z. Wu (✉)
Fujian Agriculture and Forestry University, Fuzhou, People's Republic of China
e-mail: wuzt@fafu.edu.cn

S. Chen (✉)
State Key Laboratory of Advanced Fiber Materials, College of Materials Science and Engineering, Donghua University, Shanghai, People's Republic of China
e-mail: chensy@dhu.edu.cn

F. Dourado et al. (eds.), *Advances in Bacterial Cellulose*,
https://doi.org/10.1007/978-3-032-07042-5_7

role in developing next-generation sustainable and high-performance devices across these interdisciplinary fields.

Keywords Bacterial cellulose · Flexible electronics · Optoelectronics · Photonics · Energy storage · Sustainable materials

1 Introduction

In today's society, the demand for high-performance, sustainable electronic devices, optical equipment, and energy storage and conversion technologies is increasing (Wu et al. 2023a; Ko et al. 2023; Poisson and Zhang 2024). Although traditional materials exhibit excellent performance, their manufacturing processes are often accompanied by high energy consumption and environmental pollution, and some material resources are limited and nonrenewable. With the increasing severity of global energy crises and environmental pollution, the development of novel, renewable, and environmentally friendly materials have become a significant focus of scientific research.

Bacterial cellulose (BC), a biomaterial synthesized by microorganisms, has attracted growing attention due to its abundant sources, green and environmentally friendly production process, and unique structure (Dhali et al. 2021). Characterized by a highly crystalline nanofibrous network, excellent mechanical properties, biocompatibility, and customizable physicochemical properties, as well as versatile structural modifications, BC has emerged as a promising research material in electronics and optoelectronics, energy storage, and diverse green energy conversion devices (Fig. 7.1) (Girard et al. 2024; Pan et al. 2023).

In the face of the global urgent demand for green energy and advanced materials, the widespread application of BC undoubtedly provides new solutions for future sustainable technologies and clean energy development. This chapter aims to

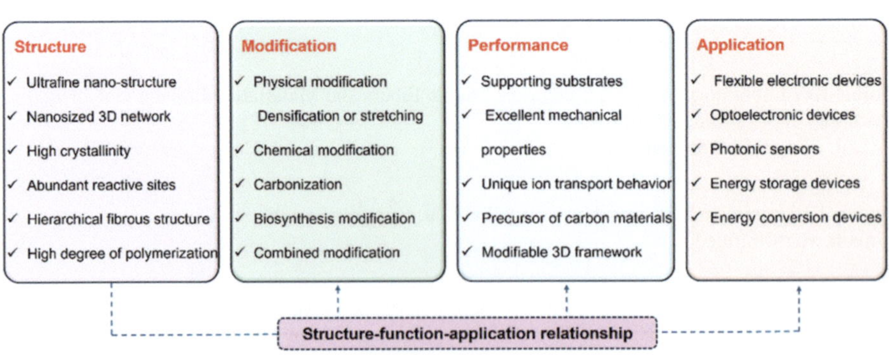

Fig. 7.1 Structure-function-application relationship of BC electronic, optoelectronic, photonics, and energy applications

provide a comprehensive overview of BC's emerging role in advanced material applications, particularly in electronic, optoelectronic, photonic, and energy-related technologies. This chapter aims to bridge the gap between fundamental BC science and its practical implementation in next-generation sustainable devices. From the perspective of sustainable development, it reveals the core value of BC materials in driving the green technology revolution, clearly positioning BC as a strategic functional material for breakthroughs in next-generation environmentally friendly technologies.

2 Bacterial Cellulose in Electronic Applications

BC is a high-purity, nontoxic, biodegradable, and sustainable natural polymer with superior mechanical strength and intricate fibrillar porous structure. For these unique biological and physical properties, BC has been widely employed in food, biomedical, textile, and cosmetic applications (Girard et al. 2024). While BC is an outstanding insulator by nature, it has been explored in electronics as a flexible substrate material. Conductive materials have been successfully embedded or polymerized onto BC surfaces or internally to create complexes with high conductivity, thus used in electronics, electromagnetic shielding, electronic printing, and other applications.

2.1 Conductive Composites

The incorporation of some conducting additives through in situ polymerization, chemical modification, and solution blending techniques can successfully transform non-conductive BC into conductive composites (Saleh et al. 2024). Electron-conductive materials (metals and their oxides, MXene, carbon-based materials, conducting polymers) and ion-conductive materials (like ionic liquids, inorganic salts) are the most commonly employed in BC-conductive composites, all demonstrating favorable conductivity covering multiple applications (Table 7.1).

The conductivity of BC-based composites is closely connected to the inherent conductivity of the applied fillers and their compatibility within BC (Saleh et al. 2024). Metal and its oxide-based materials possess various morphologies, such as nanoparticles (Au, TiO_2, ZnO, SnO_2), nanorods/nanowires (Ag), and thin films (Ag and indium-tin oxide) (Chen et al. 2018). Their incorporation into BC-based films or hydrogels forming a BC-particle network can significantly improve the electrical conductivity and mechanical properties of the composite. MXene is an emerging two-dimensional (2D) transition metal carbide, nitride, or carbon-nitride with high metallic conductivity, hydrophilicity, good mechanical properties, high aspect-ratio, and abundant surface functional groups (−OH, −O, −F). The composites that incorporate the unique properties of MXene and the intrinsic polymeric characteristics of

Table 7.1 BC-based composites, their conductivity and applications

Conductive materials	Preparation method	Conductivity S/cm	Application	References
PANI	In situ polymerization	1.08	Conductive composite paper	Fei et al. (2019)
PPy	In situ polymerization	0.01–1.2	Enhanced electrical conductivity	Muller et al. (2013)
MXene (Ti_3C_2Tx)	Solution blending and vacuum-assisted filtration	306.7	Electromagnetic interference shielding and photothermal application	Luo et al. (2023)
PEDOT	In situ polymerization	1.8×10^{-2}	Improve ionic conductivity for peripheral nerve repair	Liu et al. (2022a)
Carbon	Pyrolysis	0.20–0.41	Stretchable and foldable electronics	Liang et al. (2012)
$Li_7La_3Zr_2O_{12}$	Freeze drying and soaking	1.12×10^{-4}	Solid-state electrolytes for lithium-ion battery	Xie et al. (2018)
Cl^-	In situ molecularization	0.063	Electronic skin	Jiang et al. (2022)

BC are not only endowed with excellent mechanic performance, conductivity, and biodegradability in nature, but also provide great potential for surface modification and functionalization by virtue of their plentiful surface-encapsulation functional groups (Talipova et al. 2023).

Carbon-based materials, such as carbon nanotubes (CNTs), graphene, and graphene oxide (GO), are widely used for conductive modification. Their agglomeration, caused by strong van der Waals interactions, can be effectively mitigated through the use of highly hydrophilic BC (Li et al. 2017a; Song et al. 2024). Thus, this can guarantee the homogeneous dispersion for carbon-based conductive materials within the cellulose nanofiber-network pores and achieving the excellent conductivity of BC composites.

Conducting polymers, known as conjugated π-polymers, are introduced into BC to avoid uneven distribution or damage of the BC's mechanical properties caused by large amounts of nanofillers (such as metal- and carbon-based materials) in the composite mentioned above. Due to their compatibility with cellulose, poly(3,4-ethylenedioxythiophene)/poly(styrenesulfonate) (PEDOT: PSS), polyacetylene (PA), polypyrrole (PPy), and polyaniline (PANI), used alone or combined with metal- or carbon-based materials, are the most popular conductive polymers to create high electrical conductivity in BC-based bioelectronic and energy storage devices (Prilepskii et al. 2023). Ion-conductive materials, mainly referring to free ions of Na^+, K^+, Li^+, Fe^{3+}, Al^{3+}, ionic liquids, and polyelectrolytes, have been incorporated into BC hydrogel systems to construct ionic conducting hydrogels. The outstanding conductivity of the BC hydrogels is attained by the ionic free movement in the three-dimensional (3D) hydrogel network (Chen et al. 2024).

2.2 *Flexible Electronic Devices*

BC composites obtained after encapsulation of conductive filler materials have attracted significant attention for flexible electronic devices given the intrinsic properties of BC with 3D network structure, high specific surface area, and excellent mechanical characteristics, which mainly include sensors, actuators, electromagnetic interference (EMI) shielding, electronic skins (E-skin), electrode sheets, electrochromic (EC), supercapacitors and batteries, and so forth (Poddar and Dikshit 2021). In addition, supercapacitors and batteries are further described in the following section regarding energy applications.

The 3D network structure of BC allows for the uniform surface attachment of stimuli-responsive functional materials, enabling the successful construction of BC hydrogel-based flexible sensors for pH, electroactive, humidity, magnetic and thermal applications (Pan et al. 2023; Huang et al. 2022; de Souza et al. 2023). A variety of specifically responsive BC-based electronic sensors distributed on a single uniform surface or stacked on multiple surfaces can be assembled into a biodegradable E-skin, which possesses the three fundamental functions of human skin-like sensation, protection, and thermoregulation (Yang et al. 2024). To better mimic the biological skin, it is essential that E-skin is also endowed with properties such as breathability, self-healing, washability, and conformability. Consequently, the functional building materials and device structures in E-skin must be carefully engineered (Lv et al. 2023). The multilayer integrity and device conformability of flexible electronics highly depend on flexible conductive electrodes to enable multiple single-electronic device integration. The strong intermolecular forces among interconnected BC nanofibers create the structural support for the formation of free-standing multilayer and multi-material electrodes with strong interlayer bonding. Therefore, its main application directions include BC film electrodes with composite conductive materials (Luo et al. 2024), patterned electrodes with conductive materials/BC inks (Lv et al. 2022), or directly acting as a supporting base for electrodes (Shah and Malcolm Brown 2005).

Meanwhile, BC suffers from inhomogeneous deformation in response to external stimuli (electricity, temperature, magnetic field, and moisture) as a result of changing conformations and physicochemical properties. Subsequently, the difficulty in achieving complex and well-defined deformations for BC materials due to the nonuniform implementation of external stimuli can be overcome through the manufacturing of anisotropic structures (monolayers, bilayers, and multilayers). Hence, they are extensively used for the development of advanced soft actuators (Chen et al. 2022). Furthermore, benefiting from BC porous architecture and good dielectric properties, BC for EMI shielding tends to block the electromagnetic wave transmission via more effective scattering, reflection, or absorption of electromagnetic radiation. Therefore, the incorporation of conductive/magnetic nanoparticles into BC-based materials through surface modification, encapsulation, and coating can achieve the desired properties for effective EMI shielding. The most representative one is MXene, its BC-composite film demonstrating a specific EMI shielding efficiency of 29,141 dB cm^2g^{-1} at 4 μm thickness, which surpasses those of most previously reported carbon-based polymer composites (Talipova et al. 2023; Wan et al. 2021).

2.3 Printed Electronics

Considering self-supporting films, BC allows for smoother and less porous films compared to the conventional cellulosic papers, having also a certain level of transparency (light transmission), allowing its adoption in printed electronics (Inácio et al. 2020; Avila et al. 2018). BC has recently emerged as a promising material for conductive ink formulations, owing to its exceptional aspect ratio. This application specifically targets two fundamental challenges in conductive ink development: (1) precise control over the synthesis of conductive raw materials and (2) optimization of suspension stability, dispersion characteristics, and physicochemical properties (Hoeng et al. 2016). The utilization of BC in conductive inks follows two principal methodologies: first is to synthesize conductive ink particles (metal nanoparticles or flakes, carbon particles, and conductive polymers) by composite conductive fillers, as shown in (1) the bio-template using BC with high aspect ratios to synthesize tubular conductive particles and (2) BC with surface charged groups serving as a capping and nucleating agent to synthesize metal particles by controlling the their synthesis and growth. The second is that BC is utilized as a dispersing and stabilizing agent or a binding agent in ink formulations.

3 Bacterial Cellulose in Optoelectronic Applications

The unique growing behavior of BC-producing bacteria promotes densely packed nanofibers, reducing greatly the voids inside BC, and consequently eliminating light scattering in the visible region. Ultimately, BC films present optically transparency. Furthermore, the highly pure BC is free from hemicellulose (degrading at lower temperatures) and lignin (decomposing over a wide range of temperatures), thus it exhibits good thermal stability (>300 °C). For these reasons, BC-based films are extremely attractive for applications in optoelectronic devices when composited with conductive materials, especially for the transparent conductive films and photoelectric sensors.

3.1 Transparent Conductive Films

The transparency and nanostructure of BC allow it as an ideal candidate for transparent conductive films (TCF). The introduction of conductive materials (such as indium tin oxide and graphene) into transparent BC not only yields a transparent conductive film that simultaneously fulfills the dual functions with surface light transmission and along-surface electron transmission, but also brings in new features like recyclability, reinforced printability, and enhanced flexibility, that can even facilitate bending and easy folding (Xu et al. 2017; Khan et al. 2015). Research on BC-based TCFs are expanding and deepening due to the superior flexibility, degradability, and sustainability of BC-based composite films over that of conventional plastic-, silicone-, or glass-based ones, thus they are successfully applied in touchscreens, organic light-emitting diodes (OLEDs), solar cells, field effect transistors (FETs), and electrochromic (EC) devices.

The touchscreens used in various electronic devices (such as smartwatches, mobile phones, tablets, or TVs) perform their sensing function by receiving voltage or specific signal changes from screen sensors when someone touches them. Manuspiya's research group has successfully developed a type of flexible and transparent BC-based composite for electronic substrates, in which the BC-based TCFs can be prepared for touchscreen applications in electronic devices via incorporating an electrically active polymer, such as CNT or poly(vinylidene fluoride) (Ummartyotin et al. 2017; O-Rak et al. 2013). OLED operates through the conversion of electrical energy into light, also known as electroluminescence. Conductive BC films replace traditional rigid glass sheets as promising OLED substrates besides necessarily meeting some basic requirements: for example, they must have as uniform and smooth surface as possible and excellent chemical, thermal, and mechanical resistance. Above all, they must be transparent in the visible region (Fresta et al. 2018). Currently, research on utilizing BC as a substrate for OLED development is mainly concentrated on the surface morphology modification, thus ameliorating the damage caused by its high hygroscopicity and surface roughness (~120 nm) for proper functioning (Faraco et al. 2023). As opposed to the operating principle of OLED, solar cells employ the photovoltaic effect to directly convert light into electricity. The high optical haze of BC-based TCFs makes them extremely attractive for high-performance solar cells with advanced light management (Pleumphon et al. 2017). The first report about n-channel oxide-based thin-film FETs with cellulose paper as a dielectric layer opens the cellulose FETs era (Fortunato et al. 2008). Moreover, FETs using BC simultaneously as the dielectric layer and the substrate not only exhibited I_{ON}/I_{OFF} modulation ratio above 10^4, but the small surface roughness of BC films also did not affect the electronic performance of the FET (Barquinha et al. 2007). Also, benefiting from good biocompatibility and bio-integration, BC-based FETs demonstrate great prospects for high-sensitivity and high-speed *in vivo* sensors and electronic applications (Son et al. 2017).

EC refers to a characteristic optical phenomenon wherein materials exhibit stable and reversible changes in their optical properties—including reflectance, transmittance, and absorptivity—in response to an applied electric field. This unique behavior manifests as controllable alterations in both coloration and transparency states. Such electrochromic materials have found extensive applications across multiple advanced technological domains, particularly in smart window systems for energy-efficient buildings, dynamic information display devices, adaptive camouflage technologies for wearable applications, and real-time medical monitoring systems. In order to promote ECs to more perfectly and rapidly express the color changes, total embedding inside the support material or uniform and stable adsorption on the surface is extremely critical. Fixing EC dyes within BC microstructures in situ, during the microbial growth in static culture, can build up high-resolution dynamic display devices based on BC substrates, which can demonstrate a reversible color change down to a standard pixel-sized area (100 μm^2) upon application of switching potentials (2 ~ 5 V) (Shah and Malcolm Brown 2005). Additionally, Tajima research group produced nano-fibrillated BC in a carboxymethylcellulose (CMC)-supplemented medium, utilized as display device binders, which exploits strong interactions between negatively charged BC fibers and quaternized EC molecules, to achieve a tight and homogeneous adsorption of the EC material. The uniform and reversible color changes are successfully observed across the electric field, indicating that BC can be employed as binder materials to reach uniform surface adsorption (Tajima et al. 2020).

3.2 Photoelectric Sensors

Photoelectric sensors are devices that convert optical signals into electrical signals based on the photoelectric effect, which opens promising directions for on-site qualitative/quantitative detection in various fields like healthcare, environmental monitoring, food safety, and so on (Sharifi et al. 2022). BC with high specific surface area and active groups is widely employed in the photoelectric sensor, such as photoluminescent (PL) sensors, colorimetric sensors, and opto-electrochemical (OEC) sensors, which can achieve sensitive response to different wavelengths through loading photoelectric materials.

Given the key position of photoluminescence spectroscopy in fulfilling fast, nondestructive, easy-to-operate, and highly sensitive sensing, coupled with the ubiquitous and charming advantages conferred by BC, BC-based PL sensors have been subjected to manifold sensing needs, especially in biomedicine and environmental monitoring (Sharifi et al. 2022). Liu et al. has created a bioluminescent nano-paper for the rapid and in situ optical toxicity screening of environmental pollutants including polybrominated, tributyltin, diphenyl ether, and diuron (Liu et al. 2018). In another study, Golmohammadi group successfully performed early point-of-care diagnosis in infants and continuous therapy monitoring of phenylketonuria using a sensing mechanism, in which the L-phenylalanine contributes to the subsequent selective recovery of the fluorescent-bursting carbon dot/BC-Fe(III) detection paper (Golmohammadi et al. 2020). Interestingly, the group has also carried out a series of research work on transparent BC nano-paper-based colorimetric sensors doped with curcumin, as a fascinating phytochemical with known extraordinary chemical and optical properties, which accomplished the screening of various (bio)molecules only relying on an established color change by means of visually or a smartphone's camera as well as a spectrophotometer (Naghdi et al. 2019; Faham et al. 2019).

Opto-electrochemical (OEC) sensors are optoelectronic devices that utilize charge transfer processes resulting from chemical reactions or photophysical interactions to measure photocurrents. They are highly popular in the point-of-care testing (POCT) because they show high sensitivity at low application potential and can suppress side reactions without damaging photosensitive materials and biomolecules (Wang et al. 2023). Considering BC nano-paper's excellent optical transparency, biocompatibility, printability, and flexibility, an OEC sensing platform based on it was first established by transferring a screen-printed three-electrode system onto a pre-printed BC nano-paper film (Eynaki et al. 2020). The interaction between Prussian blue with OEC tunable properties and acetaminophen verified that OEC sensing platform possesses dual electrochemical and optical sensing. The developed smartphone-based OEC probe can be used for quantitative detection of H_2O_2. Given the remarkable characteristic of the proposed BC-based OEC sensors as a promising sensing bio-platform, it is expected to provide inspiration for the design and fabrication of dual OEC sensors, which will subsequently be practically applied in the determination of a wide variety of (bio)chemicals (Eynaki et al. 2020).

4 Bacterial Cellulose in Photonics Applications

Currently, the study of biopolymer materials used in photonics applications (e.g., polynucleotides, polypeptides, and polysaccharides) has been focused on generating a linear optical color response by diffraction or reflection of light from ordered or chiral assemblies of biopolymer nanostructures (Bharmoria et al. 2024). In this regard, cellulose has attracted a remarkable interest as the most abundant biopolymer. In 1959, Marchessault et al. first discovered the chiral liquid crystal properties of cellulose nanocrystals (CNCs), and this chiral nematic structure of CNC-based films showed magical photonic properties due to its photonic crystalline features, which opened the door to the study of cellulose for photonics applications (Marchessault et al. 1959). The 3D network and BC protofibrils formed during the biosynthesis process, in static culture, represent a bottom-up generated hierarchical structure (Fig. 7.2). The multilevel and multilayer structural organization (micro-, nano-, and molecular scales) inspired researchers to utilize BC as a template to construct luminescent composites or photonic crystals with structural colors for

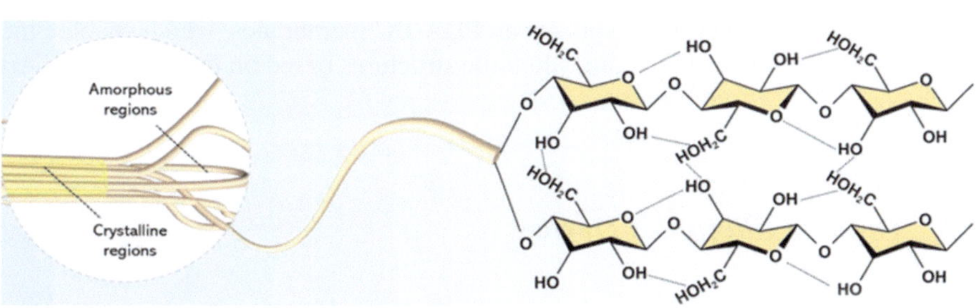

Fig. 7.2 BC structure with configurations of the crystalline and amorphous regions (Reproduced with permission (Prilepskii et al. 2023) Copyright 2023, Elsevier)

various photonic applications related to visual enhancement, camouflage, energy, communication, and photonic sensors.

4.1 Photonic Crystals

Photonic crystals are organized optical structures that periodically modulate the refractive index on the submicron scale, enabling controlled light propagation, which can be used in photonic bandgap materials, optical filters, and optical waveguides (Xiong et al. 2020). The fabrication of artificial photonic crystals has faced significant challenges, primarily owing to insufficient structural design precision and limited availability of efficient bio-fabrication techniques. In this context, cellulose nanofibrils have emerged as promising biomaterials, typically functioning as either substrates or microfluidic channels in photonic/electronic devices. Particularly noteworthy are CNCs, which demonstrate unique photonic crystal characteristics when their nanoscale crystalline structures are precisely controlled. This distinctive behavior arises from their ability to form periodically ordered arrangements under specific conditions, enabling photon bandgap modulation. Hydroxypropyl cellulose (HPC) is interesting because it not only has a liquid crystal phase, but also possesses the potential to act as a photonic architecture, adopting various shapes, thus making it extremely popular for biopolymer-based photonic platforms (Lee et al. 2022). Following a similar approach to synthesize 2D photonic crystals, using HPC as a photonic framework (Espinha et al. 2018), Bharmoria et al. employed a simple pre-programmable strategy of “photon upconversion crystals doped biopolymers” for the crystallization of nonlinear optically responsive photonic chromophores within bacterial cellulose-gelatin (BC-G) composite membranes (Bharmoria et al. 2024). Remarkably, the components of this material are physically disassembled to recover $66 \pm 1\%$ of annihilator at the end-of-service life, which demonstrates the sustainability of the developed photonics material.

In addition, BC was first applied as a substrate for the construction of optical biosensors based on structural colors (Suleimenova et al. 2023). As illustrated in Fig. 7.3, a colloidal silica particle was utilized to manufacture polydopamine (PDA) shells with imprinted sites for protein targets, followed by self-assembly of molecularly imprinted polymer (MIP) spheres on PDA-BC membranes, which enabled the sensing applications of short-range photonic structures based on BC, in a biological matrix, on a clinically relevant level.

4.2 Tunable Photonics Devices

BC presents enormous potential for tunable photonic device applications, given its excellent optical properties, the capability of facile functionalization and modification, multi-scale network structure and biocompatibility. Using a flexible BC

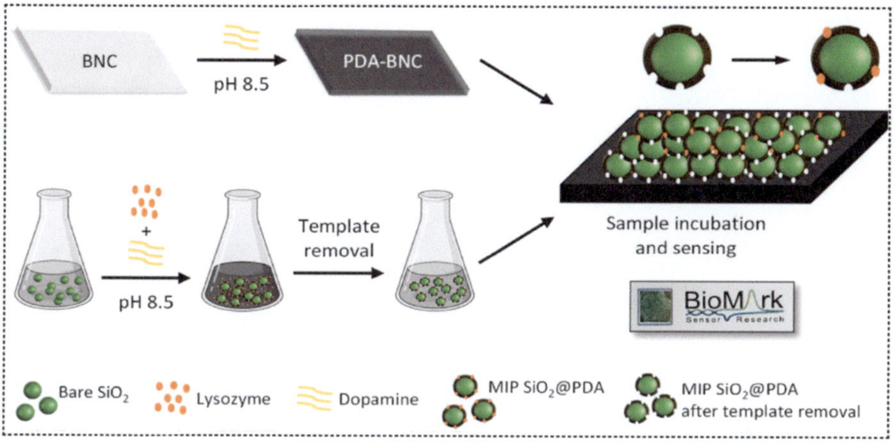

Fig. 7.3 Schematic representation of the sensor design and assembly (Suleimenova et al. 2023). (Reproduced with permission (Suleimenova et al. 2023) Copyright 2023, Elsevier)

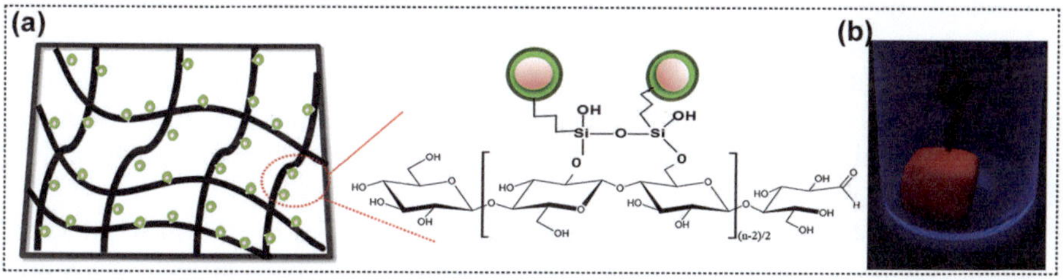

Fig. 7.4 (**a**) A schematic diagram of the BC fibers grafted with QD565 and (**b**) photos of the BC-ZCIS/ZnS QD565 aerogels under a UV lamp made via chemical binding (**a**, **b** Reproduced with permission (Wang et al. 2019) Copyright 2019, Elsevier)

membrane as the host substrate and gain medium, random lasers (RLs) eliminating the need for mirror-based optical cavities to confine and amplify light can be obtained by incorporating light-scattering particles (silver nanoparticles) into conventional lasers doped with laser-dye molecules (Rhodamine 6G) (dos Santos et al. 2014). The deeply investigated RLs based on kombucha BC show a significant shift toward longer wavelengths of the lasing emission and a remarkable decrease in the lasing threshold, suggesting that they are a promising laser light source for future applications in photonic barcoding, chemical and biological sensing (Mai et al. 2023). Wang et al. used the hydroxyl groups on BC nanofibers surface to graft ZCIS/ZnS core-shell quantum dots (QDs) via chemical covalent bonding and supercritical CO_2 drying for creating 3D network QD aerogels with tunable fluorescence emission properties (Fig. 7.4) (Wang et al. 2019).

BC is a high-molecular-weight homopolymer of β-1,4-linked anhydro-D-glucose units, where every glucose unit is twisted 180° with respect to its neighbors. While BC presents as non-birefringent upon drying because random orientations of BC

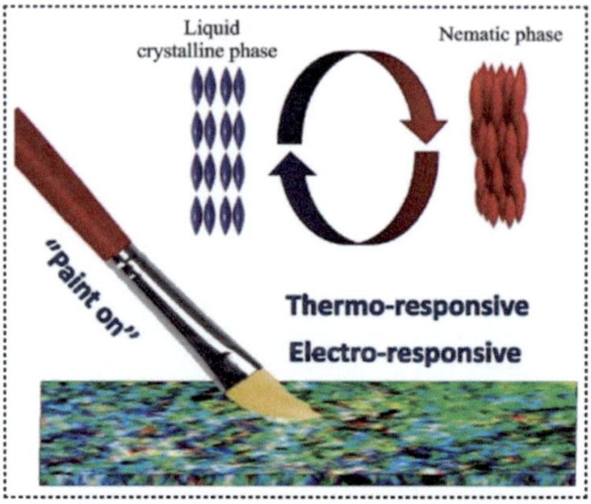

Fig. 7.5 Optical sensors (reproduced with permission (Santos et al. 2017) Copyright 2017, Elsevier)

nanofibers cancel out the intrinsic optical anisotropies of individual cellulose nanofibers, it can be compensated through mechanical stretching to induce unidirectional alignment of the nanofibers (Yu et al. 2023). The inherent stable right-hand twist of BC makes it a potential chiral host to confer luminophores with circularly polarized luminescence (CPL). Xu group exploited this structural feature to fabricate CPL-activated BC composite films with a dissymmetry factor of up to 0.03, prepared simply impregnating BC pellicles with dilute aqueous solutions of luminophores (rhodamine B, carbon dots, polymer dots) and drying under ambient conditions (Zou et al. 2019).

The study of optically active functionalities of CNCs is the most prevalent because they possess optical chirality directly relaying on the nanoscale crystalline structure (Lee et al. 2022). CNCs produced from acid-treated BC hydrogels have an unusually high aspect ratio, abbreviated as BCNCs. To precisely control the orientation in large-area photonic organization, it is crucial to achieve defect-free optical grade materials with long-range ordered and uniform optical response. The magnetic field-induced alignment is able to transform BCNC helical cholesteric (chiral nematic) decorated with superparamagnetic nanoparticles to uniaxial nematic phases, in a strongly magnetized state, thus exhibiting optical qualities such as translucency, high optical birefringence, and anisotropic light diffraction (Zhang et al. 2022). Besides, the most effective, economical, and convenient strategy adopted by researchers is the physical or chemical composite between BCNCs and various functional materials, which can be applied to photonic devices like optical sensors (Fig. 7.5) (Santos et al. 2017), tunable Bragg reflectors (De La Cruz et al. 2018), tunable filters (Santos et al. 2021), and surface-enhanced Raman spectroscopy (Rahman et al. 2022).

5 Bacterial Cellulose in Energy Application

In the current era of environmental sustainability, the development of clean and renewable energy sources coupled with advanced storage technologies has emerged as a critical research priority. This paradigm shift has driven energy storage and conversion technologies toward environmentally benign, cost-effective materials that exhibit superior electrochemical and ionic transport properties, among which bacterial cellulose (BC) has shown particular promise. BC distinguishes itself among sustainable materials through its unique three-dimensional interconnected nanofibrillar network, which confers exceptional material properties including: (1) unparalleled water retention capacity, (2) high specific surface area, (3) remarkable chemical stability, (4) outstanding mechanical strength, and (5) inherent environmental compatibility.

These characteristics collectively position BC as an ideal candidate for next-generation energy applications. Particularly noteworthy is BC's performance when incorporated into conductive matrices, where its three-dimensional architecture facilitates: enhanced electrochemical activity, rapid charge/discharge kinetics, and efficient ion extraction and transport. Recent advances have demonstrated the successful implementation of BC-derived materials—including mesoporous structures, flexible thin films, and fibrous networks—in diverse energy applications ranging from supercapacitors and batteries to energy harvesting systems. These developments underscore BC's significant potential to address contemporary challenges in sustainable energy technology.

5.1 *Energy Storage Devices*

Owing to its exceptional mechanical robustness and superior electrochemical characteristics including remarkable chemical stability across diverse solvents and an extensive electrochemical stability window, BC has emerged as a versatile material for energy storage applications. Specifically, BC has been extensively utilized as: (1) high-performance separators, (2) solid/gel polymer electrolytes, (3) bio-based adhesives, and (4) flexible substrates for advanced energy storage devices. Furthermore, BC-derived carbonaceous electrode materials have garnered significant research attention in sustainable energy storage systems, owing to their cost-effectiveness, enhanced electrical conductivity, and tunable electrochemical properties. The inherent sustainability, eco-friendliness, and facile processability of BC and its derivatives position BC-based energy storage systems as promising candidates for next-generation green electronics and renewable energy technologies.

5.1.1 Battery Separators

The separator is one of the four primary components of a battery, accounting for approximately 10–20% of the total battery cost (Lagadec et al. 2019). It serves two critical functions: (1) physically isolating the positive and negative electrodes to prevent short circuits, and (2) providing a porous structure that facilitates ion transport between the electrodes. As a result, the separator must exhibit electrical insulation, appropriate porosity, and sufficient wettability. BC is particularly promising for this application due to its natural 3D ultrafine nanofiber-network structure. This unique architecture endows BC with high porosity and a large specific surface area, which enhance electrolyte uptake and reduce interfacial resistance (Xie et al. 2019). Two main strategies have been developed to construct BC-based battery separators with low resistance and high safety (Fig. 7.6). The first strategy involves combining mass fermentation with a roll-to-roll manufacturing process, enabling large-scale production of BC membranes. However, pure BC separators are susceptible to structural degradation caused by strong Lewis acids (such as LiF and PF) released from electrolytes containing lithium hexafluorophosphate, $LiPF_6$. To address this issue, Gwon et al. introduced Lewis bases, such as polyethylene glycol (PEG), during the bacterial cultivation stage, to preserve the BC structure and enhance its thermal stability (Gwon et al. 2019). The second strategy, focused on structural modification, such as grafting various functional groups (carboxyl, aldehyde) or functional polymers (polyethyleneimine, chitosan, polydopamine), followed by vacuum filtration, may yield multiple functionalities or synergistically enhance performance compared to pure BC membranes (Huang et al. 2020; Li et al. 2021a). For instance, Wang's group reported a heat-resistant polyphenylene sulfide/BC membrane that exhibited exceptional stability and zero shrinkage at 200 °C (Zhu et al. 2021). When used as a separator in lithium-sulfur batteries, polyethyleneimine-grafted BC membranes demonstrated dendrite-free lithium deposition and superior

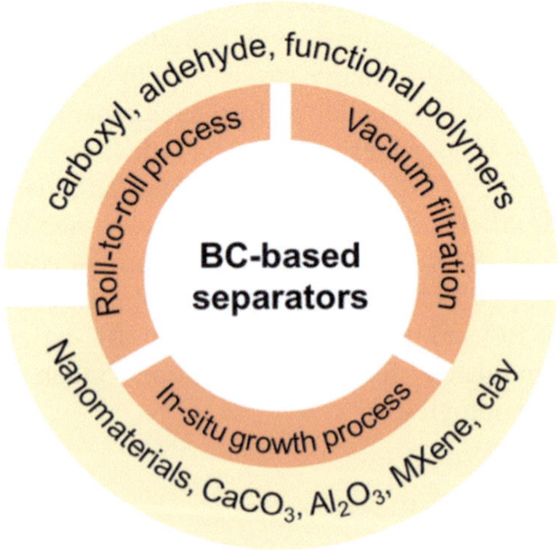

Fig. 7.6 BC-based separator obtained by nanoparticle modification, chemical modification of BC, combined with roll-to-roll process, in situ growth process, and vacuum filtration

electrochemical performance compared to batteries with commercial polyethylene (PP) and pure BC separators.

However, composite membranes prepared by directly mixing two materials often suffer from poor mechanical properties due to nanoparticle aggregation, highlighting the need for novel synthesis routes and in situ growth fabrication processes. Various nanomaterials, including $CaCO_3$, Al_2O_3, MXene, natural clay, and hydroxyapatite, can be used to modulate the pore properties, microstructure, and surface chemistry of BC membranes (Duan et al. 2023; Long et al. 2023). These modifications can be achieved through in situ growth during BC membrane synthesis or by physically blending BC nanofibers. Inorganic nanoparticles with high mechanical strength and flame retardancy are particularly effective for designing heat-resistant BC-based separators. Additionally, the tunable pore structure of metal-organic frameworks (MOFs) can enhance electrolyte absorption and ion-selective permeability in BC-based separators (Cheung et al. 2022). Certain nanomaterials, such as MXenes, possess multiple active sites, enabling the design of BC-based separators that homogenize lithium-ion flux, inhibit lithium dendrite formation, and prevent polysulfide shuttling (Wu et al. 2022a). The performance of the composite membrane is largely determined by the interfacial interactions between the nanomaterials and BC fibers. The structural manipulation and surface functionalization of BC nanofibers offer significant potential for advancing their application in electrochemical energy storage. Novel synthesis techniques, such as the in situ growth of nanoparticles on BC chain backbones or the in situ culture of BC molecules around nanoparticles, provide innovative pathways for membrane fabrication (Zheng et al. 2024).

Overall, BC exhibits superior physicochemical properties, including high tensile strength and thermal stability, which effectively suppress the formation and growth of dendrites, thereby improving battery safety. Furthermore, BC nanofibers are composed of glucose macromolecular chains that assemble into a network rich in active hydroxyl groups on their surface. These hydroxyl groups promote lithium-ion conduction while inhibiting the free diffusion of anions, significantly alleviating the polarization effects that occur during charge and discharge cycles.

5.1.2 Polymer Electrolyte (PE)

BC is a PE substrate due to its controllable ion transport channels, good ductility and flexibility, as well as excellent biocompatibility, degradability, and non-toxicity. BC-based hydrogel electrolytes can be fabricated by embedding electrolyte salt/acid/base solutions into a 3D entanglement network that anchors water molecules to BC as bound water. BC can also be mixed with other polymers to transport ions in aqueous batteries. The molecular spacing of the cellulose linear chain is expanded by grafting polar groups or metal ions with strong coordination ability on cellulose molecules, allowing ions to easily dissociate from salts and form fast single-ion conductors, which ensures fast ion transport kinetics. In addition, grafting functional groups onto BC cellulose chains can confer specific functions on BC-based electrolytes, such as non-flammability and interfacial wettability.

The first preparation of BC-based PE was obtained by directly absorbing the electrolyte solution using dried BC. For example, combining freeze-drying with impregnation methods enables large-scale fabrication of dehydrated BC hydrogel electrolytes from BC hydrogels (Li et al. 2021b). Since water molecules are bound to the hydroxyl group of BC nanofibers by hydrogen bonding, they exist in the BC hydrogel electrolyte as bound rather than free water. The presence of bound water and high mechanical strength allow Zn atoms to be deposited uniformly and densely on the electrolyte, which is superior to the flake dendrites of ordinary glass fibers and liquid electrolytes. These properties contribute to high rate performance and extended cycle life in waterborne zinc-ion batteries (Yang et al. 2023). The second strategy is to use BC as a template to synthesize a hybrid electrolyte membrane. Xie et al. synthesized an LLZO-PEO-based hybrid electrolyte using BC as a template (Xie et al. 2018). The three-dimensional nanofibrillar network architecture of BC serves to elongate and interconnect lithium-ion diffusion channels within the composite electrolyte system. Moreover, comparative mechanical analysis reveals that the hybrid electrolyte exhibits substantially greater flexibility compared to its pure electrolyte counterpart.

Overall, the high water-holding capacity of BC allows water molecules to be bound at the interface of BC, giving it satisfactory ion transport kinetics and cycling stability when used in aqueous batteries. BC can reduce the presence of water-related problems in charge storage devices, especially those that require high power and energy density, such as supercapacitors and batteries. BC has a high Young's modulus and flexibility to withstand a wide range of mechanical deformations during long-term cycling, while its interwoven network provides a 3D ion transport pathway. Anchoring the large anions of lithium salts to linear BC cellulose chains or expanding the molecular spacing of linear cellulose chains by chelating metal ions can be used to fabricate BC-based electrolytes with single-ion transport pathways. Therefore, artificial intelligence should be used to screen lithium salts and chelated metal ions. In addition, the water content of BC-based electrolytes needs to be optimized to meet a variety of application scenarios. The side reaction between the BC and the oxygen-containing functional groups in the metal electrodes should also be studied. In the future, BC-based electrolytes with specific properties such as flame retardancy and a wide operating temperature range should be prepared. This can be achieved by chemically grafting flame retardant functional groups, such as borate and phosphonate compounds, onto the cellulose chain to reduce flammability, or by mixing with non-flammable inorganic electrolytes or flame retardants.

5.1.3 Electrodes

In energy storage systems, conductivity is a critical performance metric. Given the inherently insulating nature of nanocellulose, comprehensive pretreatment through various functionalization techniques is essential to transform BC into viable conductive electrode materials. Common modification strategies include carbonization, conductive material coating, controlled deposition, composite blending, and in situ polymerization (Naghdi et al. 2019). The primary approach involves combining nanocellulose with conductive components, such as carbon-based nanomaterials, metal oxides, or conductive polymers, via two main pathways: (1) surface modification through conductive material coating or deposition and (2) bulk modification by directly blending with conductive additives or in situ polymerization.

Compared with carbon-based materials (rGO and CNT), the 3D porous network structure of BC has the advantage of accommodating more active materials for the production of flexible electrodes. Additionally, the high mechanical properties of BC offer the potential to achieve BC-based electrodes with robust structural integrity when subjected to bending or folding deformations. For instance, a 3D cross-linked structural scaffold electrode for robust supercapacitors was developed through a one-step esterification process using the covalent embedding of GO with BC fibers. Furthermore, the biosynthetic assembly strategy enables seamless integration of conductive active materials (e.g., active carbon (Guan et al. 2021), graphene (Luo et al. 2018a), graphene nanosheets (Luo et al. 2018b), nanotubes (Guan et al. 2019a), holey GO (Guan et al. 2019b)) with BC nanofibers, yielding composites with superior electron transport kinetics and exceptional potential for supercapacitor applications. A notable advancement was achieved by Guan et al., who engineered a bio-inspired spider-web structural binder for composite electrodes with high areal mass loading via biosynthetic processing (Guan et al. 2021). The resulting 3D hierarchical architecture enhances ion diffusion kinetics within the electrode matrix while preserving the porous framework of active material particles.

To develop high-performance flexible electrodes with enhanced specific capacity and cycling stability, 3D porous BC architectures offer distinct advantages: (1) increased ion storage capacity, (2) shortened ion/electron transport pathways, and (3) expanded interfacial contact between active materials and electrolytes. Numerous studies have documented the successful implementation of BC-derived carbon nanofibers and their hybrid counterparts in rechargeable battery systems. This approach addresses critical limitations of conventional synthesis techniques (e.g., CVD, PVD), including equipment complexity, high operational costs, and environmental concerns. BC and its derivatives can be thermally converted into three-dimensional conductive carbon aerogels through high-temperature pyrolysis, with material properties precisely tunable via optimization of pyrolysis parameters (e.g., temperature, dwell time, and atmosphere) (Lei et al. 2020; Ma et al. 2020). Furthermore, by leveraging the hierarchical 3D nanostructure of BC, composite electrode materials with tailored architectures have been fabricated through in situ incorporation of inorganic phases onto BC-derived carbon substrates. Owing to their inherited morphology and layered structure from pristine BC, BC-derived carbon nanofibers have demonstrated promising applicability in diverse battery systems, including lithium-oxygen (Li-O_2), lithium-sulfur (Li-S), lithium-ion (LIB), sodium-ion (NIB), and potassium-ion (PIB) batteries.

The first BC-derived carbon aerogel-based LIB anode was developed by the Titirici group (Wang et al. 2014). Due to the continuous electron and lithium-ion transport networks within the carbon aerogel, the BC-derived carbon anode exhibited a high specific capacity and superior rate performance compared to previously reported mesoporous carbon materials. It has also been reported that carbonized BC (CBC) nanofibers can serve as both anodes and cathodes in hybrid sodium-ion capacitors (Xu et al. 2020). Furthermore, BC can function as a high-quality reinforcing material for anodes in sodium-ion batteries (SIBs). A freestanding and flexible CBC/MoS_2 film was synthesized through a straightforward process involving the adsorption of ammonium tetrathiomolybdate (ATM), freeze-drying, and pyrolysis. The synthesized CBC/MoS_2 exhibited a core-shell structure with fine MoS_2 nanoparticles anchored on the surface of CBC nanofibers. The 3D interconnected structure and highly conductive framework of the CBC/MoS_2 film make it an ideal freestanding anode for SIBs (Yang et al. 2018). Additionally, the use of CBC films as sulfur hosts or multifunctional interlayers in lithium-sulfur batteries (LSBs) demonstrates their potential as durable and high-performance candidates for next-generation energy storage devices. The introduction of adsorption/catalytic conductive hosts is an effective strategy for mitigating polysulfide shuttling in LSBs, and this novel battery configuration represents an alternative approach to meeting the requirements of multi-electron reactions (Huang et al. 2015; Li et al. 2017b; Zhang et al. 2020).

5.2 *Energy Conversion Devices*

5.2.1 Salinity Gradient Energy

Traditional fossil energy resources are increasingly depleted, so it is imperative to develop and utilize new clean energy. As a sustainable alternative to fossil fuels, the salinity gradient energy that exists between seawater and river water is one of the "blue energy" sources available (Xiao et al. 2019), with a total global reserve of 2.6 billion kilowatts, which has not yet been effectively exploited (Logan and Elimelech 2012). However, the high cost, complex manufacturing process, and poor mechanical strength of basic ion exchange membranes have greatly hindered the development of reverse electrodialysis (Güler et al. 2013).

Native BC forms a three-dimensional nanoporous hydrogel network through the hierarchical assembly of nanofibers. At the nanoscale level, these ribbon-like nanofibers (20–80 nm in diameter) are composed of bundled elementary fibrils with diameters of 3–5 nm, creating an interconnected system of nanochannels (2–10 nm in width) throughout the matrix. Surface characterization reveals that the abundant hydroxyl groups present on the nanofiber surfaces impart an anionic character to the fibrils; this unique structural organization enables native BC hydrogels to function as biomimetic nanofluidic membranes, exhibiting selective ion transport properties. The nanochannel network facilitates size-dependent molecular transport while the surface charges enable electrostatic interactions with ionic species. However, due to the large number of hydroxyl groups on the BC surface, its charge density is relatively low. Additionally, the secretion process of *Acetobacter xylinum* is isotropic in the growth plan (horizontal in static culture), resulting in the disordered arrangement of BC nanofibers. These structures are unfavorable for the ion transport process. Therefore, it is necessary to appropriately modify the structure of BC to meet the regulation of ion transport behavior. Specifically, methods such as changing the culture medium and adding additives can be used to alter the morphology of nanofibers (such as porosity, pore size, and the arrangement of fibrils) (Sun et al. 2022; Shi et al. 2022). Different chemical modifications (such as TEMPO oxidation (Wang et al. 2018), quaternization (Zhang et al. 2016), and sulfonation (Yue et al. 2017)), composites (1D (Wang et al. 2022a) or 2D (Wang et al. 2022b)), physical modifications (such as hot pressing (Figueiredo et al. 2015) and stretching (Rahman and Netravali 2016)), and the construction of heterojunction structures can be employed to change the BC structure and surface properties (Xu et al. 2021), thereby realizing the regulation of ion transport performance. For example, Wu et al. successfully synthesized positively charged BC membranes in situ by introducing quaternized chitosan into the BC culture medium (Wu et al. 2021). These membranes were then applied in salinity gradient energy conversion devices, where they demonstrated excellent energy density.

Moreover, BC nanofibers can also be used to modify 2D nanochannel materials such as GO, strongly binding to the surface and tuning the ion transport. The combination of one-dimensional (1D) BC nanofibers and 2D nanosheets holds promise as a novel strategy to balance ion selectivity and permeability and enhance the output power density (Zhou et al. 2019). Thus, a negatively charged BC/GO composite membrane (NGO/NBC) and a positively charged BC/layered double hydroxide (PLDH/PBC) are constructed as cation and anion selective membranes in osmotic energy conversion. Under the concentration gradient of simulated seawater and river water, a pair of energy harvesting systems can achieve an output power density as high as 0.70 W m^{-2}. This work demonstrates the practicality and feasibility of a multilayer composite membrane that combines nanofiber confinement, coupled surface charge, and size effects as a high-performance osmotic power generator (Sheng et al. 2022). In addition to the combination of 2D materials such as GO and PLDH with BC, an asymmetric porous structure can also be constructed. Xu et al. constructed a bilayer heterostructured MXene/BC Janus membrane and demonstrated the asymmetric cation transport phenomenon in asymmetrically confined channels (Xu et al. 2022). Under the concentration gradient of simulated seawater and river water, the MXene/BC Janus membrane can obtain an output power density of 0.91 W m^{-2}. It is proposed that selective ion transport can be achieved in any confined channels (slits or networks) with surface charges.

5.2.2 Thermoelectric Conversion

Thermoelectric materials have found widespread application in thermoelectric power generation, eco-friendly cooling systems, and wearable electronics due to their ability to directly convert heat into electricity. Cellulose-based materials can be engineered into flexible ion/electron-conducting ionogels possessing strong mechanical performance using molecular-scale physical processing, chemical group substitution, and topological network regulation. Compared with lignocellulose, BC exhibits superior mechanical properties and adaptability. When integrated into flexible thermoelectric materials, BC and its derivatives can effectively address the challenging issues of tensile strength and flexibility inherent in thermoelectric materials. For instance, upon incorporating conductive materials such as graphene, carbon nanotubes, and silver arsenide into BC, the abundant pores allow these conductive materials to be uniformly distributed across the gel. Under the temperature difference effect, the migration of charge carriers is promoted, thus achieving the optimal thermoelectric voltage. For example, Han et al. employed a simple processing method to print an ink formulated with BC nanofibers and a thermoelectric material (Bi_2Te_3) onto the surface of paper (Zhao et al. 2019). Subsequently, the fabricated paper was further folded and bonded to prepare a honeycomb-structured thermoelectric generator (TEG). In this process, the main function of BC is that the strong hydrogen bonds of hydroxyl groups between BC nanofibers and paper fibers facilitate the fixation of thermoelectric particles on the paper surface. Moreover, leveraging the network-porous structure of BC, an aerogel thermoelectric generator composed of poly (3,4-ethylenedioxythiophene)/poly(styrenesulfonate) (PEDOT/PSS) and single-walled carbon nanotubes can achieve thermoelectric energy conversion in a directional manner along the BC framework under a vertical temperature difference (Chen et al. 2023).

The implementation of inorganic thermoelectric materials and organic polymer materials within the thermoelectric conversion domain is severely circumscribed by their inherently low thermoelectric coefficients. In contrast, BC ionogels, characterized by their remarkable flexibility and tailorable ionic conductivity, present substantial potential for applications in wearable electronics. The fabrication of hydrogel-based electrolytes is accomplished by introducing redox pairs into the BC hydrogels via a straightforward impregnation procedure, capitalizing on the intrinsic porous architecture of BC. Leveraging the mesoporous structure of BC, Liu et al. (2022b) engineered an ionic liquid gel (BCIG) through a simple modified cosolvent evaporation technique. This work demonstrated the mesoscale-confined ion transport phenomenon in BCIG, which is underpinned by the Soret effect. During this process, the synergistic interplay between ion-cellulose interactions and ion-ion interactions facilitates the selective conveyance of ions under a thermal gradient and concomitantly elevates the charge carrier concentration in BCIGs.

The research group of Wang et al. has embarked on comprehensive investigations into the multifunctional and multimorphological applications of BC hydrogels with their distinctive multidimensional configurations. A novel BC hydrogel featuring a calcium-ion-coordinated structure (CaBC) was synthesized using an in situ cultivation approach. The Ca^{2+} ions effectively expand the inter-chain distance of BC molecular chains, thereby enabling ions to migrate along the cellulose molecular chains and ultimately attaining enhanced thermoelectric performance (Wu et al. 2022b). Notwithstanding, due to the robust hydrogen-bond interactions among BC nanofibers, procuring a homogeneous BC nanofiber suspension remains a formidable challenge. Subsequently, by disrupting these strong hydrogen-bond interactions within a deep eutectic solvent system and incorporating acrylamide monomers, a BC/PAM gel thermogalvanic cell with a macromolecule-nanofiber entanglement architecture was fabricated (Wu et al. 2023b). BC nanofibers function as "ion-high-speed transport conduits" within the pores. This not only expedites ion transport within the pores but also substantially bolsters the effective ionic conductivity of the thermogalvanic cell.

In addition, a series of high-performance natural BC-based thermoelectric gel matrices were fabricated by introducing inorganic salts or organic solvents into gel-based electrolytes. This was achieved by harnessing the synergistic effect of thermal diffusion and the thermocouple effect, along with the crystallization effect induced by organic solvents. As a result, these matrices are capable of generating high open-circuit voltages even under small temperature differences. This accomplishment offers an effective strategy for realizing human thermal energy harvesting and its application in powering flexible wearable electronic devices (Li et al. 2023). Concurrently, an effective strategy to enhance mass transfer in thermal-battery electrolytes was devised by leveraging the reversible phase transition behavior of thermally responsive materials that can induce redox ions. Specifically, this strategy involves introducing BC hydrogels with a 3D network structure to induce the in situ confined phase transition of methyl cellulose, thereby facilitating the formation of high-entropy electrolytes (Li et al. 2024).

5.2.3 Piezoelectric and Triboelectric Conversion

Over the past two decades, researchers have developed piezoelectric, triboelectric-based energy conversion devices that can directly harvest mechanical energy and convert it into electrical energy (these conversion devices are known as nanogenerators). BC exhibits high crystallinity, with abundant hydroxyl polar groups arranged along the asymmetric monoclinic and triclinic regions of its structure, facilitating the formation of asymmetric dipoles with piezoelectric activity (Moon et al. 2011). The hydroxyl groups also provide numerous lone electron pairs, rendering cellulose a high electron-donor material that readily loses electrons upon contact with other materials, thereby demonstrating excellent triboelectric properties. Furthermore, driven by the demand for environmentally friendly materials, BC and its derivatives offer a range of advantages, including abundance, renewability, sustainability, and cost-effectiveness, making them among the most promising natural materials for energy harvesting applications in fields such as piezoelectricity and triboelectricity.

In the fabrication of BC-based piezoelectric nanogenerators, the three-dimensional network structure of BC and its adsorption properties for inorganic nanoions can be leveraged to grow vanadium-doped ZnO nanoparticles in situ via a solution immersion method. This process yields a freestanding BC/vanadium-doped ZnO composite structure, which can be utilized to construct flexible piezoelectric nanogenerators (Fig. 7.7) (Zhang et al. 2018). The incorporation of

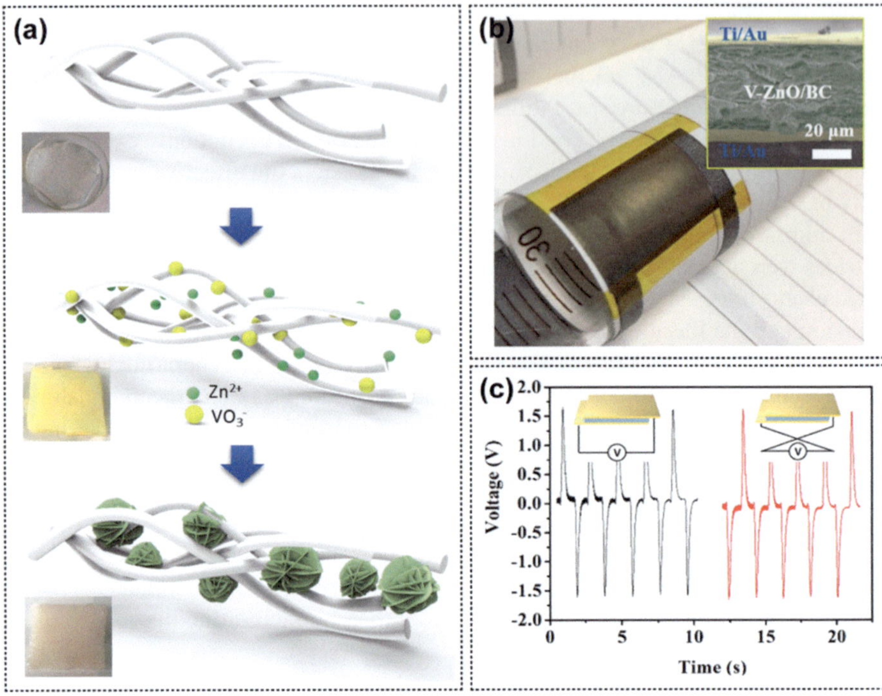

Fig. 7.7 (**a**) Schematic fabrication process of the V-ZnO/BC hybrid paper. (**b**) Optical image of the device rolled on a cylinder demonstrating its excellent flexibility. The inset shows the cross-sectional structure of the device. (**c**) Output voltage and (c) current density of the V-ZnO/BC hybrid paper-based flexible piezoelectric nanogenerators (Reproduced with permission (Zhang et al. 2018) Copyright 2018, Elsevier)

vanadium enhances the piezoelectric coefficient of ZnO, significantly improving the output performance of the piezoelectric generators. Beyond the in situ growth of piezoelectric nanoparticles, imidazole perchlorate ($ImClO_4$)/BC hydrogels have been prepared through a simple soaking and drying process, exploiting the interplay between the crystallization of ferroelectric materials and the swelling behavior of BC hydrogels (Lu et al. 2022). When applied in self-powered high-performance piezoelectric sensors, these hydrogels exhibit sensitivity that surpasses most piezoelectric sensors composed of ferroelectric polymers and rivals that of ceramic-polymer composites. Notably, the BC membrane in the hybrid piezoelectric sensor can be enzymatically degraded by cellulase, while the embedded $ImClO_4$ crystals can be dissolved and recycled using water, leaving behind no environmentally harmful electronic waste.

BC is widely utilized as a triboelectric material owing to its advantages such as less polymorphic confinement, favorable processing performance, and outstanding physical and chemical properties (Liu et al. 2022b). Due to the abundant polar hydroxyl groups, cellulose has a strong electron-donating ability and is regarded as an efficient positive triboelectric material (Yao et al. 2017). When cellulose and polytetrafluoroethylene (PTFE) come into contact, their electron clouds overlap. The lone electron from the low-electronegative atom (O) migrates to the high-electronegative atom (F), leading to electron transfer between the two materials (Liu et al. 2021). Additionally, the surface roughness of the cellulose material, which is created by the criss-crossing of the fibers, can be further tailored through mechanical or chemical treatment to enhance contact electrification (CE). For example, a highly conductive ferroelectric BC composite paper incorporating silver nanowires (Ag NWs) and barium titanate nanoparticles (BTO NPs) was fabricated using a straightforward filtration method and successfully applied in the development of large-area, high-performance triboelectric nanogenerator (TENG) devices (Oh et al. 2019). In the resulting BC composite paper, Ag NWs provided high electrical conductivity, while BTO NPs contributed ferroelectric properties. The intrinsic weak electron-donor tendency of cellulose was effectively addressed through a simple forward polarization process, which enhanced triboelectric charge generation on the ferroelectric BC composite paper during TENG operation. This improvement significantly boosted the performance of the BC-based TENG.

5.2.4 Other Energy Conversion

Hydroelectric installations, which harness energy from water to generate electricity, offer a prospective solution to green energy capture. Hydroelectric generators utilizing diverse novel nanostructured materials have demonstrated significant potential in hydropower generation. However, the advancement of high-performance and practical hydropower systems remains constrained by challenges such as low power output, high precursor material costs, and complex fabrication processes. In this work, Yun et al. (2023) developed a newly designed metal-coated BC nanofiber bilayer membrane (MBCBM) for enhanced hydroelectric power generation. The top layer of the MBCBM consists of metallic BC nanofibers, which serve as conductive electrodes for rapid charge carrier collection, while the bottom layer comprises BC nanofibers that function as aqueous voltaic materials for efficient power generation (Yun et al. 2023). In addition, an environmentally friendly BC/reduced graphene oxide (RGO) bilayer membrane (B-BRG bf) with a heterostructure was prepared using a simple vacuum filtration method (Li et al. 2025). This membrane features a gradient distribution of oxygen-containing functional groups between the top and bottom layers, enabling the maintenance of moisture differences. The BC layer, with its nano-microstructure, enhances water accessibility to the bilayer membrane while simultaneously establishing a BC/RGO interface characterized by a significant proton concentration gradient. As a result, the heterogeneous structure of the bilayer membrane enables consistently high and stable power output.

6 Summary and Prospect

BC, as an emerging biomaterial, has shown great potential in the fields of electricity, optoelectronics, photonics, and energy applications due to its unique nanostructure and excellent physical and chemical properties. However, it should be noted that the application of BC in these areas remains in the early research stage. The widespread application of BC in this field is restricted by the following challenges: firstly, the large -scale production. Currently, it is difficult to obtain large quantities of BC with uniformity, high quality and a high aspect ratio; secondly, traditional methods such as chemical modification, solution impregnation, or physical coating can cause environmental pollution. These methods may also lead to the loss of functional components and are not conducive to large-scale production.

Future research should further explore the functional regulation of BC in these fields, promoting the large-scale application of BC materials in green energy, sustainable electronic devices and intelligent optical devices. From what was described in this chapter, it is clear that BC has received considerable attention in recent years and rapid development in the fabrication of advanced electronic, optoelectronic, and photonic devices using BC-based composites may be expected, due to its mechanical stability, thermostability, high crystallinity, high purity, high porosity, modifiability, and biocompatibility. Although various green electronic, optoelectronic, and photonics devices utilizing BC as a substrate have been evidenced to present high transmittance, excellent flexibility, and bendability, there are still several challenges to further practical and extensive application for BC:

1. Given its strong hydrophilicity, the unsatisfactory humidity stability can be addressed through surface modification or introducing hydrophobic components, while avoiding aggregation phenomenon caused by surface/interfacial properties.
2. BC-based conductive composites predominantly exhibit a black coloration, which is attributed to the inherent optical properties of the constituent conductive materials, including metals, metal oxides, MXenes, carbon-based materials, and conducting polymers. Achieving an optimal equilibrium between high electrical conductivity and optical transparency in these composites can be realized through strategic modifications of the BC matrix or the advancement of novel conductive materials with tailored properties.
3. Insufficient study about the internal structure of modified BC or BC introduced into functional materials may result in inaccurate characterization of BC-based complexes, and to obtain uniform samples across the entire volume (surface area/internal volume), the moment of modification must be combined with the process of BC growth to minimize post-doping.

Overcoming these challenges will bring BC-based composites to fulfill their potential, which could produce significant improvements in sustainable and high-performance electronic, optoelectronic, and photonics devices.

References

Avila OI, Santos MV, Shimizu FM, Almeida GFB, Siqueir JP, Andrade MB, Balogh DT, Ribeiro SJL, Mendonca CR (2018) Direct femtosecond laser printing of PPV on bacterial cellulose-based paper for flexible organic devices. Macromol Mater Eng 303(10):1800265

Barquinha P, Gonçalves G, Pereira L, Martins R, Fortunato E (2007) Effect of annealing temperature on the properties of IZO films and IZO based transparent TFTs. Thin Solid Films 515(24):8450–8454

Bharmoria P, Naimovičius L, Abol-Fotouh D, Miroshnichenko M, Lekavičius J, De Luca G, Saeed U, Kazlauskas K, Candau N, Baronas P, Roig A, Moth-Poulsen K (2024) Photon upconversion crystals doped bacterial cellulose composite films as recyclable photonic bioplastics. Commun Mater 5(1):200

Chen G, Matsuhisa N, Liu Z, Qi D, Cai P, Jiang Y, Wan C, Cui Y, Leow WR, Liu Z, Gong S, Zhang K-Q, Cheng Y, Chen X (2018) Plasticizing silk protein for on-skin stretchable electrodes. Adv Mater 30(21):1800129

Chen W, Sun B, Biehl P, Zhang K (2022) Cellulose-based soft actuators. Macromol Mater Eng 307(6):2200072

Chen LZ, Lou J, Zong YD, Liu ZQ, Jiang YF, Han WJ (2023) Wood-like aerogel for thermoelectric generators based on BC/PEDOT/SWCNT. Cellulose 30(5):3141–3152

Chen J, Liu F, Abdiryim T, Liu X (2024) An overview of conductive composite hydrogels for flexible electronic devices. Adv Compos Hybrid Mater 7(2):35

Cheung YH, Ma K, Wasson MC, Wang X, Idrees KB, Islamoglu T, Mahle J, Peterson GW, Xin JH, Farha OK (2022) Environmentally benign biosynthesis of hierarchical MOF/bacterial cellulose composite sponge for nerve agent protection. Angew Chem Int Ed 61(19):e202202207

De La Cruz JA, Liu Q, Senyuk B, Frazier AW, Peddireddy K, Smalyukh II (2018) Cellulose-based reflective liquid crystal films as optical filters and solar gain regulators. ACS Photonics 5(6):2468–2477

de Souza TC, Amorim JD, Silva Junior CJ, de Medeiros AD, Santana Costa AF, Vinhas GM, Sarubbo LA (2023) Magnetic bacterial cellulose biopolymers: production and potential applications in the electronics sector. Polymers (Basel) 15(4):853

Dhali K, Ghasemlou M, Daver F, Cass P, Adhikari B (2021) A review of nanocellulose as a new material towards environmental sustainability. Sci Total Environ 775:145871

dos Santos MV, Dominguez CT, Schiavon JV, Barud HS, de Melo LSA, Ribeiro SJL, Gomes ASL, de Araújo CB (2014) Random laser action from flexible biocellulose-based device. J Appl Phys 115(8):083108

Duan QS, Zhang ZJ, Zhao JM, He JX, Peng WQ, Zhang Y, Liu T, Wang SF, Nie SX (2023) Fire-retardant hydroxyapatite/cellulosic triboelectric materials for energy harvesting and sensing at extreme conditions. Nano Energy 117:108851

Espinha A, Dore C, Matricardi C, Alonso MI, Goñi AR, Mihi A (2018) Hydroxypropyl cellulose photonic architectures by soft nanoimprinting lithography. Nat Photonics 12(6):343–348

Eynaki H, Kiani MA, Golmohammadi H (2020) Nanopaper-based screen-printed electrodes: a hybrid sensing bioplatform for dual opto-electrochemical sensing applications. Nanoscale 12(35):18409–18417

Faham S, Golmohammadi H, Ghavami R, Khayatian G (2019) A nanocellulose-based colorimetric assay kit for smartphone sensing of iron and iron-chelating deferoxamine drug in biofluids. Anal Chim Acta 1087:104–112

Faraco TA, Fontes MD, Paschoalin RT, Claro AM, Gonçalves IS, Cavicchioli M, Farias RL, Cremona M, Ribeiro SJ, Barud HD, Legnani C (2023) Review of bacterial nanocellulose as suitable substrate for conformable and flexible organic light-emitting diodes. Polymers 15(3):479

Fei G, Wang Y, Wang H, Ma Y, Guo Q, Huang W, Yang D, Shao Y, Ni Y (2019) Fabrication of bacterial cellulose/polyaniline nanocomposite paper with excellent conductivity, strength, and flexibility. ACS Sustain Chem Eng 7(9):8215–8225

Figueiredo ARP, Silvestre AJD, Neto CP, Freire CSR (2015) In situ synthesis of bacterial cellulose/polycaprolactone blends for hot pressing nanocomposite films production. Carbohydr Polym 132:400–408

Fortunato E, Correia N, Barquinha P, Pereira L, Goncalves G, Martins R (2008) High-performance flexible hybrid field-effect transistors based on cellulose fiber paper. IEEE Electron Device Lett 29(9):988–990

Fresta E, Fernández-Luna V, Coto PB, Costa RD (2018) Merging biology and solid-state lighting: recent advances in light-emitting diodes based on biological materials. Adv Funct Mater 28(24):1707011

Girard V-D, Chaussé J, Vermette P (2024) Bacterial cellulose: a comprehensive review. J Appl Polym Sci 141(15):e55163

Golmohammadi H, Hamzei Z, Hosseinifard M, Ahmadi SH (2020) Smart fully integrated lab: a smartphone-based compact miniaturized analytical/diagnostic device. Adv. Mater. Technol 5(12):2000742

Guan Q-F, Han Z-M, Luo T-T, Yang H-B, Liang H-W, Chen S-M, Wang G-S, Yu S-H (2019a) A general aerosol-assisted biosynthesis of functional bulk nanocomposites. Natl Sci Rev 6(1):64–73

Guan F, Chen S, Sheng N, Chen Y, Yao J, Pei Q, Wang H (2019b) Mechanically robust reduced graphene oxide/bacterial cellulose film obtained via biosynthesis for flexible supercapacitor. Chem Eng J 360:829–837

Guan QF, Ling ZC, Han ZM, Luo TT, Yang HB, Yang KP, Yin CH, Yu SH (2021) Sustainable 3D structural binder for high-performance supercapacitor by biosynthesis process. Adv Funct Mater 31(42):2105070

Güler E, Elizen R, Vermaas DA, Saakes M, Nijmeijer K (2013) Performance-determining membrane properties in reverse electrodialysis. J Membr Sci 446:266–276

Gwon H, Park K, Chung S-C, Kim R-H, Kang Jin K, Ji Sang M, Kim N-J, Lee S, Ku J-H, Do Eun C, Park S, Kim M, Shim Woo Y, Rhee Hong S, Kim J-Y, Kim J, Kim Tae Y, Yamaguchi Y, Iwamuro R, Saito S, Kim G, Jung I-S, Park H, Lee C, Lee S, Jeon Woo S, Jang Woo D, Kim Hyun U, Lee Sang Y, Im D, Doo S-G, Lee Sang Y, Lee Hyun C, Park Jin H (2019) A safe and sustainable bacterial cellulose nanofiber separator for lithium rechargeable batteries. Proc Natl Acad Sci U S A 116(39):19288–19293

Hoeng F, Denneulin A, Bras J (2016) Use of nanocellulose in printed electronics: a review. Nanoscale 8(27):13131–13154

Huang Y, Zheng MB, Lin ZX, Zhao B, Zhang ST, Yang JZ, Zhu CL, Zhang H, Sun DP, Shi Y (2015) Flexible cathodes and multifunctional interlayers based on carbonized bacterial cellulose for high-performance lithium-sulfur batteries. J Mater Chem A 3(20):10910–10918

Huang C, Ji H, Yang Y, Guo B, Luo L, Meng Z, Fan L, Xu J (2020) TEMPO-oxidized bacterial cellulose nanofiber membranes as high-performance separators for lithium-ion batteries. Carbohydr Polym 230:115570

Huang J, Zhao M, Hao Y, Wei Q (2022) Recent advances in functional bacterial cellulose for wearable physical sensing applications. Adv Mater Technol 7(1):2100617

Inácio PMC, Medeiros MCR, Carvalho T, Félix RC, Mestre A, Hubbard PC, Ferreira Q, Morgado J, Charas A, Freire CSR, Biscarini F, Power DM, Gomes HL (2020) Ultra-low noise PEDOT:PSS electrodes on bacterial cellulose: a sensor to access bioelectrical signals in non-electrogenic cells. Org Electron 85:105882

Jiang G, Wang G, Zhu Y, Cheng W, Cao K, Xu G, Zhao D, Yu H (2022) A scalable bacterial cellulose ionogel for multisensory electronic skin. Research 2022:9814767

Khan S, Ul-Islam M, Khattak WA, Ullah MW, Park JK (2015) Bacterial cellulose–poly(3,4-ethylenedioxythiophene)–poly(styrenesulfonate) composites for optoelectronic applications. Carbohydr Polym 127:86–93

Ko Y, Kwon G, Choi H, Lee K, Jeon Y, Lee S, Kim J, You J (2023) Cutting edge use of conductive patterns in nanocellulose-based green electronics. Adv Funct Mater 33(37):2302785

Lagadec MF, Zahn R, Wood V (2019) Characterization and performance evaluation of lithium-ion battery separators. Nat Energy 4(1):16–25

Lee C, Kim S, Cho Y-H (2022) Silk and paper: progress and prospects in green photonics and electronics. Adv Sustain Syst 6(2):2000216
Lei W, Jin D, Liu H, Tong Z, Zhang H (2020) An overview of bacterial cellulose in flexible electrochemical energy storage. ChemSusChem 13(15):3731–3753
Li C, Wu ZY, Liang HW, Chen JF, Yu SH (2017a) Ultralight multifunctional carbon-based aerogels by combining graphene oxide and bacterial cellulose. Small 13(25):1700453
Li SQ, Warzywoda J, Wang S, Ren GF, Fan ZY (2017b) Bacterial cellulose derived carbon nanofiber aerogel with lithium polysulfide catholyte for lithium-sulfur batteries. Carbon 124:212–218
Li W, Wang S, Fan Z, Li S, Bernussi A, Newman N (2021a) Functionalized bacterial cellulose as a separator to address polysulfides shuttling in lithium-sulfur batteries. Mater Today Energy 21:100813
Li XL, Lin ZF, Jin N, Yang XJ, Du YB, Lei L, Rozier P, Simon P, Liu Y (2021b) Perovskite-type $SrVO_3$ as high-performance anode materials for lithium-ion batteries. Adv Mater 34(46):2107262
Li J, Chen S, Han Z, Qu X, Jin M, Deng L, Liang Q, Jia Y, Wang H (2023) High performance bacterial cellulose organogel-based thermoelectrochemical cells by organic solvent-driven crystallization for body heat harvest and self-powered wearable strain sensors. Adv Funct Mater 33(46):2306509
Li J, Chen S, Qu X, Han Z, Zhou Z, Deng L, Jia Y, Zhang S, Xie R, Wang H (2024) Confined phase transition triggering a high-performance energy storage thermo-battery. Energy Environ Sci 17(18):6606–6615
Li X, Zhang R, Ai X, Tang P, Wang H, Bin Y (2025) Bacterial cellulose/reduced graphene oxide bilayer films for moist-electric power generation. J Mater Chem A 13(12):8571–8579
Liang H-W, Guan Q-F, Zhu Z, Song L-T, Yao H-B, Lei X, Yu S-H (2012) Highly conductive and stretchable conductors fabricated from bacterial cellulose. NPG Asia Mater 4(6):e19
Liu J, Morales-Narváez E, Orozco J, Vicent T, Zhong G, Merkoçi A (2018) Bioluminescent nanopaper for rapid screening of toxic substances. Nano Res 11(1):114–125
Liu YH, Fu Q, Mo JL, Lu YX, Cai CC, Luo B, Nie SX (2021) Chemically tailored molecular surface modification of cellulose nanofibrils for manipulating the charge density of triboelectric nanogenerators. Nano Energy 89:106369
Liu G, Ma M, Meng H, Liu J, Zheng Y, Peng J, Wei S, Sun Y, Wang Y, Xie Y, Li J (2022a) In-situ self-assembly of bacterial cellulose/poly(3,4-ethylenedioxythiophene)-sulfonated nanofibers for peripheral nerve repair. Carbohydr Polym 281:119044
Liu K, Lv J, Fan G, Wang B, Mao Z, Sui X, Feng X (2022b) Flexible and robust bacterial cellulose-based ionogels with high thermoelectric properties for low-grade heat harvesting. Adv Funct Mater 32(6):2107105
Logan BE, Elimelech M (2012) Membrane-based processes for sustainable power generation using water. Nature 488(7411):313–319
Long TY, Zhu C, Li L, Shao L, Zhu SQ, Rueping M, Chu LL (2023) Ligand-controlled stereodivergent alkenylation of alkynes to access functionalized trans- and cis-1,3-dienes. Nat Commun 14(1):55
Lu J, Hu S, Li W, Wang X, Mo X, Gong X, Liu H, Luo W, Dong W, Sima C, Wang Y, Yang G, Luo J-T, Jiang S, Shi Z, Zhang G (2022) A biodegradable and recyclable piezoelectric sensor based on a molecular ferroelectric embedded in a bacterial cellulose hydrogel. ACS Nano 16(3):3744–3755
Luo H, Dong J, Zhang Y, Li G, Guo R, Zuo G, Ye M, Wang Z, Yang Z, Wan Y (2018a) Constructing 3D bacterial cellulose/graphene/polyaniline nanocomposites by novel layer-by-layer in situ culture toward mechanically robust and highly flexible freestanding electrodes for supercapacitors. Chem Eng J 334:1148–1158
Luo H, Xiong P, Xie J, Yang Z, Huang Y, Hu J, Wan Y, Xu Y (2018b) Uniformly dispersed freestanding carbon nanofiber/graphene electrodes made by a scalable biological method for high-performance flexible supercapacitors. Adv Funct Mater 28(48):1803075

Luo S, Li Q, Xue Y, Zhou B, Feng Y, Liu C (2023) Reinforcing and toughening bacterial cellulose/MXene films assisted by interfacial multiple cross-linking for electromagnetic interference shielding and photothermal response. J Colloid Interface Sci 652:1645–1652

Luo Y, Que W, Tang Y, Kang Y, Bin X, Wu Z, Yuliarto B, Gao B, Henzie J, Yamauchi Y (2024) Regulating functional groups enhances the performance of flexible microporous MXene/bacterial cellulose electrodes in supercapacitors. ACS Nano 18(18):11675–11687

Lv Z, Wang C, Wan C, Wang R, Dai X, Wei J, Xia H, Li W, Zhang W, Cao S, Zhang F, Yang H, Loh XJ, Chen X (2022) Strain-driven auto-detachable patterning of flexible electrodes. Adv Mater 34(30):2202877

Lv X, Liu Y, Yu J, Li Z, Ding B (2023) Smart fibers for self-powered electronic skins. Adv Fiber Mater 5(2):401–428

Ma L, Bi Z, Xue Y, Zhang W, Huang Q, Zhang L, Huang Y (2020) Bacterial cellulose: an encouraging eco-friendly nano-candidate for energy storage and energy conversion. J Mater Chem A 8(12):5812–5842

Mai HH, Dao QK, Nguyen MN, Pham HM (2023) Random lasing from kombucha bacterial cellulose—ZnO bionanocomposite foam. J Phys D Appl Phys 56(49):495107

Marchessault RH, Morehead FF, Walter NM (1959) Liquid crystal systems from fibrillar polysaccharides. Nature 184(4686):632–633

Moon RJ, Martini A, Nairn J, Simonsen J, Youngblood J (2011) Cellulose nanomaterials review: structure, properties and nanocomposites. Chem Soc Rev 40(7):3941–3994

Muller D, Rambo CR, Porto LM, Schreiner WH, Barra GMO (2013) Structure and properties of polypyrrole/bacterial cellulose nanocomposites. Carbohydr Polym 94(1):655–662

Naghdi T, Golmohammadi H, Vosough M, Atashi M, Saeedi I, Maghsoudi MT (2019) Lab-on-nanopaper: an optical sensing bioplatform based on curcumin embedded in bacterial nanocellulose as an albumin assay kit. Anal Chim Acta 1070:104–111

Oh H, Kwak SS, Kim B, Han E, Lim G-H, Kim S-W, Lim B (2019) Highly conductive ferroelectric cellulose composite papers for efficient triboelectric nanogenerators. Adv Funct Mater 29(37):1904066

O-Rak K, Ummartyotin S, Sain M, Manuspiya H (2013) Covalently grafted carbon nanotube on bacterial cellulose composite for flexible touch screen application. Mater Lett 107:247–250

Pan X, Li J, Ma N, Ma X, Gao M (2023) Bacterial cellulose hydrogel for sensors. Chem Eng J 461:142062

Pleumphon C, Thiangtham S, Pechyen C, Manuspiya H, Ummartyotin S (2017) Development of conductive bacterial cellulose composites: an approach to bio-based substrates for solar cells. J Biobased Mater Bioenergy 11(4):321–329

Poddar MK, Dikshit PK (2021) Recent development in bacterial cellulose production and synthesis of cellulose based conductive polymer nanocomposites. Nano Select 2(9):1605–1628

Poisson J, Zhang K (2024) Unique optical properties of cellulosic materials. Acc Mater Res 5(8):920–932

Prilepskii A, Nikolaev V, Klaving A (2023) Conductive bacterial cellulose: from drug delivery to flexible electronics. Carbohydr Polym 313:120850

Rahman MM, Netravali AN (2016) Aligned bacterial cellulose arrays as "green" nanofibers for composite materials. ACS Macro Lett 5(9):1070–1074

Rahman A, Kang S, Wang W, Huang Q, Kim I, Vikesland PJ (2022) Lectin-modified bacterial cellulose nanocrystals decorated with au nanoparticles for selective detection of bacteria using surface-enhanced Raman scattering coupled with machine learning. ACS Appl Nano Mater 5(1):259–268

Saleh AK, El-Sayed MH, El-Sakhawy MA, Alshareef SA, Omer N, Abdelaziz MA, Jame R, Zheng H, Gao M, Du H (2024) Cellulose-based conductive materials for bioelectronics. ChemSusChem 6(18):e202401762

Santos MV, Tercjak A, Gutierrez J, Barud HS, Napoli M, Nalin M, Ribeiro SJL (2017) Optical sensor platform based on cellulose nanocrystals (CNC)—4′-(hexyloxy)-4-biphenylcarbonitrile (HOBC) bi-phase nematic liquid crystal composite films. Carbohydr Polym 168:346–355

Santos MV, Maturi FE, Pecoraro É, Barud HS, Lima LR, Ferreira RAS, Carlos LD, Ribeiro SJL (2021) Cellulose based photonic materials displaying direction modulated photoluminescence. Front Bioeng Biotechnol 9:617328

Shah J, Malcolm Brown R (2005) Towards electronic paper displays made from microbial cellulose. Appl Microbiol Biotechnol 66(4):352–355

Sharifi AR, Naghdi T, Yousefi H, Kiani MA, Golmohammadi H (2022) Nanopapers toward green photonic and optical applications. ACS Sustain Chem Eng 10(51):16995–17026

Sheng N, Zhang M, Song Q, Zhang H, Chen S, Wang H, Zhang K (2022) Enhanced salinity gradient energy harvesting with oppositely charged bacterial cellulose-based composite membranes. Nano Energy 101:107548

Shi L, Wang T, Yang L, Chen C, Dou R, Yang X, Sun B, Zhou B, Zhang L, Sun D (2022) Enhanced mechanical properties and biocompatibility on BC/HAp composite through calcium gluconate fortified bacterial. Carbohydr Polym 281:119085

Son KA, Yang B, Seo HC, Wong D, Moon JS, Hussain T (2017) High-speed graphene field effect transistors on microbial cellulose biomembrane. IEEE Trans Nanotechnol 16(2):239–244

Song J, Fan M, Zhang R, Qu M, Tang P, Wang H, Bin Y (2024) Highly sensitive humidity sensor based on composite film of partially reduced graphene oxide and bacterial cellulose. Biosens Bioelectron 257:116296

Suleimenova A, Frasco MF, Soares da Silva FAG, Gama M, Fortunato E, Sales MGF (2023) Bacterial nanocellulose membrane as novel substrate for biomimetic structural color materials: application to lysozyme sensing. Biosens Bioelectron: X 13:100310

Sun B, Lin J, Liu M, Li W, Yang L, Zhang L, Chen C, Sun D (2022) In situ biosynthesis of biodegradable functional bacterial cellulose for high-efficiency particulate air filtration. ACS Sustain Chem Eng 10(4):1644–1652

Tajima K, Tahara K, Ohba J, Kusumoto R, Kose R, Kono H, Matsushima T, Fushimi K, Isono T, Yamamoto T, Satoh T (2020) Detailed structural analyses of nanofibrillated bacterial cellulose and its application as binder material for a display device. Biomacromolecules 21(2):581–588

Talipova AB, Buranych VV, Savitskaya IS, Bondar OV, Turlybekuly A, Pogrebnjak AD (2023) Synthesis, properties, and applications of nanocomposite materials based on bacterial cellulose and MXene. Polymers 15(20):4067

Ummartyotin S, Thiangtham S, Manuspiya H (2017) Strontium-modified bacterial cellulose and a polyvinylidene fluoride composite as an electroactive material. Forest Prod J 67(3–4):288–296

Wan Y, Xiong P, Liu J, Feng F, Xun X, Gama FM, Zhang Q, Yao F, Yang Z, Luo H, Xu Y (2021) Ultrathin, strong, and highly flexible Ti3C2Tx MXene/bacterial cellulose composite films for high-performance electromagnetic interference shielding. ACS Nano 15(5):8439–8449

Wang LP, Schütz C, Salazar-Alvarez G, Titirici MM (2014) Carbon aerogels from bacterial nanocellulose as anodes for lithium ion batteries. RSC Adv 4(34):17549–17554

Wang C, Wang S, Chen G, Kong W, Ping W, Dai J, Pastel G, Xie H, He S, Das S, Hu L (2018) Flexible, bio-compatible nanofluidic ion conductor. Chem Mater 30(21):7707–7713

Wang H, Qian H, Luo Z, Zhang K, Shen X, Zhang Y, Zhang M, Liebner F (2019) ZCIS/ZnS QDs fluorescent aerogels with tunable emission prepared from porous 3D nanofibrillar bacterial cellulose. Carbohydr Polym 224:115173

Wang Y, Wang F, Kong Y, Wang L, Li Q (2022a) Novel ionic bioartificial muscles based on ionically crosslinked multi-walled carbon nanotubes-mediated bacterial cellulose membranes and PEDOT: PSS electrodes. Smart Mater Struct 31(2):025023

Wang B, Li J, Wu Z, Sheng N, Zhang M, Han Z, Jin M, Li J, Lv X, Ou K, Wang H, Chen S (2022b) Salinity power generation based biocompatible bacterial cellulose/MXene membrane for biological power source. Nano Energy 102:107702

Wang Y, Rong Y, Ma T, Li L, Li X, Zhu P, Zhou S, Yu J, Zhang Y (2023) Photoelectrochemical sensors based on paper and their emerging applications in point-of-care testing. Biosens Bioelectron 236:115400

Wu Z, Zhang T, Wang B, Ji P, Sheng N, Zhang M, Liang Q, Chen S, Wang H (2021) Scalable bacterial cellulose biofilms with improved ion transport for high osmotic power generation. Nano Energy 88:106275

Wu S, Shi J, Nie X, Yu Z, Huang F (2022a) Multi-duties for one post: biodegradable bacterial cellulose-based separator for lithium sulfur batteries. Carbohydr Polym 285:119201

Wu Z, Wang B, Li J, Wu R, Jin M, Zhao H, Chen S, Wang H (2022b) Advanced bacterial cellulose ionic conductors with gigantic thermopower for low-grade heat harvesting. Nano Lett 22(20):8152–8160

Wu Z, Chen S, Li J, Wang B, Jin M, Liang Q, Zhang D, Han Z, Deng L, Qu X, Wang H (2023a) Insights into hierarchical structure–property–application relationships of advanced bacterial cellulose materials. Adv Funct Mater 33(12):2214327

Wu Z, Wang B, Li J, Jia Y, Chen S, Wang H, Chen L, Shuai L (2023b) Stretchable and durable bacterial cellulose-based thermocell with improved thermopower density for low-grade heat harvesting. Nano Lett 23(22):10297–10304

Xiao K, Jiang L, Antonietti M (2019) Ion transport in nanofluidic devices for energy harvesting. Joule 3(10):2364–2380

Xie H, Yang C, Fu K, Yao Y, Jiang F, Hitz E, Liu B, Wang S, Hu L (2018) Flexible, scalable, and highly conductive garnet-polymer solid electrolyte templated by bacterial cellulose. Adv Energy Mater 8(18):1703474

Xie W, Liu W, Dang Y, Tang A, Deng T, Qiu W (2019) Investigation on electrolyte-immersed properties of lithium-ion battery cellulose separator through multi-scale method. J Power Sources 417:150–158

Xiong R, Luan J, Kang S, Ye C, Singamaneni S, Tsukruk VV (2020) Biopolymeric photonic structures: design, fabrication, and emerging applications. Chem Soc Rev 49(3):983–1031

Xu X, Zhou J, Xin Y, Lubineau G, Ma Q, Jiang L (2017) Alcohol recognition by flexible, transparent and highly sensitive graphene-based thin-film sensors. Sci Rep 7(1):4317

Xu JX, Liu ZY, Zhang F, Tao J, Shen LF, Zhang XG (2020) Bacterial cellulose-derived carbon nanofibers as both anode and cathode for hybrid sodium ion capacitor. RSC Adv 10(13):7780–7790

Xu Y, Song Y, Xu F (2021) TEMPO oxidized cellulose nanofibers-based heterogenous membrane employed for concentration-gradient-driven energy harvesting. Nano Energy 79:105468

Xu Y, Zhang K, Chen S, Zhang X, Chen Y, Li D, Xu F (2022) Two-dimensional lamellar MXene/three-dimensional network bacterial nanocellulose nanofiber composite Janus membranes as nanofluidic osmotic power generators. Electrochim Acta 412:140162

Yang H, Wang M, Liu X, Jiang Y, Yu Y (2018) MoS2 embedded in 3D interconnected carbon nanofiber film as a free-standing anode for sodium-ion batteries. Nano Res 11(7):3844–3853

Yang X, Zhang Z, Wu M, Guo Z-P, Zheng Z-J (2023) Reshaping zinc plating/stripping behavior by interfacial water bonding for high-utilization-rate zinc batteries. Adv Mater 35(49):2303550

Yang Y, Tang J, Guo H, Pan F, Jiang H, Wu Y, Chen C, Li X, Yuan B, Lu W (2024) Robust and environmentally friendly MXene-based electronic skin enabling the three essential functions of natural skin: perception, protection, and thermoregulation. Nano Lett 24(35):10883–10891

Yao CH, Yin X, Yu YH, Cai ZY, Wang XD (2017) Chemically functionalized natural cellulose materials for effective triboelectric nanogenerator development. Adv Funct Mater 27(30):1700794

Yu X, Zhang D, Xu M, Song G, Xu Y (2023) Bacterial cellulose wave plates for polarization-encryption applications. J Mater Chem C 11(5):1684–1689

Yue L, Xie Y, Zheng Y, He W, Guo S, Sun Y, Zhang T, Liu S (2017) Sulfonated bacterial cellulose/polyaniline composite membrane for use as gel polymer electrolyte. Compos Sci Technol 145:122–131

Yun YJ, Yoon OJ, Son DI, Jun Y (2023) Metal/bacteria cellulose nanofiber bilayer membranes for high-performance hydrovoltaic electric power generation. Nano Energy 118:108934

Zhang H, Guo H, Wang B, Shi S, Xiong L, Chen X (2016) Synthesis and characterization of quaternized bacterial cellulose prepared in homogeneous aqueous solution. Carbohydr Polym 136:171–176

Zhang G, Liao Q, Ma M, Gao F, Zhang Z, Kang Z, Zhang Y (2018) Uniformly assembled vanadium doped ZnO microflowers/bacterial cellulose hybrid paper for flexible piezoelectric nanogenerators and self-powered sensors. Nano Energy 52:501–509

Zhang M, Chen W, Xue LX, Jiao Y, Lei TY, Chu JW, Huang JW, Gong CH, Yan CY, Yan YC, Hu Y, Wang XF, Xiong J (2020) Adsorption-catalysis design in the lithium-sulfur battery. Adv Energy Mater 10(2):1903008

Zhang X, Kang S, Adstedt K, Kim M, Xiong R, Yu J, Chen X, Zhao X, Ye C, Tsukruk VV (2022) Uniformly aligned flexible magnetic films from bacterial nanocelluloses for fast actuating optical materials. Nat Commun 13(1):5804

Zhao X, Han WJ, Jiang YF, Zhao CS, Ji XX, Kong FG, Xu WY, Zhang X (2019) A honeycomb-like paper-based thermoelectric generator based on a Bi_2Te_3/bacterial cellulose nanofiber coating. Nanoscale 11(38):17725–17735

Zheng Z-J, Ye H, Guo Z-P (2024) Bacterial cellulose applications in electrochemical energy storage devices. Adv Mater 37(22):2412908

Zhou Y, Chen C, Zhang X, Liu D, Xu L, Dai J, Liou S-C, Wang Y, Li C, Xie H, Wu Q, Foster B, Li T, Briber RM, Hu L (2019) Decoupling ionic and electronic pathways in low-dimensional hybrid conductors. J Am Chem Soc 141(44):17830–17837

Zhu CQ, Zhang JX, Qiu SY, Jia YB, Wang LX, Wang H (2021) Tailoring the pore size of polyphenylene sulfide nonwoven with bacterial cellulose (BC) for heat-resistant and high-wettability separator in lithium-ion battery. Compos Commun 24:100659

Zou C, Qu D, Jiang H, Lu D, Ma X, Zhao Z, Xu Y (2019) Bacterial cellulose: a versatile chiral host for circularly polarized luminescence. Molecules 24(6):1008

Chapter 8
Bacterial Cellulose in Foods: Toxicology and Regulatory Landscape

Ricardo Silva-Carvalho, Daniela Martins, L. Costa, Ana C. Rodrigues, Fernando Dourado, and Miguel Gama

Abstract Nanocellulose, particularly bacterial cellulose (BC), has gained increasing attention for its diverse applications in several fields, notability in the food industry. Despite its long history of safe human consumption in Asian countries as "nata de coco", with no reported adverse health effects, its nanoscalar dimension raises regulatory concerns regarding its safety upon oral ingestion. Consequently, global regulatory bodies, such as the European Commission (through the European Food Safety Authority—EFSA) and the FDA (U.S. Food and Drug Administration), have implemented pre-market safety assessment frameworks to evaluate the potential toxicity of nanomaterials like BC in food applications.

This chapter presents a comprehensive overview of the toxicological profile and regulatory landscape of BC, with a focus on its safety for human consumption. The chapter also addresses the existing challenges in the characterizing of BC using conventional nanomaterial assessment techniques, as recommended by EFSA, which are inadequate for its unique structure and highlights the importance of further research to bridge knowledge gaps.

Briefly, although the existing studies suggest that nanocellulose, in particular BC, generally exhibits low toxicity and may even confer prebiotic benefits, critical knowledge gaps persist. Further, there is a limited understanding of the long-term fate of nanocellulose in the gastrointestinal tract (GIT), particularly concerning degradation, absorption, and systemic distribution.

While current regulatory frameworks emphasize the need for robust risk assessment strategies, challenges remain in adapting these methodologies to structured biopolymers like BC. The chapter concludes with recommendations for future research directions aimed at addressing uncertainties in nanocellulose toxicology and ensuring safe and sustainable integration into food systems.

R. Silva-Carvalho · D. Martins · L. Costa · A. C. Rodrigues · F. Dourado · M. Gama (✉)
CEB—Centre of Biological Engineering, University of Minho, Campus de Gualtar, Braga, Portugal

LABBELS–Associate Laboratory in Biotechnology and Bioengineering and Microelectromechanical Systems, Braga, Guimarães, Portugal
e-mail: fdourado@ceb.uminho.pt; fmgama@deb.uminho.pt

F. Dourado et al. (eds.), *Advances in Bacterial Cellulose*,
https://doi.org/10.1007/978-3-032-07042-5_8

Keywords Nanocellulose · Bacterial cellulose · Risk assessment · Human gastrointestinal tract · Toxicity

1 Introduction

The advent of nanotechnology has revolutionized various industries, including food science, where nanoscale materials (size between 1 and 100 nm) are integrated to enhance texture, functionality, and preservation (Onyeaka et al. 2022). Among these materials, nanocelluloses (NC) have emerged as promising functional natural additives, serving as emulsifiers, texturing agents, and stabilizers (Wüstenberg 2014). Their use may enable the reduction of synthetic compounds in food products, thus aligning with the growing demand for natural and sustainable ingredients (Li et al. 2021; Nile et al. 2020; Pasupuleti 2022).

Nanocelluloses include nanocrystals (CNC, 6–25 nm in width and 50–1100 nm in length (Mota et al. 2022)), nanofibres (CNF, 3–100 nm in width and up to 100 μm in length (Moriyama et al. 2023)), and bacterial cellulose (BC, with fibre widths up to 100 nm (Babi et al. 2022)). CNC and CNF are usually obtained from wood, cotton, or other cellulose fibres, by physical and/or chemical processes, whereas BC is synthesized by certain strains of acetic acid bacteria, notably those of the *Komagataeibacter* genus (Brand et al. 2022). Although chemically identical to plant-derived nanocelluloses, BC offers superior purity (free of lignin, hemicelluloses, and other biogenic compounds) and unique structural, physicochemical, and mechanical properties, thereby enhancing the textural and organoleptic qualities of food (Dourado et al. 2016a; Nsor-Atindana et al. 2020). BC has been widely recognized for its significant potential as a gelling agent, thickener and suspension, and foaming stabilizer (Dourado et al. 2016a; Martins et al. 2020b, 2024). In this review, we address the toxicology of BC, a discussion related to NC being also made whenever relevant. BC is naturally nano, and therefore cannot be considered an "engineered nanomaterial", differently from CNC and CNF that can be obtained only through physical/chemical processing. Although BC has special features that make it a different material compared to other nanocelluloses, the regulatory assessment of BC is made more complex due to the perception of some risks related to wood-derived nanocelluloses.

Notably, BC has been consumed for decades in Asian countries (marketed under the trade name "nata de coco") without reported cases of health-related issues. This fact alone would be expected to support the conclusion that BC is safe and could be adopted by the food industry in a generalized way. However, according to European legislation, any product absent from the food market before 1995 is considered a novel food and mandatorily needs to be assessed by the European Food Safety Authority (EFSA) before it can be marketed (EC 2015). The fact that BC is nano further complicates this assessment, as discussed ahead. Consequently, unlike commercial micron-sized forms of cellulose and many of its derivatives (e.g. methyl cellulose and carboxymethyl cellulose, among others) which have been safely used for decades in food products, BC and other NC are not yet approved by the regulatory bodies for commercial use in food (Dourado et al. 2017). Several *in vitro*, ex vivo, and *in vivo* studies have been published regarding the safety of plant (Andrade et al. 2015; Chen et al. 2020; DeLoid et al. 2019; Lin et al. 2021; Mackie et al. 2019; Moriyama et al. 2023; Müller et al. 2023; Ong et al. 2020) and bacterial (Hagiwara et al. 2010; Rodrigues et al. 2023; Schmitt et al. 1991) nanocelluloses. Other studies have shown limited toxicological risk under realistic doses of NC and CNF and exposure scenarios (Cañas-Gutiérrez et al. 2024; Endes et al. 2016; Nel et al. 2006). The existing literature regarding the potential hazards posed by the various forms of NC, especially towards human and environmental health, is thus inconsistent. This is pivotal considering the increasing focus on food quality and safety by policymakers, industries, and consumers (Carocho et al. 2015). Nanomaterials exhibit very different properties (e.g. particle size and surface area, agglomeration, surface functionalization, and crystallinity) and interactions with biological systems compared to their micron- and larger scale counterparts. These differences raise concerns about their potentially hazardous effects and uncertain fate once ingested (Schoonjans et al. 2023). In response to recent legislation on nanomaterials and (cellulosic) fibres (Cañas-Gutiérrez et al. 2024), further investigations into the impact of NC (particularly of BC) on human health are essential to demonstrate their safety and responsible commercialization. Globally, the risk assessment of nanocelluloses in food products falls within the remit of regulatory bodies such as the European Food Safety Authority (EFSA) and the U.S. Food and Drug Administration (FDA). Both agencies have developed toxicological frameworks for assessing the risks associated with nanomaterials, yet these frameworks are continually evolving as more data on nanotoxicology becomes available (EFSA 2021a; FDA 2014b).

In previous work, a comprehensive overview on the safety and potential health benefits of BC when used in food applications was made. In that work, the safety and toxicology, the dietetic properties, the biocompatibility and inflammatory response, as well as regulatory considerations were reviewed, supporting the conclusion that BC not only is safe for human consumption, but also that it has beneficial dietary properties (Dourado et al. 2016b). In the present chapter, we provide an overview of the current regulatory landscape, with a focus on the European Union, and the potential risks associated with the use of BC in foods. An in-depth assessment of the recent literature on the toxicology of BC is presented. The *in vitro* and *in vivo* risk assessment studies, the impact on gut microbiota and highlights on the structural and functional properties of BC are also covered.

2 Regulatory Landscape for Nanomaterials in Food

The rapid advancement of nanotechnology in the food industry led the regulatory bodies worldwide to develop frameworks for assessing the safety of nanomaterials, which includes BC. The methodological guidance provided aims to ensure the thorough evaluation of nanomaterials used in food applications for their potential toxicological effects, before they are approved for use. Thus, organizations such as EFSA and FDA have established guidelines for risk assessment of nanomaterials in food and food contact materials (EFSA 2021a; FDA 2014b). However, the regulatory approaches of these organizations differ in scope and specificity, reflecting the evolving understanding of nanomaterials and their potential risks resulting in frequent updates to their respective guidelines.

In the European Union (EU), the regulation of nanomaterials in food is governed by a comprehensive legislative framework. The General Food Law Regulation (EC) No 178/2002 establishes the bottom lines of food safety and risk assessment (EC 2002), making sure that any potential risks associated with emerging technologies, including nanomaterials, are thoroughly evaluated before they are allowed in the food chain. The General Food Law also determines the legal basis for the European Food Safety Authority (EFSA) to conduct scientific risk assessments (Dourado et al. 2016b).

Based on the European Commission Recommendation 2011/696/EU (EC 2011), the Novel Food Regulation (EU) 2015/2283 specifically addresses nanomaterials, defined as "any intentionally produced material that has one or more dimensions of the order of 100 nm or less, or that is composed of discrete functional parts, either internally or at the surface, many of which have one or more dimensions of the order of 100 nm or less" (EC 2015). This regulation determines that any engineered nanomaterial used as a novel food or food ingredient undergoes a pre-market safety assessment and authorization by EFSA. This includes a thorough evaluation of the nanomaterial's safety, bioavailability, persistence, potential accumulation in tissues, nutritional impact, and potential allergenicity. This regulation makes mandatory the labelling of nanomaterials in food products, ensuring consumer transparency. Thus, food products containing engineered nanomaterials must be explicitly labelled, regardless of their intended function or concentration (The European Parliament and the Council of the European Union 2015).

EFSA plays a pivotal role in evaluating the safety of nanomaterials in food and feed. In 2021, EFSA issued an updated guidance on risk assessment of nanomaterials to be applied in the food and feed chain: human and animal health (EFSA 2021a). This guidance provides a detailed orientation for assessing the safety of nanomaterials on a case-by-case basis, considering their physicochemical properties, potential exposure routes, and toxicological profiles. It addresses key threats related to nanomaterials, such as the potential ability to cross biological barriers, accumulate in tissues and induce adverse effects such as cytotoxicity, genotoxicity, and immunotoxicity. The importance of evaluating the stability and biopersistence of nanomaterials in biological systems, as well as their potential impact on gut microbiota and other physiological processes, is also highlighted. In response to a mandate from the European Commission (EFSA 2021b), EFSA has also developed the Guidance on Technical Requirements for Regulated Food and Feed Product Applications to Establish the Presence of Small Particles Including Nanoparticles (Guidance on Particle-TR). This complementary guidance aims to provide specific technical requirements for identifying and characterizing small particles, accurately detecting and quantifying nanoparticles in food and feed products. It establishes criteria for distinguishing between intentionally engineered nanomaterials and incidental nanoscale particles that may be present as a result of processing. Together, these EFSA documents create a structured framework for evaluating the safety of nanomaterials, integrating both biological risk assessments and analytical methodologies to ensure a robust regulatory approach (Fig. 8.1).

In the United States, the regulation of nanomaterials in food falls under the jurisdiction of the FDA. Unlike the EU, the FDA does not have a specific regulatory definition for nanomaterials but evaluates them within the broader category of food ingredients or food contact substances on a case-by-case basis. The FDA's approach to nanomaterials is outlined in its Guidance for Industry: Assessing the Effects of Significant Manufacturing Process Changes, Including Emerging Technologies, on the Safety and Regulatory Status of Food Ingredients and Food Contact Substances, Including Food Ingredients that are Colour Additives. This guidance emphasizes the need for manufacturers to assess the safety of nanomaterials used in food, particularly in cases where the material properties (e.g. size, surface area, and reactivity) differ significantly from their macro-sized counterparts (FDA 2014a, b). The FDA encourages the use of existing toxicological data and risk assessment methodologies while acknowledging that additional studies may be required to address the unique properties of nanomaterials. For instance, the FDA highlights the importance of evaluating whether nanomaterials may accumulate in the body, cross biological barriers and induce long-term health effects such as immune system modulation or organ toxicity (FDA 2014b). Manufacturers are encouraged to consult with the FDA early in the development process to address any scientific or regulatory issues related to their nanotechnology products (FDA 2014b).

In addition to the FDA guidance, the FDA participates in the National Nanotechnology Initiative (NNI) (https://www.nano.gov/about-nni), a multi-agency effort established in 2000 to coordinate federal research and policy development on nanotechnology. Within the NNI, the Nanotechnology Environmental and Health Implications (NEHI) Working Group (https://www.nano.gov/about-nni/working-groups/nehi) serves as a platform for interagency collaboration on the safety assessment of nanomaterials. The NNI supports research on the environmental, health, and safety (EHS) implications of nanomaterials, which informs regulatory decision-making across agencies, including the FDA. Furthermore, the FDA's Nanotechnology Task Force, established in 2006, plays a key role in addressing regulatory and scientific issues related to nanotechnology, ensuring that the agency's approach remains science-based and adaptive as new data emerges (FDA 2024).

Beyond the EU and the USA, international organizations such as the World Health Organization (WHO) and Food and the Agriculture Organization (FAO) have contributed to the development of global guidelines for the safety assessment of nanomaterials in food. The WHO/FAO Joint Expert Committee on Food Additives (JECFA) has published reports on the nanomaterials safety, emphasizing the need for a comprehensive evaluation of their toxicological properties and potential risks to human health (FAO/WHO 2013; Takeuchi et al. 2014). These guidelines

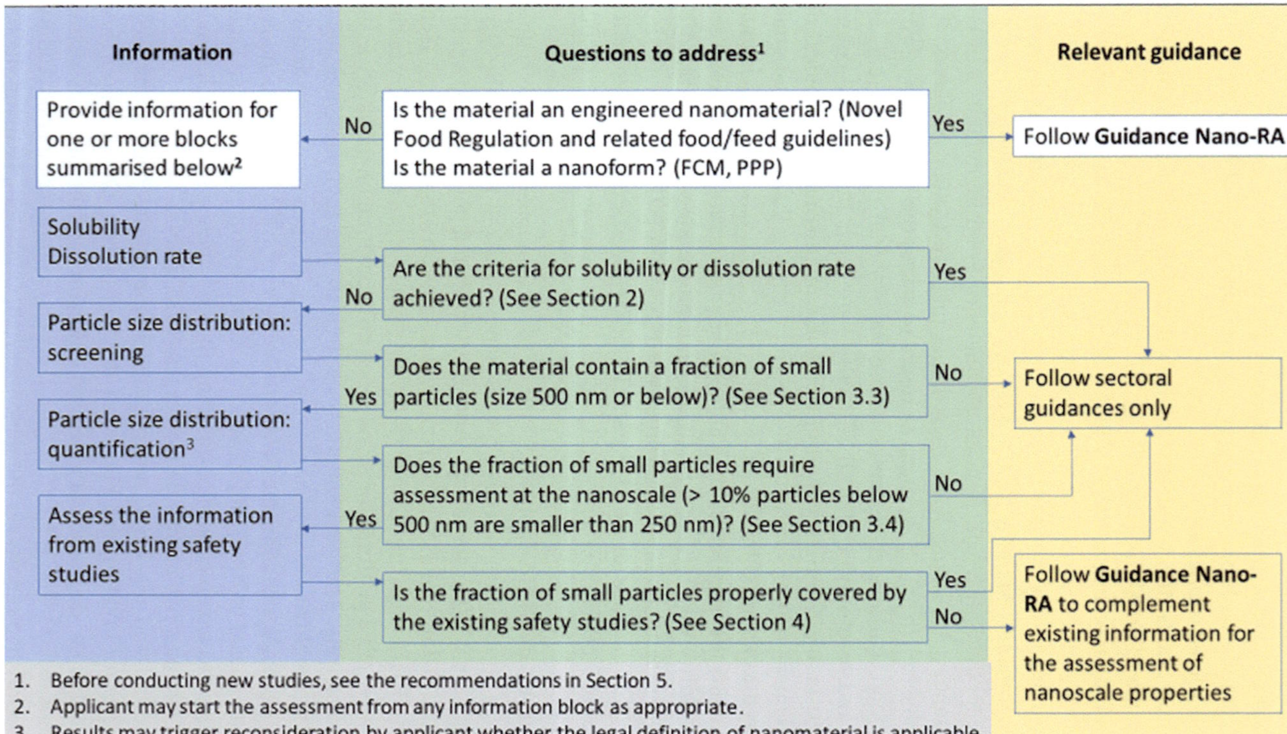

Fig. 8.1 Decision process for selecting the applicable guidance document(s) to be used for the risk assessment of the material regarding the assessment of small particles (Figure reproduced from EFSA Journal 2021;19(8):6769 under the terms of the Creative Commons Attribution-NoDerivatives 4.0 International (CC BY-ND 4.0) licence. https://doi.org/10.2903/j.efsa.2021.6769 (EFSA 2021a))

represent a reference for countries developing their own regulatory frameworks for the assessment of nanomaterials in food.

Despite the regulatory frameworks and technical recommendations, significant challenges persist in the assessment of nanomaterials, such as BC. A primary concern regards the attempt to develop standardized methods for universal use in the characterization of nanomaterials and assessing their toxicological effects. While the availability of a roadmap to characterize the unique features of nanomaterials, including their size, shape, surface charge, and colloidal behaviour (e.g. aggregation), is very desirable, the current scenario simply fails to properly characterize BC. It is well known that nanoparticles with high surface energy tend to aggregate. Thus, measuring the size distribution using light scattering may overestimate the size due to aggregation. Consequently, the EFSA Guidance on Particle-TR (EFSA 2021b) advises methodologies to disperse nanoparticles (e.g. by sonication) and recommends the use of descriptive Transmission Electron Microscopy (TEM), setting aside the use of light scattering. While this approach is suitable for, for example, metallic nanoparticles, it does not provide the full picture of structured nanomaterials such as BC, which naturally bears a complex structure involving crosslinks and entanglements. The characterization of BC using descriptive TEM is a bit like characterizing a forest by the size of the trees. In the experience of the authors, EFSA is not willing to adapt their methodological orthodoxy to special materials like BC. Therefore, by following the guidance of EFSA, (1) it is not possible to properly disaggregate BC without changing its natural morphology; (2) it is not possible to properly characterize the size of BC using TEM only. This matter is discussed ahead in more detail in Sect. 3.

Another challenge concerns the understanding of the fate of nanomaterials in the human body, particularly in the context of chronic exposure through food consumption (El-Kalliny et al. 2023). Many studies have investigated the absorption, distribution, metabolism, and excretion (ADME) of nanomaterials; however, there is still limited information on how materials like BC behave in the gastrointestinal tract and whether they can accumulate in tissues over time (Haripriyaa and Suthindhiran 2023; Xue et al. 2023). Studies on cellulose-based nanomaterials have demonstrated that their behaviour in the GIT is governed by factors such as the particle size, surface charge, and degree of crystallinity (Chumchoochart et al. 2023). Moreover, the ability of CNCs to cross the intestinal mucus layer—but not the intestinal tissue barrier—was demonstrated (Müller et al. 2023). A potential DNA damage to human intestinal cells is yet another concern (Vital et al. 2025). The surface charge of cellulose nanocrystals has also been shown to affect their interactions with digestive system components (e.g. enzymes, gastrointestinal fluid, mucosa, microbiota), potentially influencing their bioavailability (Koshani and Madadlou 2018). Furthermore, the aspect ratio of cellulose nanomaterials also influences their interactions with biological systems, including the innate immune response (Wang et al. 2019). Research on other types of nanomaterials, such as silica and titanium dioxide nanoparticles (both used in the food industry), has demonstrated their potential to cross the intestinal barrier and accumulate in organs, such as the liver and spleen (Böhmert et al. 2014; McClements et al. 2017), generating the hypothesis that the same may occur with CNC and CNF. Therefore, data from cellulosic nanomaterials like CNC and CNF are sometimes generalized to "nanocelluloses", even though they are substantially different from BC with regard to, for example, size, charge, interaction, and organization of fibres in the 3D space. It must be stressed that BC fibres are nano in one dimension (thickness), but are always observed as very large particles composed of entangled fibres. This is discussed in more detail in Part III.

The gut microbiome may influence the biodegradability of the material and its potential impact on gut health. On the other hand, emerging evidence suggests that nanomaterials can alter the composition and function of gut microbiota, potentially leading to dysbiosis and associated health effects, such as immune dysfunctions ((Lamas et al. 2023) and increased susceptibility to chronic diseases like inflammatory bowel diseases (IBD), obesity, and colorectal cancer (Ghebretatios et al. 2021; Lamas et al. 2020). Such alterations may also induce changes in nutrient digestion, absorption, and overall gut health (Campos et al. 2022). Indeed, it has been shown that inorganic nanoparticles can inhibit probiotic reproduction, stimulate the inflammatory response of the gut immune system, increase opportunistic infections and disrupt microbial diversity (Ma et al. 2023), reduce the abundance of beneficial bacteria such as *Lactobacillus* and *Bifidobacterium* (Chen et al. 2017; Pinget et al. 2019). These changes in the gut microbiome may compromise the intestinal barrier function and increase the susceptibility to IBD (Mu et al. 2017). Although humans do not produce cellulases and therefore are not capable of degrading celluloses, BC and other NC may be processed by microbial enzymes in the gut, potentially affecting their absorption, persistence, and potential for bioaccumulation (Bher et al. 2022). Studies have shown that ingested nanocellulose materials may interact with the gut microbiome, potentially modulating microbial populations and intestinal function, which could have significant health implications (Khare et al. 2020). However, the specific mechanisms by which nanomaterials interact with the gut microbiome and any potential long-term consequences remain poorly understood, underscoring the need for further research (McClements et al. 2017).

3 BC: A Structured Nanocellulose

Bacterial cellulose (BC) is a unique biopolymer characterized by its nanoscale dimensions, high purity, exceptional physicochemical properties, high water retention capacity, and biocompatibility, making it an attractive material for various applications in the food industry (Dufresne 2013). BC is synthesized extracellularly by certain bacterial species, most notably *Komagataeibacter xylinus* (formerly known as *Gluconacetobacter xylinus*), through a process that involves the polymerization of glucose into β-1,4-glucan chains (Iguchi et al. 2000; Ross et al. 1991). The resulting material is composed of a three-dimensional (3D) network of nanofibres, typically ranging from 20 to 100 nm in diameter, which are highly crystalline and exhibit a large surface area (Klemm et al. 2005). The length of the fibres spans from 2 to 20 μm (Tagawa et al. 2024).

The guidance on risk assessment of the application of nanoscience and nanotechnologies in the food and feed chain: Part I, provides detailed recommendations for the physicochemical characterization of nanomaterials. It outlines how to establish whether a material is a nanomaterial, identifies key parameters that should be measured, and describes methods and techniques that can be used for the characterization of nanomaterials and their determination in complex matrices (EFSA 2018). However, applying these guidelines to the characterization of BC presents specific challenges such as the characterization, stability assessment, biological interactions, and regulatory compliance. EFSA's document establishes a comprehensive methodological framework for nanomaterial risk assessment, covering everything from physicochemical characterization to exposure assessment and hazard characterization. In particular, nanospecific considerations relating to *in vivo*/*in vitro* toxicological studies are discussed and a tiered framework for toxicological testing is outlined. It describes *in vitro* degradation, toxicokinetics, genotoxicity as well as general issues relating to testing of nanomaterials. Depending on the initial tier results, further studies may be needed to investigate the reproductive and developmental toxicity, immunotoxicity, allergenicity, neurotoxicity, effects on gut microbiome, and endocrine activity.

The characterization of the nanomaterial deserves naturally a detailed analysis, in particular with regard to the characterization of the particle size (EFSA 2018). When the guidance on risk assessment was published, in 2018, EFSA advised on the collection of data on primary and secondary (agglomerates and aggregates) particle size. The number-based size distribution and mass-based size distribution of the material should be provided as measured by more than one independent technique, one being electron microscopy (EM), for assessing the presence of particles smaller than 500 nm. Data obtained with other techniques (such as centrifugal liquid sedimentation (CLS), particle tracking analysis (PTA), or dynamic light scattering (DLS)) should be provided in the original metrics as produced by the technique (e.g. intensity, volume- or mass-based) (EFSA 2018). In a paper by Schoonjans et al. (2023), the authors discuss the regulatory safety assessment of nanoparticles in the food chain in Europe and reference the EFSA guidance, highlighting EM as the recommended method for determining particle size distribution in regulated food and feed products. According to this work, as also elaborated in the practical cases in dossiers that have so far been assessed by EFSA, often applicants have used techniques, such as DLS, that are no longer considered suitable for screening for the presence of small particles or characterizing the particle size distribution (EFSA 2021b).

The use of EM, alone or complemented with CSL or PTA, to characterize BC provides an incomplete picture of the fibres and of their relevant dimensions. The recent work of Babi et al. (2022) showed a hierarchical model of the BC ribbons based on TEM measurements (Fig. 8.2). In that work, the so-called BC ribbons were shown to have a width of 30–60 nm and height of 4–16 nm. These ribbons are composed of several sheets, each with 4–6 nm height. In turn, each sheet has several laterally assembled bundles, each bundle having 10–14 nm wide and 4–6 nm height. Finally, each bundle is composed of microfibrils, 1.6 nm wide and 2.6 nm height,

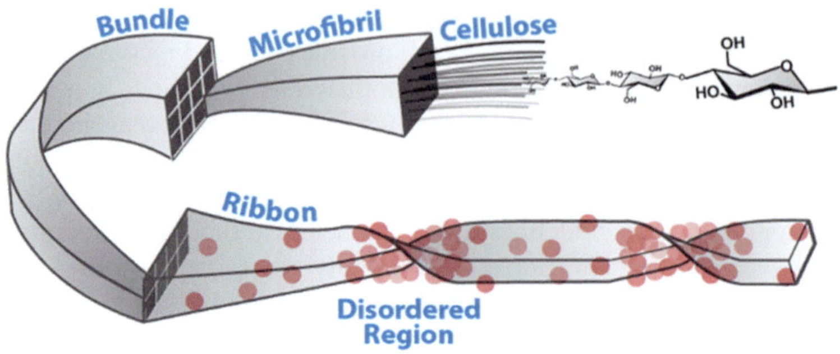

Fig. 8.2 Ultrastructure of BC (Reproduced from (Babi et al. 2022) (https://doi.org/10.1021/acs.biomac.2c01108))

which is the elementary structural unit according to this model, well supported on the AFM and TEM observations. Similar models have been proposed by other authors, based on information gathered from BC samples produced by different strains, such that it translates (with slight differences on the dimensions of each structural unit) a generally observed BC structure.

Furthermore, as obtained in static culture, BC is a 3D nanofibrillar network of entangled fibres. In a recent work, Takayama and Kondo (2023) investigated the 3D structure of BC. Based on experimental evidence, these authors proposed that the BC fibres' crosslinking points (defined as points close to each other), have two types of bonding based on the nature of the crosslinks: those with branching points and those with contact points. At a branching point, one fibre branches into two or more fibres (Fig. 8.3). This is a characteristic of BC and results in strong and permanent physical crosslinking due to crystalline junction zones. A contact point occurs when there is literal contact between fibres, and it is supposed to be a transient crosslink based on the weak interaction due to hydrogen bonding and/or van der Waals forces. The crosslinked nature of BC produced in static culture is shown in that work to be a common feature of seven studied strains (and is probably shared by every BC membranes obtained by static fermentation), its density reaching 10^9 crosslinks per mm^3 (Takayama and Kondo 2023). The crosslinked nature of BC, which brings together a large number of fibres resulting in macroscopic membranes, cannot be confused with the formation of aggregates, as discussed in more detail below. In addition to the mentioned physical crosslinking mechanisms, mechanical interlocking or physical entanglement is probably yet another major contributor to the internal cohesion of the BC's 3D network.

The BC produced in static culture may be ground into a pulp containing fleeces of variable size. These fleeces have a star-resembling structure, with loose fibre branches emerging from a deeply entangled core, which preserve the crosslinks naturally present in the original membrane obtained by static culture. The size of these fleeces, as measured by laser diffraction, commonly spans from a few tenths of μm to a few mm, depending on the mechanical technology used for comminution (e.g. blend grinder, high pressure homogenizer). Interestingly, as observed in the

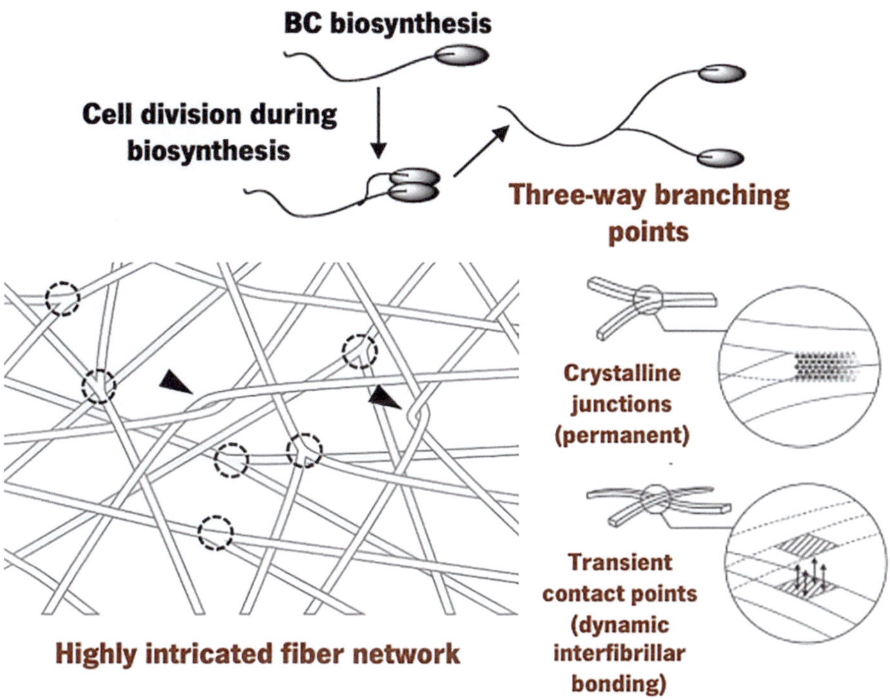

Fig. 8.3 3D organization of BC displaying the interactions between fibres that hold together the BC membrane (arrow—fibre entanglements) (Adapted from Takayama and Kondo (2023). https://doi.org/10.1016/j.carbpol.2023.121311)

work by Martins et al. (2020a), no particles are detected below 10 μm in a number-based distribution (Fig. 8.4).

Thus, BC has a complex structure, comprising the organization of microfibrils into bundles, in turn organized in sheets, furthermore organized into ribbons (commonly known as the BC fibre), which finally are connected to other ribbons/fibres through several bridging and stabilizing mechanisms (mechanical interlocking or physical entanglements, van der Waals contacts, fibre ramifications), which preserve the 3D organization of the BC hydrogel. Characterizing this multidimensional organization through descriptive TEM alone, as recommended by EFSA, is not possible. TEM does not allow a proper characterization of the organization at a supra-ribbon level. Yet, this level is highly relevant with regard to the topic discussed ahead.

One central concern related to the use of materials in food regards its potential absorption in the gut. As a rule, the chances of particle translocation increase with a decrease in the particle size. Studies so far have suggested that specific cellular uptake mechanisms for particles extend up to approximately 250 nm with a preferred uptake size of about 50 nm (EFSA 2021c; Frohlich and Roblegg 2012; Howe et al. 2014; Macierzanka et al. 2014; Powell et al. 2010). This means that particles larger than 250 nm are unlikely to have biodistribution or accumulation routes other

	Percentiles (μm)				Mean Diameters (μm)		
	Dv(10)	Dv(50)	Dv(90)	Span	D[4,3]	D[3,2]	
HB	254	1228	2299	2	1205	514	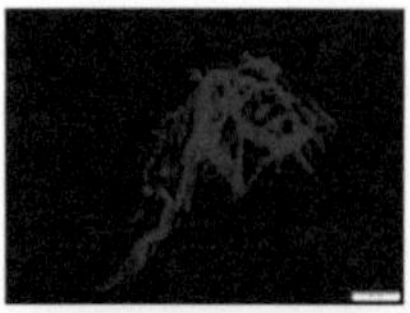
MHI	108	575	1700	3	757	254	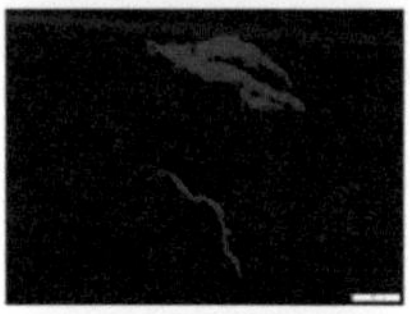
HPH-1	24	79	229	3	107	52	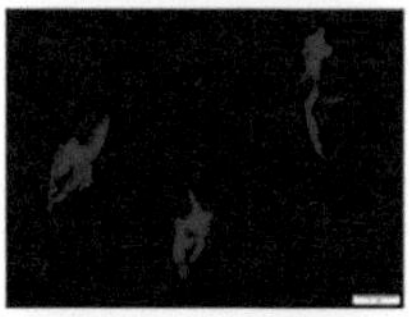
HPH-2	24	71	187	2	91	50	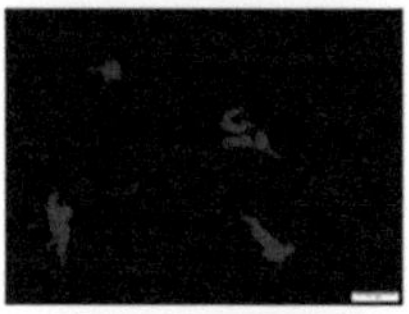
HPH-6	22	55	139	2	70	43	

Fig. 8.4 Percentiles and Mean Diameters of the particle size distribution (by volume) of BC samples wet-ground by different methods (HB—hand blender; MHI—microhead impeller; HPH—high pressure homogenization (different cycles)), and respective micrographs of the cellulose stained with calcofluor (scale bars correspond to 100 μm) (Reproduced with permission from: Martins, D. et al. (2020). Dry Bacterial Cellulose and Carboxymethyl Cellulose formulations with interfacial-active performance: processing conditions and redispersion. *Cellulose*. Springer Nature (Martins et al. 2020a))

than the corresponding conventional material. For regulatory purposes, EFSA has included a screening option. If all particles have a minimum external dimension of 500 nm or larger, a conventional risk assessment is sufficient (Schoonjans et al. 2023).

Taking into account the above discussion on the structure of BC, it is highly unlikely that it can be absorbed in the intestine. The fibres (ribbons) are typically a few times longer than 500 nm, and they are naturally interconnected to many other fibres, generating particles well over 10 μm. As a matter of fact, as discussed bekiw in more detail, studies conducted on the absorption of BC by the epithelium, both *in vivo* and *in vitro*, failed to detect any absorption (Rodrigues et al. 2023; Silva-Carvalho et al. 2024). However, contradictory results were obtained by Vincentini et al. (2023), highlighting the complexity of the toxicological assessment of structured materials, and also the inadequacy of the methodologies developed by EFSA, which are inappropriate to analyse BC.

4 Toxicology

Although *in vitro* tests provide valuable preliminary insights, they are insufficient to determine the systemic toxicity of NCs upon consumption. Whenever hazard assessment requires, *in vivo* assays using relevant animal models should be conducted, while carefully integrating *in vitro* data and adhering to the principles of the 3Rs (replacement, reduction, and refinement). Despite their limitations due to differences in digestive physiology, microbiome composition and fibre degradation rates compared to humans (Vincentini et al. 2023), *in vivo* systems, such as mice and rats, are commonly used to investigate the behaviour of NC in the gastrointestinal tract and their health effects. In fact, these systems are complex and allow the evaluation of the interactions of nanostructures with biological components (e.g. proteins and cells) that may promote distinctive patterns of biodistribution, clearance, immune response, and metabolism (Fischer and Chan 2007). Data on the acute oral toxicity, subchronic toxicity, and genotoxicity (Schmitt et al. 1991) have been reviewed elsewhere (Dourado et al. 2017) and will not be discussed here again.

4.1 Translocation of BC Across the Epithelium

4.1.1 *In Vitro* Assays

For decades, micron-sized cellulose and its derivatives (e.g. methylcellulose and carboxymethylcellulose) have been designated "generally regarded as safe" (GRAS) for ingestion and used in foods as fillers and thickeners. However, NC, which reached industrial production in recent years, are still not approved by regulatory entities for use in food products. To date, there have been limited studies on its behaviour in the gastrointestinal tract. It is still controversial whether NCs can cross the intestinal barriers, such as the mucus layer or epithelium cells, potentially distributing systemically, generating toxicological issues (Silva-Carvalho et al. 2024; Vincentini et al. 2023).

Because of the apparent difficulty in translating the results from the animal models to humans and the increasing need for human-relevant information (Leist et al. 2014), a variety of *in vitro* methods that mimic different human biological targets, including epithelial barriers, are often required to meet regulatory requirements.

The EFSA guidance on risk assessment of nanomaterials (EFSA Guidance on Nano-RA) suggests that nano-specific risk assessment is best achieved through Integrated Approaches to Testing and Assessment (IATAs) with new approach methodologies (NAMs) as the first choice. Integrating NAMs in risk assessment promises several advantages such as a better human focus, more detailed insights into molecular mechanisms and a higher efficacy, while minimizing animal trials. Therefore, some researchers have investigated the *in vitro* biological properties of BC and plant NC and their potential implications for human health (Chen et al. 2020; Lin et al. 2021; Lopes et al. 2020; Mejía-Jaramillo et al. 2023; Mortensen et al. 2022). Caco-2 cells are the most commonly used cellular model to mimic the intestinal barrier because cell monolayer permeability correlates well with *in vivo* absorption (Fröhlich 2018).

In vitro models based on a single cell line, such as Caco-2 cell monolayer, lack physiological accuracy because they do not properly represent the complex gut environment. Therefore, to enhance the physiological relevance of *in vitro* intestinal models, a triculture model incorporating different human-derived cell lines, such as microfold (M) cells and mucus-secreting cells, into Caco-2 cell cultures has been adopted in various studies (DeLoid et al. 2019; Ede et al. 2020; Silva-Carvalho et al. 2024; Vincentini et al. 2023).

According to Vincentini et al. (2023), and in agreement with the recommendations from EFSA, *in vitro* studies using exposure concentrations above 100 μg/mL require proper dispersion of the sample to prevent agglomeration. Recently, the authors demonstrated the *in vitro* cell uptake of 30 μg/mL BC, CNF, and CNC subjected to a sonication dispersion protocol (energy delivered: ±7 kJ) in a triculture model. The internalization rates were 0.9–3.9% after 24 h and 3.3–4.7% after 72 h, which were higher compared to Caco-2 monolayers, where the internalization rates were 1.4–2.4% after 24 h and increased to 8.7% for CNC after 72 h. However, in the case of BC, the dispersion protocol appears to completely disrupt the fibres into nanocrystals since the authors reported an estimated median diameter ranging from 5 to 9 nm for all the materials. As discussed above, the hierarchical organization of BC comprises fibres with a width between 20 and 80 nm (Babi et al. 2022). Therefore, the results show that the triculture intestinal model was capable of internalizing the thin fibres released from BC, but no conclusions can be reached with regard to the native (thicker and longer) fibres, the pristine BC. In fact, recently Mota et al. (2022) and Silva-Carvalho et al. (2024) demonstrated using a simple, sensitive, and specific detection method (carbohydrate-binding module coupled to a green fluorescent protein (CBM-GFP) capable of detecting single cellulose nanocrystals (Fig. 8.5) that cellulose at a concentration of 50 μg/mL—whether derived from BC or Avicel LM310—does not penetrate an *in vitro* triculture intestinal model and does not affect the integrity of the cell monolayer. Furthermore, no internalization or significant changes in cell metabolic activity were observed in the intestinal co-culture after 24 h of exposure to 10, 50, and 100 μg/mL BC, CNC, and Avicel LM310. Similar results for CNC were obtained by Müller et al. (2023). The authors showed that although CNC could cross the ex vivo porcine intestinal mucus layer, they could not pass the tissue barrier ex vivo or *in vitro* (in a triple co-culture model,

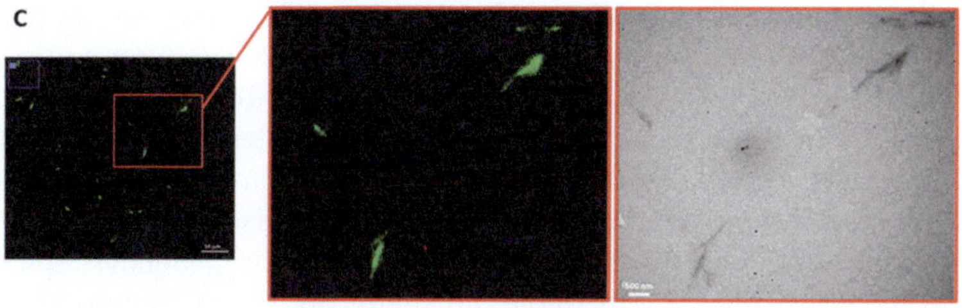

Fig. 8.5 Example of CLEM images of BC-derived nanocrystals labelled with GFP-CBM. Small green dots in fluorescence microscopy (FM) correspond to single BC in TEM (highlighted in the red boxes). Scale bar: 10 μm for FM; 500 nm for TEM. These images demonstrate the detections of isolated nanocrystals using GFP-CBM, taking into account the colocalization of the signals obtained by FM and TEM. (Reproduced from *Nanomaterials*, *12*(15), 2605. https://doi.org/10.3390/nano12152605)

Caco-2, HT29-MTX, and THP1 cells). Finally, no *in vitro* toxicity was observed after 2 and 24 h of treatment with 100 μg/mL of CNCs.

The findings of recent studies therefore present inconsistent results regarding the assessment of hazards potentially associated with NC oral exposure. While limited permeability and low toxicity of BC is observed in some studies, suggesting their potential safety when used in food applications, opposite results are shown by other authors. Nevertheless, standard experimental protocols to enhance reproducibility and reliability are required. A key aspect includes the use of appropriate dispersion techniques and physiologically relevant exposure concentrations to ensure meaningful and consistent results, while making sure that the natural morphology of the fibres is not changed. Although there is no significant *in vitro* penetration or toxicity, a cautious yet optimistic perspective on the safety of BC is warranted. Nevertheless, further research is crucial to support the responsible and safe incorporation of BC into consumer products, in alignment with regulatory standards and sustainability objectives.

4.1.2 *In Vivo* Assays

Overall, the studies described in the previous section provide good indications for the absence of adverse effects during exposure to BC. However, *in vivo* absorption data are needed for further risk assessment. It is hypothesized that because of their smaller size and other physicochemical properties, BC and other NC may be able to cross the intestinal epithelium through, for example, endocytosis or through the M-cells in the Peyer's patches (Koshani and Madadlou 2018), once they pass through the mucus layer. However, in the case of BC, although the nanometric thickness of the fibres cannot be disregarded but the large size of the fibre bundles is certainly decisive with regard to *in vivo* absorption; this appears indeed to be limited or negligible, in addition to the large size probably owing to mucoadhesion and restricted transport capacities (Brand et al. 2022). Recently, this aspect was evaluated in female and male Wistar Han IGS rats after being fed daily by oral gavage with an aqueous suspension of 1% (w/v) BC or its counterpart of plant origin (Avicel LM310) for 21 days (Rodrigues et al. 2023). To the best of our knowledge, this is the first *in vivo* study to track potentially internalized intact cellulose fibres. The results suggest that BC did not exert any systemic toxicity since no significant differences were detected in the serum biochemistry analysis and did not affect the global architecture and morphology of the liver, kidneys, spleen, and small and large intestines. Furthermore, no significant histopathological changes, including inflammation-related alterations, were observed. Importantly, no evidence of BC persorption was found in the histological analysis of the gut using GFP-CBM. BC was not detected in the Peyer's patches, colonic patches, or mesenteric lymphatic nodules. Although no evidence of BC translocation or accumulation in gut lymphoid tissue was observed, the authors speculated that if a low-efficiency translocation process does occur, it would likely apply to any cellulose source in food undergoing degradation. Therefore, this study supports the safety of BC as a food additive (Rodrigues et al. 2023). Notwithstanding the promising *in vivo* results, regulatory bodies still consider the need for further in-depth studies to address the existing data gaps that hamper conclusions regarding the potential hazards and risks of oral exposure to NC in humans.

5 Biodegradability, Impact on Human Gut Microbiota and Excretion Pathways: The Fate of BC in the Gastrointestinal Tract

Micron-scale plant cellulose and its derivatives are allowed in the EU market as food additives. Studies have shown that even high daily doses of up to 6000 mg/person for 8 months cause no adverse effects (Younes et al. 2018). However, given the reduced size dimensions and possibly altered properties due to surface modifications, the behaviour and effects of NCs along the GIT are not fully understood and cannot be predicted simply from the chemical nature of cellulose. Regulatory bodies continue to raise concerns about the possible degradation/digestion of NC in the GIT, as it may release smaller particles that can cross the intestinal barrier and trigger pathological effects in organs and tissues. Given the relevance with regard to the potential biodegradation of BC in the GIT (and the lack of data available), the results obtained with other NC are here reviewed, together with the information available concerning BC.

Dietary fibre plays a vital role in maintaining gut microbiome homeostasis and diversity, which significantly affects human health (Makki et al. 2018). In the human gut, fermentation of dietary fibre generates energy and produces metabolites, such as short-chain fatty acids (SCFA), which are essential not only for colonic health, glucose and cholesterol homeostasis, and regulation of appetite (Blaak et al. 2020; den Besten et al. 2013; Wong et al. 2006), but also for digestive transit regulation, prevention of obesity and diabetes, and reduction of cardiovascular diseases and cancer (Makki et al. 2018; Robert and Bernalier-Donadille 2003).

As a key component of the plant cell wall, cellulose is an important source of insoluble, non-digestible dietary fibre in human nutrition, resisting digestion by gastric acid and digestive enzymes in the stomach and remaining intact upon reaching the small intestine (He et al. 2023). Once in the colon, digestion or degradation of cellulose by the human microbiome is very limited, in contrast to that in ruminants and other herbivores (Moraïs et al. 2024). Evidence of microcrystalline cellulose fermentability in the human colon was first reported in 1984 (Kelleher et al. 1984) when ^{14}C-labelled cellulose, orally administered to 10 healthy volunteers, was metabolized and detected as $^{14}CO_2$ in expired air. A subsequent study in 2003 (Robert and Bernalier-Donadille 2003) demonstrated cellulose degradation by a colonic cellulolytic microbial community in faecal samples of methane excretors. However, this process is generally rare or absent in the human gut, despite the presence of putative cellulolytic enzymes produced by various bacterial species in the colon (e.g. *Ruminococcus, Enterococcus, and Bacteroides*) (Chassard et al. 2010; Robert and Bernalier-Donadille 2003). Notably, some bacteria isolated from the human microbiome such as *Ruminococcus champanellensis* (Chassard et al. 2012) and *Bacteroides cellulosilyticus* sp. nov. (Robert et al. 2007) have been found to contain functional multi-enzymatic cellulosome structures, but the highly ordered nature of cellulose, high crystallinity, and high degree of polymerization, especially when considering BC, limits the accessibility of cellulosic fibres and glucosidic β-1,4 bonds for enzymatic breakdown (Thielemans et al. 2023). Nevertheless, studies on the potential digestion or degradation of NC by the human microbiome, which could result in smaller fibres, remain scarce, leaving this topic open for further exploration. To date, there have been few studies on the use of NC as a prebiotic and its effects on the microbiota in the human colon. In these studies, the authors improved plant cellulose fermentation in the human colon by changing some of its properties. Nsor-Atindana et al. (2020) improved the fermentability of microcrystalline cellulose by reducing its hydrodynamic size to the nanometric scale (125.01 ± 25.43 nm in length and 3.24 ± 0.75 nm in diameter) through ultrasonication. Although the safety of NCs remains a concern for EFSA, the study demonstrated that the resulting NC could provide beneficial prebiotic effects on the gut microbiota. This was evidenced by a significant increase in Short Chain Fatty Acids (SCFA) production (such as acetate, butyrate, and propionate) and a notable rise in *Bifidobacterium* counts. In a subsequent study, Thielemans et al. (2023) combined mechanical treatment and acid hydrolysis to produce amorphized and depolymerized cellulose with an average degree of polymerization below 100 anhydroglucose units and a crystallinity index under 30%. The modified cellulose demonstrated improved fermentability of human faecal microbiota. After 48 h, the minimal fermentability using pooled human faecal microbiota increased from 5 to 45%, accompanied by an eight-fold increase in SCFA production. The fermentability of modified NC has also been studied. An *in vitro* colonic fermentation study demonstrated that TEMPO-modified CNC and CNF exhibited higher fermentability than microcrystalline cellulose and carboxymethyl cellulose, resulting in increased total production of SCFAs. Furthermore, *Bacteroides cellulosilyticus* showed high specificity for utilizing TEMPO-modified CNC and CNF, as the relative abundance increased by at least dozens of folds after fermentation (He et al. 2022).

The effects of NCs on the gut microbiota and other health benefits have also been studied using animal models. Nsor-Atindana et al. (2020) demonstrated that oral administration of NC (size = 125 nm) to rats via gavage significantly increased SCFA yields and *Bifidobacterium* counts compared with both the control group and the group receiving microsized cellulose. In mice models of constipation (Wang et al. 2022) and ulcerative colitis (Wang et al. 2023), the administration of CNC significantly enhanced the richness and evenness of colonic microbe species. For constipation, a 5-day CNC treatment increased the relative abundance of seven bacterial species, including *Bacteroides*, which play key roles in promoting SCFA production and maintaining bile acid homeostasis (Wang et al. 2022). In ulcerative colitis, 7-day CNC treatment improved body weight, reduced colonic inflammation, decreased pro-inflammatory cytokines, and elevated anti-inflammatory cytokines in the serum. The treatment also promoted mucin secretion to restore the intestinal chemical barrier, while enhancing gut microbiota diversity and increasing the abundance of *Akkermansia* and *Odoribacter*, contributing to the restoration of the intestinal biological barrier (Wang et al. 2023). In contrast, 5 weeks of biweekly gavage treatment with water or cream, with or without 1% CNF resulted in changes in microbial diversity, reducing the populations of *Coprococcus catus* and *Bacteroides acidifaciens*, both of which are known for producing significant amounts of SCFA. However, no associated pathological changes were detected within the duration of this study (Khare et al. 2020). Nagano and Yano (2020) proposed that CNF had an inhibitory effect on obesity by controlling the gut microbiota balance. In fact, high-fat diet-fed mice that were provided with 0.2% CNF dispersions via drinking water for 7 weeks presented suppression in weight gain and a lower accumulation of epididymal and subcutaneous fat. In addition, the treatment decreased the abundance of *Streptococcaceae* and *Rikenellaceae*, which are obesity-related bacteria. Han et al. (2024) evaluated the potential of BC in preventing obesity. After 9 weeks of feeding mice a diet supplemented with 15% BC, significant health benefits were observed. More specifically, body and liver weights decreased, along with fat accumulation in the kidneys and epididymis. Additionally, glucose, total triglycerides, total cholesterol, high-density lipoprotein cholesterol, and low-density lipoprotein cholesterol levels were restored to control levels. BC consumption also enhances liver fat metabolism, improves intestinal function, and reduces inflammation in the ileum and epididymis. In another study, 30-days of BC-containing diet also significantly reduced serum triglycerides, total cholesterol, low-density lipoprotein cholesterol, total lipids, and cholesterol in the liver of hamsters. Furthermore, animals showed enhanced faecal excretion of lipids, cholesterol, and bile acids (Chau et al. 2008).

However, the behaviour and fate of NC in the colon remain largely unexplored in all of these studies. Consequently, the potential absorption of NC in the gastrointestinal tract, particularly in humans, represents a significant knowledge gap that affects their safety assessment. To the best of our knowledge, only one study has assessed the possible digestion or degradation of NC by the human microbiome, with the formation of (potentially more toxic) smaller particles. Recently, Vincentini et al. (2023) investigated the potential degradation of ultra-sonicated CNF (2.7 μm length × 5 nm diameter) and CNCs (176 nm length × 7 nm diameter) at a concentration of 2 g/L by human microbiome using the ARCOL (ARtificial COLon) system. In contrast to the above-described studies and aside from the impact on gut microbiota β-diversity, their findings suggest that no significant degradation occurred, as gas production, composition, and SCFA profiles remained unchanged following the treatments. Importantly, TEM analysis of the fermented samples revealed that their size and shape remained similar to those of the pristine materials. Consequently, despite some degree of uncertainty, the results suggest that no smaller NC particles capable of entering systemic circulation via the intestinal route were detected. Notwithstanding that, further studies are needed to fully understand the GIT fate and potential impact of the NCs on human health.

6 Conclusion

The assessment of the safety of BC in the food industry remains an unsolved matter even though it is being commonly consumed in different geographies as Nata de Coco. This puzzling situation remains a challenge for a number of reasons: (1) the parallel with other nanocelluloses (unlike BC, engineered nanomaterials) and the generalization of concerns to "nanocelluloses"; (2) the methodologies for the disaggregation and characterization of nanomaterials advised by EFSA are not appropriate for complex structures like BC; (3) a more comprehensive understanding of the organization of BC obtained in static culture is required. Most results in the literature support the assumption that BC is safe, but some controversial results suggest the opposite. Additional studies will eventually shed light on this matter.

References

Andrade DR, Mendonça MH, Helm CV, Magalhães WL, de Muniz GI, Kestur SG (2015) Assessment of nano cellulose from peach palm residue as potential food additive: part II: preliminary studies. J Food Sci Technol 52(9):5641–5650. https://doi.org/10.1007/s13197-014-1684-0

Babi M, Williams A, Reid M, Grandfield K, Bassim ND, Moran-Mirabal JM (2022) Unraveling the supramolecular structure and nanoscale dislocations of bacterial cellulose ribbons using correlative super-resolution light and electron microscopy. Biomacromolecules 24(1):258–268. https://doi.org/10.1021/acs.biomac.2c01108

Bher A, Mayekar PC, Auras RA, Schvezov CE (2022) Biodegradation of biodegradable polymers in mesophilic aerobic environments. Int J Mol Sci 23(20):12165. https://doi.org/10.3390/ijms232012165

Blaak EE, Canfora EE, Theis S, Frost G, Groen AK, Mithieux G, Nauta A, Scott K, Stahl B, van Harsselaar J, van Tol R, Vaughan EE, Verbeke K (2020) Short chain fatty acids in human gut and metabolic health. Benefic Microbes 11(5):411–455. https://doi.org/10.3920/BM2020.0057

Böhmert L, Girod M, Hansen U, Maul R, Knappe P, Niemann B, Weidner SM, Thünemann AF, Lampen A (2014) Analytically monitored digestion of silver nanoparticles and their toxicity on human intestinal cells. Nanotoxicology 8(6):631–642. https://doi.org/10.3109/17435390.2013.815284

Brand W, van Kesteren PCE, Swart E, Oomen AG (2022) Overview of potential adverse health effects of oral exposure to nanocellulose. Nanotoxicology 16(2):217–246. https://doi.org/10.1080/17435390.2022.2069057

Campos D, Goméz-García R, Oliveira D, Madureira AR (2022) Intake of nanoparticles and impact on gut microbiota: in vitro and animal models available for testing. Gut Microbiome 3:e1. https://doi.org/10.1017/gmb.2021.5

Cañas-Gutiérrez A, Gómez Hoyos C, Velásquez-Cock J, Gañán P, Triana O, Cogollo-Flórez J, Romero-Sáez M, Correa-Hincapié N, Zuluaga R (2024) Health and toxicological effects of nanocellulose when used as a food ingredient: a review. Carbohydr Polym 323:121382. https://doi.org/10.1016/j.carbpol.2023.121382

Carocho M, Morales P, Ferreira ICFR (2015) Natural food additives: quo vadis? Trends Food Sci Technol 45(2):284–295. https://doi.org/10.1016/j.tifs.2015.06.007

Chassard C, Delmas E, Robert C, Bernalier-Donadille A (2010) The cellulose-degrading microbial community of the human gut varies according to the presence or absence of methanogens. FEMS Microbiol Ecol 74(1):205–213. https://doi.org/10.1111/j.1574-6941.2010.00941.x

Chassard C, Delmas E, Robert C, Lawson PA, Bernalier-Donadille A (2012) Ruminococcus champanellensis sp. nov., a cellulose-degrading bacterium from human gut microbiota. Int J Syst Evol Microbiol 62(1):138–143. https://doi.org/10.1099/ijs.0.027375-0

Chau C-F, Yang P, Yu C-M, Yen G-C (2008) Investigation on the lipid- and cholesterol-lowering abilities of biocellulose. J Agric Food Chem 56(6):2291–2295. https://doi.org/10.1021/jf7035802

Chen H, Zhao R, Wang B, Cai C, Zheng L, Wang H, Wang M, Ouyang H, Zhou X, Chai Z, Zhao Y, Feng W (2017) The effects of orally administered Ag, TiO2 and SiO2 nanoparticles on gut microbiota composition and colitis induction in mice. NanoImpact 8:80–88. https://doi.org/10.1016/j.impact.2017.07.005

Chen Y, Lin Y-J, Nagy T, Kong F, Guo TL (2020) Subchronic exposure to cellulose nanofibrils induces nutritional risk by non-specifically reducing the intestinal absorption. Carbohydr Polym 229:115536. https://doi.org/10.1016/j.carbpol.2019.115536

Chumchoochart W, Chandet N, Saenjum C, Tinoi J (2023) Important role and properties of granular nanocellulose particles in an in vitro simulated gastrointestinal system and in lipid digestibility and permeability. Biomolecules 13(10):1479. https://www.mdpi.com/2218-273X/13/10/1479

DeLoid GM, Cao X, Molina RM, Silva DI, Bhattacharya K, Ng KW, Loo SCJ, Brain JD, Demokritou P (2019) Toxicological effects of ingested nanocellulose in in vitro intestinal epithelium and in vivo rat models. Environ Sci Nano 6(7):2105–2115. https://doi.org/10.1039/C9EN00184K

den Besten G, van Eunen K, Groen AK, Venema K, Reijngoud D-J, Bakker BM (2013) The role of short-chain fatty acids in the interplay between diet, gut microbiota, and host energy metabolism. J Lipid Res 54(9):2325–2340. https://doi.org/10.1194/jlr.R036012

Dourado F, Leal M, Martins D, Fontão A, Cristina Rodrigues A, Gama M (2016a) Celluloses as food ingredients/additives: is there a room for BNC? In: Gama M, Dourado F, Bielecki S (eds) Bacterial nanocellulose. Elsevier, pp 123–133

Dourado F, van den Berg C, Gama M (2016b) European regulatory framework on novel foods and novel food additives. In: Gama M, Bielecky S, Dourado F (eds) Bacterial nanocellulose: from biotechnology to bio-economy. Elsevier, pp 135–144. https://doi.org/10.1016/B978-0-444-63458-0.00008-1

Dourado F, Gama M, Rodrigues AC (2017) A review on the toxicology and dietetic role of bacterial cellulose. Toxicol Rep 4:543–553. https://doi.org/10.1016/j.toxrep.2017.09.005

Dufresne A (2013) Nanocellulose: a new ageless bionanomaterial. Mater Today 16:220–227. https://doi.org/10.1016/j.mattod.2013.06.004

EC (2002) Regulation (EC) no 178/2002 of the European Parliament and of the Council of 28 January 2002 laying down the general principles and requirements of food law, establishing the European Food Safety Authority and laying down procedures in matters of food safety. Off J Eur Union L31:1–24. https://eur-lex.europa.eu/eli/reg/2002/178/oj/eng

EC (2011) Commission recommendation of 18 October 2011 on the definition of nanomaterial (Text with EEA relevance) (2011/696/EU). Off J Eur Union L275:38–40. https://eur-lex.europa.eu/eli/reco/2011/696/oj/eng

EC (2015) Regulation (EU) 2015/2283 of the European Parliament and of the Council of 25 November 2015 on novel foods, amending Regulation (EU) No 1169/2011 of the European Parliament and of the Council and repealing Regulation (EC) No 258/97 of the European Parliament and of the Council and Commission Regulation (EC) No 1852/2001. Off J Eur Union 327:1–22

Ede JD, Ong KJ, Mulenos MR, Pradhan S, Gibb M, Sayes CM, Shatkin JA (2020) Physical, chemical, and toxicological characterization of sulfated cellulose nanocrystals for food-related applications using in vivo and in vitro strategies. Toxicol Res (Camb) 9(6):808–822. https://doi.org/10.1093/toxres/tfaa082

EFSA (2018) Guidance on risk assessment of the application of nanoscience and nanotechnologies in the food and feed chain: part 1, human and animal health. EFSA J 16(7):e05327. https://doi.org/10.2903/j.efsa.2018.5327

EFSA (2021a) Guidance on risk assessment of nanomaterials to be applied in the food and feed chain: human and animal health. EFSA J 19(8):e06768. https://doi.org/10.2903/j.efsa.2021.6768

EFSA (2021b) Guidance on technical requirements for regulated food and feed product applications to establish the presence of small particles including nanoparticles. EFSA J 19(8):e06769. https://doi.org/10.2903/j.efsa.2021.6769

EFSA (2021c) Outcome of the public consultation on the draft guidance on technical requirements for regulated food and feed product applications to establish the presence of small particles including nanoparticles. EFSA Support Publ 18(8):6804E. https://doi.org/10.2903/sp.efsa.2021.EN-6804

El-Kalliny AS, Abdel-Wahed MS, El-Zahhar AA, Hamza IA, Gad-Allah TA (2023) Nanomaterials: a review of emerging contaminants with potential health or environmental impact. Discov Nano 18(1):68. https://doi.org/10.1186/s11671-023-03787-8

Endes C, Camarero-Espinosa S, Mueller S, Foster EJ, Petri-Fink A, Rothen-Rutishauser B, Weder C, Clift MJD (2016) A critical review of the current knowledge regarding the biological impact of nanocellulose. J Nanobiotechnol 14(1):78. https://doi.org/10.1186/s12951-016-0230-9

FAO/WHO (2013) State of the art on the initiatives and activities relevant to risk assessment and risk management of nanotechnologies in the food and agriculture sectors. FAO/WHO

FDA (2014a) FDA's approach to regulation of nanotechnology products. U.S. Food and Drug Administration. https://www.fda.gov/science-research/nanotechnology-programs-fda/fdas-approach-regulation-nanotechnology-products?utm_source=chatgpt.com. Accessed 16 February

FDA (2014b) Guidance for industry: assessing the effects of significant manufacturing process changes, including emerging technologies, on the safety and regulatory status of food ingredients and food contact substances, including food ingredients that are color additives. Food and Drug Administration. https://www.fda.gov/regulatory-information/search-fda-guidance-documents/guidance-industry-assessing-effects-significant-manufacturing-process-changes-including-emerging

FDA (2024) Nanotechnology task force. U.S. Food and Drug Administration. https://www.fda.gov/science-research/nanotechnology-programs-fda/nanotechnology-task-force. Accessed 16 February

Fischer HC, Chan WCW (2007) Nanotoxicity: the growing need for in vivo study. Curr Opin Biotechnol 18(6):565–571. https://doi.org/10.1016/j.copbio.2007.11.008

Fröhlich E (2018) Comparison of conventional and advanced in vitro models in the toxicity testing of nanoparticles. Artif Cells Nanomed Biotechnol 46(sup2):1091–1107. https://doi.org/10.1080/21691401.2018.1479709

Frohlich E, Roblegg E (2012) Models for oral uptake of nanoparticles in consumer products. Toxicology 291(1–3):10–17

Ghebretatios M, Schaly S, Prakash S (2021) Nanoparticles in the food industry and their impact on human gut microbiome and diseases. Int J Mol Sci 22(4):1942. https://doi.org/10.3390/ijms22041942

Hagiwara A, Imai N, Sano M, Kawabe M, Tamano S, Kitamura S, Omoto T, Asai I, Yasuhara K, Hayashi SM (2010) A 28-day oral toxicity study of fermentation-derived cellulose, produced by Acetobacter aceti subspecies xylinum, in F344 rats. J Toxicol Sci 35(3):317–325. https://doi.org/10.2131/jts.35.317

Han Y-H, Cui X-W, Li Y-X, Chen X, Zhang H, Zhang Y, Wang S-S, Li M (2024) Bacterial cellulose is a desirable biological macromolecule that can prevent obesity via modulating lipid metabolism and gut microbiota. Int J Biol Macromol 283:137522. https://doi.org/10.1016/j.ijbiomac.2024.137522

Haripriyaa M, Suthindhiran K (2023) Pharmacokinetics of nanoparticles: current knowledge, future directions and its implications in drug delivery. Future J Pharm Sci 9(1):113. https://doi.org/10.1186/s43094-023-00569-y

He X, Sun C, Fang J, Wu C, Zhang Y, Zhang X, Fang Y (2022) In vitro colonic fermentation profiles and microbial responses of cellulose derivatives with different colloidal states. J Agric Food Chem 70(30):9509–9519. https://doi.org/10.1021/acs.jafc.2c01721

He H, Teng H, An F, Wang Y, Qiu R, Chen L, Song H (2023) Nanocelluloses review: preparation, biological properties, safety, and applications in the food field. Food Front 4(1):85–99. https://doi.org/10.1002/fft2.208

Howe SE, Lickteig DJ, Plunkett KN, Ryerse JS, Konjufca V (2014) The uptake of soluble and particulate antigens by epithelial cells in the mouse small intestine. PLoS One 9(1):e86656. https://doi.org/10.1371/journal.pone.0086656

Iguchi M, Yamanaka S, Budhiono A (2000) Bacterial cellulose—a masterpiece of nature's arts. J Mater Sci 35(2):261–270. http://www.springerlink.com/index/P0620LH2326V0774.pdf

Kelleher J, Walters MP, Srinivasan TR, Hart G, Findlay JM, Losowsky MS (1984) Degradation of cellulose within the gastrointestinal tract in man. Gut 25(8):811–815. https://doi.org/10.1136/gut.25.8.811

Khare S, DeLoid GM, Molina RM, Gokulan K, Couvillion SP, Bloodsworth KJ, Eder EK, Wong AR, Hoyt DW, Bramer LM, Metz TO, Thrall BD, Brain JD, Demokritou P (2020) Effects of ingested nanocellulose on intestinal microbiota and homeostasis in Wistar Han rats. NanoImpact 18:100216. https://doi.org/10.1016/j.impact.2020.100216

Klemm D, Heublein B, Fink H-P, Bohn A (2005) Cellulose: fascinating biopolymer and sustainable raw material. Angew Chem Int Ed 44(22):3358–3393. https://doi.org/10.1002/anie.200460587

Koshani R, Madadlou A (2018) A viewpoint on the gastrointestinal fate of cellulose nanocrystals. Trends Food Sci Technol 71:268–273. https://doi.org/10.1016/j.tifs.2017.10.023

Lamas B, Martins Breyner N, Houdeau E (2020) Impacts of foodborne inorganic nanoparticles on the gut microbiota-immune axis: potential consequences for host health. Part Fibre Toxicol 17(1):19. https://doi.org/10.1186/s12989-020-00349-z

Lamas B, Evariste L, Houdeau E (2023) Dysregulation along the gut microbiota-immune system axis after oral exposure to titanium dioxide nanoparticles: a possible environmental factor promoting obesity-related metabolic disorders. Environ Pollut 330:121795. https://doi.org/10.1016/j.envpol.2023.121795

Leist M, Hasiwa N, Rovida C, Daneshian M, Basketter D, Kimber I, Clewell H, Gocht T, Goldberg A, Busquet F, Rossi A-M, Schwarz M, Stephens M, Taalman R, Knudsen TB, McKim J, Harris G, Pamies D, Hartung T (2014) Consensus report on the future of animal-free systemic toxicity testing. ALTEX 31(3):341–356. https://doi.org/10.14573/altex.1406091

Li Z, Zhang Y, Anankanbil S, Guo Z (2021) Applications of nanocellulosic products in food: manufacturing processes, structural features and multifaceted functionalities. Trends Food Sci Technol 113:277–300. https://doi.org/10.1016/j.tifs.2021.03.027

Lin Y-J, Qin Z, Paton CM, Fox DM, Kong F (2021) Influence of cellulose nanocrystals (CNC) on permeation through intestinal monolayer and mucus model in vitro. Carbohydr Polym 263:117984. https://doi.org/10.1016/j.carbpol.2021.117984

Lopes VR, Strømme M, Ferraz N (2020) In vitro biological impact of nanocellulose fibers on human gut bacteria and gastrointestinal cells. Nanomaterials (Basel) 10(6):1159. https://doi.org/10.3390/nano10061159

Ma Y, Zhang J, Yu N, Shi J, Zhang Y, Chen Z, Jia G (2023) Effect of nanomaterials on gut microbiota. Toxics 11(4):384. https://doi.org/10.3390/toxics11040384

Macierzanka A, Mackie AR, Bajka BH, Rigby NM, Nau F, Dupont D (2014) Transport of particles in intestinal mucus under simulated infant and adult physiological conditions: impact of mucus structure and extracellular DNA. PLoS One 9(4):e95274. https://doi.org/10.1371/journal.pone.0095274

Mackie A, Gourcy S, Rigby N, Moffat J, Capron I, Bajka B (2019) The fate of cellulose nanocrystal stabilised emulsions after simulated gastrointestinal digestion and exposure to intestinal mucosa. Nanoscale 11:2991–2998. https://doi.org/10.1039/C8NR05860A

Makki K, Deehan EC, Walter J, Bäckhed F (2018) The impact of dietary fiber on gut microbiota in host health and disease. Cell Host Microbe 23(6):705–715. https://doi.org/10.1016/j.chom.2018.05.012

Martins D, de Carvalho Ferreira D, Gama M, Dourado F (2020a) Dry bacterial cellulose and carboxymethyl cellulose formulations with interfacial-active performance: processing conditions and redispersion. Cellulose 27:6505–6520

Martins D, Estevinho B, Rocha F, Dourado F, Gama M (2020b) A dry and fully dispersible bacterial cellulose formulation as a stabilizer for oil-in-water emulsions. Carbohydr Polym 230:115657. https://doi.org/10.1016/j.carbpol.2019.115657

Martins D, Khodamoradi N, Silva-Carvalho R, Gama M, Moradi M, Dourado F (2024) Effect of bacterial cellulose on the foaming properties of egg white and soy proteins. Carbohydr Polymer Technol Appl 7:100517. https://doi.org/10.1016/j.carpta.2024.100517

McClements DJ, Xiao H, Demokritou P (2017) Physicochemical and colloidal aspects of food matrix effects on gastrointestinal fate of ingested inorganic nanoparticles. Adv Colloid Interface Sci 246:165–180. https://doi.org/10.1016/j.cis.2017.05.010

Mejía-Jaramillo AM, Gómez-Hoyos C, Cañas Gutierrez AI, Correa-Hincapié N, Zuluaga Gallego R, Triana-Chávez O (2023) Tackling the cytotoxicity and genotoxicity of cellulose nanofibers from the banana rachis: a new food packaging alternative. Heliyon 9(11):e21560. https://doi.org/10.1016/j.heliyon.2023.e21560

Moraïs S, Winkler S, Zorea A, Levin L, Nagies FSP, Kapust N, Lamed E, Artan-Furman A, Bolam DN, Yadav MP, Bayer EA, Martin WF, Mizrahi I (2024) Cryptic diversity of cellulose-degrading gut bacteria in industrialized humans. Science 383(6688):eadj9223. https://doi.org/10.1126/science.adj9223

Moriyama A, Ogura I, Fujita K (2023) Potential issues specific to cytotoxicity tests of cellulose nanofibrils. J Appl Toxicol 43(1):195–207. https://doi.org/10.1002/jat.4390

Mortensen NP, Moreno Caffaro M, Davis K, Aravamudhan S, Sumner SJ, Fennell TR (2022) Investigation of eight cellulose nanomaterials' impact on differentiated Caco-2 monolayer integrity and cytotoxicity. Food Chem Toxicol 166:113204. https://doi.org/10.1016/j.fct.2022.113204

Mota R, Rodrigues AC, Silva-Carvalho R, Costa L, Martins D, Sampaio P, Dourado F, Gama M (2022) Tracking bacterial nanocellulose in animal tissues by fluorescence microscopy. Nanomaterials 12(15):2605. https://doi.org/10.3390/nano12152605

Mu Q, Kirby J, Reilly CM, Luo XM (2017) Leaky gut as a danger signal for autoimmune diseases. Front Immunol 8:598. https://doi.org/10.3389/fimmu.2017.00598

Müller M, Drexel R, Burkhart M, Dähnhardt-Pfeiffer S, Wien L, Herrmann C, Knoll T, Metzger C, Briesen H, Wagner S, Meier F, Kohl Y (2023) Ex vivo models for intestinal transloca-

tion studies of cellulose nanocrystals. In Vitro Model 2(5):181–194. https://doi.org/10.1007/s44164-023-00056-x
Nagano T, Yano H (2020) Dietary cellulose nanofiber modulates obesity and gut microbiota in high-fat-fed mice. Bioact Carbohydr Diet Fibre 22:100214. https://doi.org/10.1016/j.bcdf.2020.100214
Nel A, Xia T, Mädler L, Li N (2006) Toxic potential of materials at the nanolevel. Science 311(5761):622–627. https://doi.org/10.1126/science.1114397
Nile SH, Baskar V, Selvaraj D, Nile A, Xiao J, Kai G (2020) Nanotechnologies in food science: applications, recent trends, and future perspectives. Nanomicro Lett 12(1):45. https://doi.org/10.1007/s40820-020-0383-9
Nsor-Atindana J, Zhou YX, Saqib MN, Chen M, Douglas Goff H, Ma J, Zhong F (2020) Enhancing the prebiotic effect of cellulose biopolymer in the gut by physical structuring via particle size manipulation. Food Res Int 131:108935. https://doi.org/10.1016/j.foodres.2019.108935
Ong KJ, Ede JD, Pomeroy-Carter CA, Sayes CM, Mulenos MR, Shatkin JA (2020) A 90-day dietary study with fibrillated cellulose in Sprague-Dawley rats. Toxicol Rep 7:174–182. https://doi.org/10.1016/j.toxrep.2020.01.003
Onyeaka H, Passaretti P, Miri T, Al-Sharify ZT (2022) The safety of nanomaterials in food production and packaging. Curr Res Food Sci 5:763–774. https://doi.org/10.1016/j.crfs.2022.04.005
Pasupuleti VR (2022) Nanoscience and nanotechnology advances in food industry. In: Bhat R (ed) Future foods. Academic, pp 721–732. https://doi.org/10.1016/B978-0-323-91001-9.00011-6
Pinget G, Tan J, Janac B, Kaakoush NO, Angelatos AS, O'Sullivan J, Koay YC, Sierro F, Davis J, Divakarla SK, Khanal D, Moore RJ, Stanley D, Chrzanowski W, Macia L (2019) Impact of the food additive titanium dioxide (E171) on gut microbiota-host interaction. Front Nutr 6:57. https://doi.org/10.3389/fnut.2019.00057
Powell JJ, Faria N, Thomas-McKay E, Pele LC (2010) Origin and fate of dietary nanoparticles and microparticles in the gastrointestinal tract. J Autoimmun 34(3):J226–J233. https://doi.org/10.1016/j.jaut.2009.11.006
Robert C, Bernalier-Donadille A (2003) The cellulolytic microflora of the human colon: evidence of microcrystalline cellulose-degrading bacteria in methane-excreting subjects. FEMS Microbiol Ecol 46(1):81–89. https://doi.org/10.1016/s0168-6496(03)00207-1
Robert C, Chassard C, Lawson PA, Bernalier-Donadille A (2007) Bacteroides cellulosilyticus sp. nov., a cellulolytic bacterium from the human gut microbial community. Int J Syst Evol Microbiol 57(7):1516–1520. https://doi.org/10.1099/ijs.0.64998-0
Rodrigues AC, Costa L, Silva-Carvalho R, Mota R, Duarte-Silva S, Teixeira-Castro A, Lamas N, Oliveira GNP, Wan Y, Dourado F, Gama M (2023) Assessment of the gastrointestinal fate of bacterial nanocellulose and its toxicological effects after repeated-dose oral administration. Environ Sci Nano 10(3):781–799. https://doi.org/10.1039/D2EN01019D
Ross P, Mayer R, Benziman M (1991) Cellulose biosynthesis and function in bacteria. Microbiol Rev 55(1):35–58. http://www.ncbi.nlm.nih.gov/pubmed/2030672
Schmitt DF, Frankos VH, Westland J, Zoetis T (1991) Toxicologic evaluation of Cellulon™ fiber; genotoxicity, pyrogenicity, acute and subchronic toxicity. J Am Coll Toxicol 10(5):541–554. https://doi.org/10.3109/10915819109078651
Schoonjans R, Castenmiller J, Chaudhry Q, Cubadda F, Daskaleros T, Franz R, Gott D, Mast J, Mortensen A, Oomen AG, Rauscher H, Weigel S, Astuto MC, Cattaneo I, Barthelemy E, Rincon A, Tarazona J (2023) Regulatory safety assessment of nanoparticles for the food chain in Europe. Trends Food Sci Technol 134:98–111. https://doi.org/10.1016/j.tifs.2023.01.017
Silva-Carvalho R, Rodrigues PM, Martins D, Rodrigues AC, Sampaio P, Dourado F, Gonçalves C, Gama M (2024) Bacterial cellulose in vitro uptake by macrophages, epithelial cells, and a triculture model of the gastrointestinal tract. Biomacromolecules 25(10):6748–6761. https://doi.org/10.1021/acs.biomac.4c00972
Tagawa S, Tokuyasu K, Yamagishi K, Ike M, Amano Y, Mizuno M (2024) Characterization of hybrid nanofibrils composed of xyloglucan and disintegrated bacterial cellulose. Cellulose 31(4):2239–2249. https://doi.org/10.1007/s10570-023-05712-9

Takayama G, Kondo T (2023) Quantitative evaluation of fiber network structure–property relationships in bacterial cellulose hydrogels. Carbohydr Polym 321:121311. https://doi.org/10.1016/j.carbpol.2023.121311

Takeuchi MT, Kojima M, Luetzow M (2014) State of the art on the initiatives and activities relevant to risk assessment and risk management of nanotechnologies in the food and agriculture sectors. Food Res Int 64:976–981. https://doi.org/10.1016/j.foodres.2014.03.022

The European Parliament and the Council of the European Union (2015) Regulation (EU) 2015/2283 of the European Parliament and of the Council of 25 November 2015 on novel foods, amending Regulation (EU) No 1169/2011 of the European Parliament and of the Council and repealing Regulation (EC) No 258/97 of the European Parliament and of the Council and Commission Regulation (EC) No 1852/2001 (Text with EEA relevance) Text with EEA relevance. Off J Eur Union. https://eur-lex.europa.eu/legal-content/EN/TXT/?uri=CELEX%3A02015R2283-20210327

Thielemans K, De Bondt Y, Comer L, Raes J, Everaert N, Sels BF, Courtin CM (2023) Decreasing the crystallinity and degree of polymerization of cellulose increases its susceptibility to enzymatic hydrolysis and fermentation by colon microbiota. Foods 12(5):1100. https://www.mdpi.com/2304-8158/12/5/1100

Vincentini O, Blier A-L, Bogni A, Brun M, Cecchetti S, De Battistis F, Denis S, Etienne-Mesmin L, Ferraris F, Fumagalli FS, Hogeveen K, Iacoponi F, Raggi A, Siciliani L, Stanco D, Verleysen E, Fessard V, Mast J, Blanquet-Diot S et al (2023) EFSA project on the use of new approach methodologies (NAMs) for the hazard assessment of nanofibres. Lot 1, nanocellulose oral exposure: gastrointestinal digestion, nanofibres uptake and local effects. EFSA Support Publ 20(9):8258E. https://doi.org/10.2903/sp.efsa.2023.EN-8258

Vital N, Cardoso M, Kranendonk M, Silva MJ, Louro H (2025) Evaluation of the cyto- and genotoxicity of two types of cellulose nanomaterials using human intestinal cells and in vitro digestion simulation. Arch Toxicol 99(2):575–596. https://doi.org/10.1007/s00204-024-03911-2

Wang X, Chang CH, Jiang J, Liu Q, Liao YP, Lu J, Li L, Liu X, Kim J, Ahmed A, Nel AE, Xia T (2019) The crystallinity and aspect ratio of cellulose nanomaterials determine their pro-inflammatory and immune adjuvant effects in vitro and in vivo. Small 15(42):e1901642. https://doi.org/10.1002/smll.201901642

Wang M, Cha R, Hao W, Du R, Zhang P, Hu Y, Jiang X (2022) Nanocrystalline cellulose cures constipation via gut microbiota metabolism. ACS Nano 16(10):16481–16496. https://doi.org/10.1021/acsnano.2c05809

Wang M, Cha R, Hao W, Jiang X (2023) Nanocrystalline cellulose modulates dysregulated intestinal barriers in ulcerative colitis. ACS Nano 17(19):18965–18978. https://doi.org/10.1021/acsnano.3c04569

Wong JM, de Souza R, Kendall CW, Emam A, Jenkins DJ (2006) Colonic health: fermentation and short chain fatty acids. J Clin Gastroenterol 40(3):235–243. https://doi.org/10.1097/00004836-200603000-00015

Wüstenberg T (2014) Rheology of food hydrocolloids. In: Cellulose and cellulose derivatives in the food industry. Wiley-VCH Verlag GmbH & Co. KGaA, pp 69–90. https://doi.org/10.1002/9783527682935.ch02

Xue J, Blesso C, Luo Y (2023) A comprehensive review of nanoparticles for oral delivery in food: biological fate, evaluation models, and gut microbiota influences. Annu Rev Food Sci Technol 14:1–33. https://doi.org/10.1146/annurev-food-060721-025159

Younes M, Aggett P, Aguilar F, Crebelli R, Di Domenico A, Dusemund B, Filipič M, Jose Frutos M, Galtier P, Gott D, Gundert-Remy U, Georg Kuhnle G, Lambré C, Leblanc J-C, Lillegaard IT, Moldeus P, Mortensen A, Oskarsson A, Stankovic I, … Woutersen RA (2018) Re-evaluation of celluloses E 460(i), E 460(ii), E 461, E 462, E 463, E 464, E 465, E 466, E 468 and E 469 as food additives (1831-4732). EFSA J. https://efsa.onlinelibrary.wiley.com/doi/abs/10.2903/j.efsa.2018.5047

Chapter 9
Techno-economic Analysis and Life Cycle Assessment of Bacterial Cellulose Production: A Comparative Study with Plant-Derived Micro- and Nanocelluloses

Ana Cristina Rodrigues, Francisco Garrett, Miguel Gama, and Fernando Dourado

Abstract The transition to sustainable materials is crucial for addressing global environmental challenges. Bacterial cellulose (BC), a high-purity biopolymer with unique properties, has emerged as a promising alternative to plant-derived cellulose in various applications. However, its widespread industrial adoption is limited, mainly for economic reasons. This chapter presents a comprehensive techno-economic analysis (TEA) and revises the life cycle assessment (LCA) of BC production, comparing its feasibility, cost, and sustainability with those of plant-derived micro- and nanocellulose (PMNC).

A process model using SuperPro Designer software was used to evaluate the economic viability of an industrial-scale BC production facility (264 ton/year—dry mass). Despite the high initial capital investment (€21.5 million) and energy-intensive fermentation and purification processes, economic feasibility can be achieved within a 4–5 year payback period. A market analysis reveals that BC, with an estimated price of €41/kg, may have a challenging market position against microcrystalline cellulose (MCC) and cellulose nanocrystals (CNC), advising strategic positioning in high-value applications such as biomedical materials, cosmetics, and advanced composites.

The comparative LCA, using standardized methodologies, highlighted key environmental trade-offs. While BC production shows lower environmental impacts than other nanocelluloses in some categories, its use in downstream products, such as textile fibers, significantly increases its environmental footprint owing to the high energy demand and use of resource-intensive chemicals (e.g., sodium phosphate

A. C. Rodrigues · F. Garrett · M. Gama · F. Dourado (✉)
CEB—Centre of Biological Engineering, University of Minho, Braga, Portugal

LABBELS—Associate Laboratory in Biotechnology and Bioengineering and Microelectromechanical Systems, Guimarães, Portugal
e-mail: fmgama@deb.uminho.pt; fdourado@ceb.uminho.pt

F. Dourado et al. (eds.), *Advances in Bacterial Cellulose*,
https://doi.org/10.1007/978-3-032-07042-5_9

and sodium hydroxide), as compared to the available cellulose sources (cotton, regenerated from forest biomass). Sensitivity analyses suggested that substituting conventional culture medium with agro-industrial byproducts could mitigate these impacts by over 60%, emphasizing the potential of circular economy approaches for improving BC sustainability.

This chapter underscores the need for continued process optimization, regulatory incentives, and strategic market alignment to enhance the economic and environmental viability of BC production. By addressing production bottlenecks and leveraging sustainable feedstocks, BC could contribute to the transition toward a bio-based economy.

Keywords Bacterial nanocellulose · Techno-economic analysis · Life cycle assessment · Sustainability · Circular bioeconomy · Market analysis

1 Introduction

1.1 Global Challenge: The Urgent Need for Sustainable Materials

Plastics are ubiquitous in our daily lives and are increasingly associated with environmental damage. They are responsible for pollution, global warming, and escalating consumption of non-renewable resources. Thus, a global challenge has emerged: the urgent need to transition from petrochemical-based to sustainable materials (Borrelle et al. 2020; Rosenboom et al. 2022). The United Nations has estimated that over 300 million tons of plastic waste are generated annually worldwide. Much of this waste enters marine and ecological habitats and can persist for centuries (Lau et al. 2020; MacLeod et al. 2021; United Nations 2018). Unless substantial changes are made in the usage of plastics, worldwide plastic production is predicted to double by 2050, further intensifying the ongoing environmental and public health crisis (Borrelle et al. 2020). These interconnected and global crises call for the development of novel, renewable, and sustainable materials aligned with circular economy principles (The European Environmental Bureau 2016; Karan et al. 2019). In response, governments and industries worldwide are crafting policies, such as the European Green Deal (The EU Green Deal: 2024 Edition 2024) and the Extended Producer Responsibility (EPR) (OECD 2024) to foster the transition to sustainable materials.

Cellulose, traditionally regarded as a common material, is experiencing a resurgence of attention owing to its promise as a sustainable substitute for synthetic compounds, especially in the form of micro- and nanocellulose (PMNC), which enables the development of strong and lightweight materials, for instance, by additive manufacturing (3D printing), advanced composite fabrication, and the formulation of protective surface treatments (da Cruz et al. 2024; Habibi et al. 2010; Trache et al. 2020). Moreover, PMNC can be obtained from a readily available supply chain and well-established extraction methods, significantly enhancing its potential for cost-effective, large-scale commercialization (Penloglou et al. 2023; Sharma et al. 2019; Trache et al. 2020; Vanhatalo et al. 2014). Conversely, bacterial cellulose (BC), synthesized through bacterial fermentation, presents a distinct profile of unique properties. Notably, BC is prized for its notable mechanical strength, nanofibrillar morphology, and high degree of purity. BC often exhibits crystallinity indices surpassing 80%, a value considerably greater than that observed in many plant-sourced counterparts (Cazón and Vázquez 2021; R et al. 2021). Furthermore, its remarkable hygroscopic capacity—retaining up to 99% of its weight in water—coupled with its impressive tensile properties and biocompatibility, renders it particularly well-suited for specialized applications such as wound care, bioengineered skin substitutes, and the development of flexible electronic components among others (Chandana et al. 2022; Taokaew 2024). While both PMNC and BC represent substantial progress in green chemistry, their successful market penetration will depend on the capacity to overcome economic and environmental challenges, as discussed ahead.

1.2 The Central Debate: Balancing Economic Viability and Sustainability

1.2.1 Economic Competitiveness and Environmental Impacts

In today's economy, where financial considerations often dictate the viability of emerging technologies, achieving cost-effective cellulose production has become increasingly important. Despite its exceptional properties, BC continues to face challenges in its industrialization, largely due to its high production costs. Being mainly reliant on microbial fermentation, it raises doubts regarding its economic viability given its high investment and operating costs (Behera et al. 2021; Dourado et al. 2016, 2018; Jozala et al. 2016; Kamal et al. 2022). However, recent breakthroughs in biotechnology may be capable of reshaping the landscape for BC production. Researchers are actively investigating approaches such as genetic engineering to enhance cellulose yields (Florea et al. 2016; Montenegro-Silva et al. 2024; R et al. 2021; Singh et al. 2020), as well as the use of low-cost and readily available feedstocks including fruit residues, molasses, and glycerol derived from biodiesel production (Alabbosh et al. 2021; Cazón and Vázquez 2021; Rodrigues et al. 2019; Wang et al. 2019).

In contrast, plant-based micro- and nanocellulose (PMNC) draws upon decades of industrial expertise within sectors such as forestry, agriculture, and the pulp and paper industry, where scalable processing methods are already deeply embedded. Moreover, recent advancements in PMNC extraction technologies have significantly enhanced both efficiency and yield, further lowering the barriers to commercialization, (Das et al. 2024; Hu et al. 2024). As a result, PMNC is gaining increasingly interest as a more cost-effective alternative, especially in high-volume applications such as food packaging, textile manufacturing, and various industrial sectors (Klemm et al. 2011, 2018; Sharma et al. 2019).

A key consideration when evaluating the economic viability of both types of cellulose is their equally important environmental implications. Both materials are biodegradable and derived from renewable sources, positioning them as compelling alternatives to petroleum-based polymers. Nonetheless, their environmental profiles diverge in meaningful ways as will be shown further ahead. For instance, BC production requires nutrient input during biosynthesis and is an energy-intensive process, whereas PMNC depends on biomass primarily sourced from agriculture and forestry. While BC entails a high upfront cost and high operating costs (Dourado et al. 2018), its superior properties may reduce the need for supplementary physicochemical or mechanical treatments, or reinforcing agents, potentially offsetting its financial burden. Conversely, despite the equally renewable PMNC, concerns have been raised regarding its long-term sustainability (Qureshi et al. 2024; Vanhatalo et al. 2014).

To date, a literature survey shows a significant gap regarding a systematic comparison of the techno-economic performance and life cycle impacts of BC and PMNCs. This chapter aims to fill such gap by providing a detailed techno-economic and environmental assessment comparison of the production of BC versus plant cellulose. Through this, the chapter hopes to help guiding academics and industry stakeholders on how to develop a green transition toward the circular bioeconomy in terms of more sustainable production and consumption/use of these biobased materials.

2 Process Modeling and Techno-economic Analysis of BC Production

2.1 Process Modeling

Drawing on previous research (Dourado et al. 2016, 2018), the BC production process (Table 9.1) was modeled using SuperPro Designer software (version 10, Windows OS) to facilitate the techno-economic analysis. In many fermentation processes, the cost and availability of substrates play crucial roles in determining the economic feasibility of the process. Consequently, there has been an increased focus on using agricultural and industrial waste as nutrient sources to reduce production costs (Quijano et al. 2024; Velásquez-Riaño and Bojacá 2017; Zhang et al. 2020). Thus, a culture medium (CM) containing molasses (a byproduct of the sugar-making process) as the carbon source and corn steep liquor (a byproduct of the corn wet-milling process) as the protein source was used in this simulation (Table 9.1). The production of BC, using a commercial strain ATCC 700178 (BPR2001) under static culture conditions, was optimized in a previous study (Rodrigues et al. 2019).

Table 9.1 Composition and estimated purchase prices of raw materials for BC production and washing, based on (Rodrigues et al. 2019)

	Density kg/L	Culture medium g/L	Price USD$/ton[a]
BC fermentation (upstream)			
Corn step liquor	1.2	7.00	20.0
Molasses	1.5	4.00	185.0
Ammonium sulfate	1.8	5.00	115.0
Disodium phosphate		2.70	380.0
Ethanol	0.8	1.5% (vol/vol)	940.0
Citric acid	1.7	1.26	520.0
Water	1.0		2.50
BC washing (downstream)			
NaOH	2.1	4.00	350.0

[a] An average purchase price was calculated based on values from different suppliers

The assumption here was that a similar BC yield (8.5 g/L, dry mass) could be achieved using an air-lift fermenter (ALF) operating for 72 h (Chao et al. 2001; Cheng et al. 2002; Wu and Li 2015). This corresponds to a BC productivity of 0.118 g/L/h, which is within the range reported in the literature (Dourado et al. 2018).

The modeled BC production facility was projected to produce 264 tons/year of BC (dry mass), assuming a product loss of 15% (dry basis) during downstream operations (further details are provided in Table 9.3. For practicality, a single entry containing a mixture of CM components was fed to a mixing tank with a volume of 100 m^3, which was sufficient to prepare enough CM for all fermenters. Pasteurized CM was then fed sequentially into each seed bioreactor and ALF (Table 9.3).

Inoculum propagation is usually achieved by successive propagations of biomass in fresh CM at a ratio of 1:10 (biomass:CM), which was also used in this study. Hence, three seed bioreactors, each operating for 12 h, with increasing volumes (100 L, 1000 L, and 10,000 L) were used for sequential biomass growth before inoculating each ALF. For simplicity, propagations below 100 L were omitted from the design, as they can easily be performed at laboratory scale. Fermentation occurred in a 100 m^3 ALF. Considering that each fermentation lasts for 72 h, to ferment daily 100 m^3, three ALF were used, each starting 1 day apart. A temperature of 30 °C was used for all fermentations and inoculum propagation. Upon fermentation, the resulting BC was collected, centrifuged, and washed with NaOH and water to remove bacteria and remnants of the culture medium. For extensive BC washing, typically—5 to 6 times the total volume of fermentation is used (unpublished data). The purified BC was then centrifuged, pasteurized, and packed in plastic bags and cardboard boxes. The packed BC corresponds to a wet product containing 20% (m/m) BC. This is a highly compressed cake.

Despite these nutrient-dense alternative substrates allowing high BC yields following fermentation, the wastewater (WW) generated from BC washing contains a very high organic load, which must be treated to comply with environmental regulations and local legal restrictions prior to its disposal. Therefore, integrated pretreatment systems or dedicated wastewater treatment plants are essential (da Silva et al. 2020; Rodrigues et al. 2025). Even considering initial investment and operating expenses, for large-scale, *in situ* treatment facilities (WWT) may provide long-term advantages, such as lower subcontracting expenses and opportunities for water recycling and energy production. Anaerobic digestion (AD) is an established method for treating wastewater on an industrial scale, decreasing the organic waste load, and generating biogas, a form of renewable energy (da Silva et al. 2020; Li et al. 2019; Mao et al. 2015). Anaerobic digestion is effective for managing high-organic-load waste and solid organic materials, such as industrial sludge, food scraps, forestry residues, and animal manure (Appels et al. 2008; Batstone and Virdis 2014; Mata-Alvarez et al. 2000; Wilkie 2005). In this simulation, a WWT facility using AD was included to process the high organic load of the waste streams.

Finally, a total of 39 people were estimated to be needed, distributed across different areas, from non-productive ones (management, finance, marketing, and others) to productive ones (laboratory, inoculum preparation, fermentation, downstream processing, warehouse, maintenance, and WWT facility). A typical work shift was defined as 8 h/day. For simplicity, the facility was operated for 365 days/year (more details are provided in Table 9.3).

2.2 *Techno-Economic Analysis (TEA)*

Table 9.2 offers a detailed breakdown of the estimated CAPEX associated with the different stages of the BC production process, highlighting the areas where most investments are concentrated. The total CAPEX for the modeled BC production facility amounts to 21.5 M€. The air-lift fermentation (ALF) system (where three ALF were used) represents the largest single investment, accounting for 6.2 M€ (29% of total CAPEX). The choice of ALFs, while potentially offering technological advantages (mixing and oxygen transfer), comes with high upfront costs. The anaerobic wastewater treatment plant (WWTP) is the second-largest expense, costing 6 M€ (28% of CAPEX) because of the very high effluent generated during BC

Table 9.2 Summary of the Capital Expenses (CAPEX) of the modeled BC production process

Unit operations	CAPEX (M€)	%
Storage of raw materials and finished goods	0.5	3
CM preparation and inoculum propagation	1.4	6
Fermentation	6.2	29
BC downstream processing	2.2	10
CIP unit, boiler and chiller rooms, compressed air production	1.1	5
Wastewater Treatment Plant (WWTP)	6.0	28
Building construction	4.2	19
Total without construction	17.3	
Total	**21.5**	

CM Culture medium

washing. The construction of the production facility accounted for 4 M€ (19% of CAPEX). This includes the costs of land, buildings, accessory infrastructure, and building projects.

Downstream processing includes discharge, grinding, filtration washing, and packaging.

Equipment prices were obtained from quotations provided by multiple manufacturers and suppliers; the facility design and cost estimations, including installation and auxiliary systems, were developed by a specialized engineering firm with expertise in bioprocess facilities.

Table 9.3 presents a comprehensive breakdown of the various costs, namely the operating costs, revenue projections, and financial performance metrics over a 5-year period (Year 0 to Year 4). Year zero corresponds to the year in which the CAPEX investment was made, and the production facility was built. Then, a projected gradual ramp-up in production capacity, starting at 60% in Year 1, increasing to full capacity (100%) by Year 3, and maintaining this level in Year 4 was modeled. The initial years often involve process stabilization, market penetration, and operational efficiency.

Table 9.3 Capital expenditure breakdown of the modeled BC production process

Assumptions	**Year 0**	**Year 1**	**Year 2**	**Year 3**	**Year 4**
Production days		365	365	365	365
Shifts (8 h)		3	3	3	3
% of full capacity		*60%*	*80%*	*100%*	*100%*
Production capacity (ton/year)[a]		158	211	264	264
Capex (M€)	21.5				
Property and Plant (M€)	*0.4*				
Selling price (€/kg)		**41.0**	**41.0**	**41.0**	**41.0**
Direct cost (€/kg) Culture medium and washing		4.28	5.71	7.13	7.13
Electricity (k€/year)		314	419	524	524
Personnel (number)		23	31	39	39
Mean cost per employee (€/year)		*21,449*	*21,449*	*21,449*	*21,449*
SG&A (% of turnover)		30%	25%	25%	20%
Depreciations (straight line)					
Plant	*50 years*	2%	2%	2%	2%
Equipment	*10 years*	10%	10%	10%	10%
Maintenance costs (% of fixed assets)		1%	1%	1%	1%
P&L statement (M€/year)		**Year 1**	**Year 2**	**Year 3**	**Year 4**
Turnover (gross revenue)		6.7	8.7	10.8	10.8
Cost of Goods (COGS)		0.7	1.2	1.9	1.9
Personnel		0.5	0.7	0.8	0.8
Electricity		0.3	0.4	0.5	0.5
Maintenance costs		0.2	0.2	0.2	0.2

(continued)

Table 9.3 (continued)

SG&A		2.0	2.2	2.7	2.2
EBITDA		2.8	4.0	4.7	5.2
Depreciations		1.8	1.8	1.8	1.8
EBIT		1.0	2.2	2.8	3.4
EBITDA margin		43.7%	46.0%	43.0%	48.0%
EBIT margin		15.7%	25.1%	26.3%	31.3%
Tax rate		22.5%	22.5%	22.5%	22.5%
Working capital (% sales)		10.0%	10.0%	10.0%	10.0%
Discount rate	16.0%				
Terminal value—EBITDA multiple	5.0x				
Free Cash Flow Statement (M€)	**Year 0**	**Year 1**	**Year 2**	**Year 3**	**Year 4**
EBIT		1.0	2.2	2.8	3.4
Tax		0.3	0.5	0.6	0.8
D&A		1.8	1.8	1.8	1.8
Net working capital		0.6	0.2	0.2	
Terminal value					26.0
Free cash flow	−21.5	2.0	3.3	3.8	30.4
Net Present Value	**1.9**				
MSP (NPV = 0)	**38.89**				

D&A Depreciation and Amortization Expenses, *EBIT* Earnings Before Interest and Taxes, *EBITDA* Earnings Before Interest, Taxes, Depreciation, and Amortization, *P&L* Profit and Loss, *SG&A* Selling, General, and Administrative expenses, *MSP* Minimum Selling Price (the price at which the net present value (NPV) of a company or project is zero)
[a] Dry mass

The nominal production capacity was 722.5 kg/day on a dry basis (corresponding to 3.6 ton BC cake at approximately 80% moisture). To simplify the financial model, a constant selling price of €41/kg (dry basis) was assumed throughout the projected period, emulating a stable market demand and pricing environment for BC or fixed contractual agreements.

Direct costs, also known as prime or production costs, are expenses that can be directly attributed to the production of goods or services. Because these costs, as well as personnel and electricity costs, are directly proportional to production levels, they also increase with production capacity. The costs of the culture medium and BC washing were based on the values listed in Table 9.1. The cost of washing the BC represented 31% of the total direct cost. Salary prices vary according to the specific role of human resources. To simplify the model, a mean salary was set for all 39 workers. Electricity costs were calculated based on the specific equipment data (Fig. 9.1).

Selling, General and Administrative (SG&A) expenses (indirect costs) represent the costs associated with the day-to-day operations of a business that are not directly tied to the production of goods or services. These expenses include sales team salaries, marketing and advertising costs, office supplies, administrative salaries, and other overheads. When expressed as a percentage of turnover (sales), SG&A

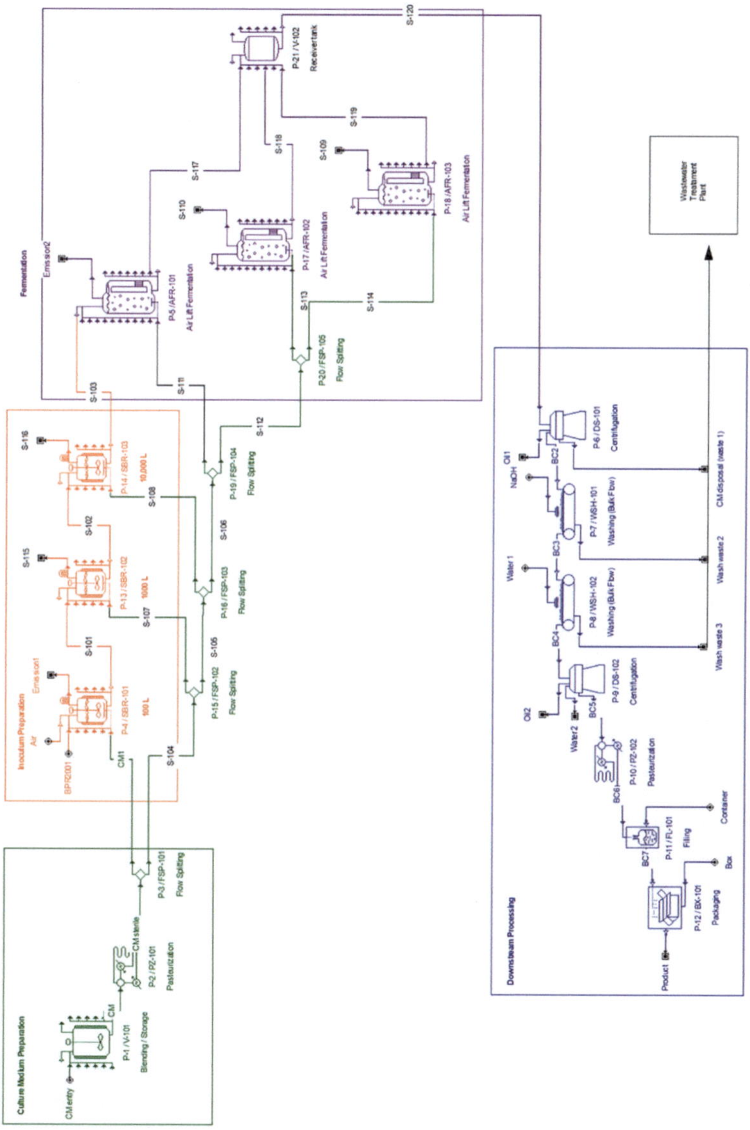

Fig. 9.1 SuperPro process flowsheet for the fermentative production of bacterial cellulose

expenses provide insights into the efficiency of a company's operational management compared to its revenues. For a biotechnology production process, the expected range of SG&A expenses as a percentage of turnover varies greatly depending on the stage of the company (e.g., startup, growth, or mature phase), scale of operations, and specific market dynamics. In the biotechnology and pharmaceutical industries, SG&A expenses often range between 20 and 40% of turnover. Startups and early stage biotech firms often have higher SG&A expenses as a percentage of turnover due to significant investments in marketing, salesforce development, and administrative infrastructure to establish a market presence (Angelis et al. 2023; Lowe 2023). Thus, the SG&A was set at 30% for the first year of BC production, followed by a 20% decrease in the last year.

Depreciation concerns the allocation of the cost of tangible assets over their useful lives. The straightline method is one of the most used depreciation methods, owing to its simplicity and consistency. This method spreads the cost of an asset evenly over its useful life, resulting in equal annual depreciation expenses (Green et al. 2002; Kırlı 2018). For plant depreciation, a period of 50 years is often chosen to reflect the long-term nature of these assets. This includes infrastructure, such as production facilities, piping systems, and structural components (Ben-Shahar et al. 2009). Equipment depreciation is often set for a 10-year period. This category includes fermenters, bioreactors, filtration systems, and other processing units. Unlike buildings, bioprocessing equipment is subject to wear and technological obsolescence, needing more frequent upgrades or replacements (Erumban 2008).

Maintenance costs represent the periodic expenses required to preserve the operational condition of fixed assets, including repairs, servicing, and preventive maintenance. The maintenance cost was set at 1% of the fixed assets, which is a standard assumption in capital-intensive industries (Thomas 2018).

The free cash flow statement provides important insights into the financial health and feasibility of the simulated BC production process. By analyzing the free cash flow over the years, it is possible to assess whether the project generates sufficient cash to cover operating expenses, reinvestment needs, and returns for investors. The project moves from a heavy cash outflow in Year 0 to a positive free cash flow from Year 1 onwards, indicating that the business can sustain itself after the initial investment. Given the positive flow, the payback period (time taken to recover the initial investment) can be estimated to be within—4 to 5 years. From these observations and the positive Net Present Value, the simulated BC production process appears to be economically viable if sold at a market price of 41€/kg (dry basis).

2.3 Market Competitiveness of BC: A Comparative Analysis with Plant Celluloses

The market positioning of BC compared to other forms of cellulose, such as microcrystalline cellulose (MCC) and nanocellulose (NC), is a critical factor in determining its economic viability and potential for widespread adoption. This price point of

41€/kg for BC warrants a comparison with the market prices of other cellulose derivatives to assess BC's competitiveness. Importantly, this discussion is restricted to putative price markets and does not consider the well-reported superior properties of BC compared with plant cellulose. Its greater technological role may influence the preference for BC over other celluloses, potentially justifying its higher price in certain market segments, as described below. Indeed, while BC and PMNC may address some similar markets (e.g., composites), BC has much better performance, such as in the cases of biomedicine, cosmetics and potentially foods, given the much larger aspect ratio, purity, and properties and presentation (e.g., the unique membranes obtained by static culture).

MCC is a well-established product, and its market price (Table 9.4) varies depending on its grade and application. Standard grades are typically priced between 4 and 5€/kg, whereas premium grades, which offer higher purity and performance, are priced between 6 and 7€/kg (Ioelovich 2022). A techno-economic analysis conducted by Vanhatalo et al. (2014) reported that the price of MCC ranged within 4–10 €/air-dry kg, depending on the particle size distribution and purity. A review by Ioelovich (2022) found that MCC typically sells for about 5.7–6.7 €/kg in the USA and Europe. Since BC is more expensive to produce, it is better suited for specialized, high-value markets like premium cosmetics, advanced wound care products and complex biomedical implants.

Within the nanocellulose category, cellulose nanocrystals (CNC) and cellulose nanofibrils (CNF) tend to have higher price tags due to their unique properties and the cutting-edge applications they support (Table 9.4). For instance, a study by Bondancia et al. (2020) on producing CNF and CNC from eucalyptus kraft pulp, calculated an average Minimum Selling Price (MSP) of 17.5 €/kg for CNC and 12.2 €/kg for CNF. Another analysis by Rosales-Calderon et al. (2021), using different acids for hydrolysis, estimated an MSP of 96 €/kg for dry CNC. Meanwhile, Yuzawa (2017) reported an MSP of 42 €/kg for dry CNF. In a 2017 study, the cost of TEMPO-oxidized CNF was estimated at 28 €/kg, considering only energy costs and minimal consumption of chemicals (Serra et al. 2017). In a market projection study, Cowie et al. (2014) stated that nanocellulose can reach a price range of 4.4–11 €/kg. Although presented here, these are unrealistically low values,

Table 9.4 Price overview (dry mass) of cellulose and cellulose derivatives

Type of cellulose		Price type	Price (€/kg)	References
MCC		Market price	€4.0–€10.0	Vanhatalo et al. (2014)
			€4.0–€7.0	Ioelovich (2022)
CNF			€10.0–€10,000.0	
MCC		Selling price	€1.6	Vanhatalo et al. (2014)
CNC	Acid hydrolysis	MSP	€9.1	Rosales-Calderon et al. (2021)
	With acid recovery		€4.7	Rajendran et al. (2023)
	Without acid recovery		€4.9	

An approximate exchange rate of 1 USD = 0.91 EUR was considered
MSP Minimum Selling Price

considering the high production costs related to the high acid consumption at the hydrolysis stage and the increased energy consumption required for hydrolysis, the disintegration of nanocrystalline aggregates, concentration by evaporation, freeze-drying, etc. For instance, the costs of the hydrolysis and evaporation steps were estimated to be 20–30 €/kg. Considering the full expenses, including disintegration, salary, equipment depreciation, overhead costs, and others, the actual production cost of CNC can reach 29–39 €/kg (Ioelovich 2022). More recently, the cost of nanocelluloses such as CNC was estimated to be above 95 €/kg (Arumugam and Gao 2022), a much higher value than that calculated for BC. Recent studies have highlighted the potential for reducing the cost of nanocellulose production through technological advancements and process optimization. For instance, waste-free technology has shown promise in reducing the production costs of MCC to between 3.50 and 4.00 €/Kg (Ioelovich 2022). Similarly, the development of more efficient hydrolysis processes and the recovery of acids and other chemicals could lower the production costs of CNC and, thus, their selling prices (Rajendran et al. 2023).

From the above, the BC market positioning at 41 €/Kg places it at a premium compared to the MCC. Although BC is pricier than cellulose nanocrystals (CNC) and cellulose nanofibrils (CNF), its unique qualities and potential for high-value applications could justify the extra cost. However, BC faces rough competition due to ongoing improvements in cost-saving techniques and advanced nanocellulose production methods.

Table 9.4 provides a price overview of cellulose and its derivatives on a dry basis.

3 Comparative Life Cycle Assessment

3.1 Introduction to Life Cycle Assessment (LCA)

Life Cycle Assessment (LCA) is a methodology that provides a systematic and rigorous approach to evaluate the environmental implications of a product throughout its entire existence, or life cycle, which encompasses all stages from the acquisition of raw materials to its fate, whether through disposal or recycling. This approach allows for the identification of the important hotspots of environmental concern, as well as areas where improvements can be made (Curran 2013; Finnveden et al. 2009; Jacquemin et al. 2012). The consistency and credibility of the LCA studies is supported by a robust methodological framework which is maintained by internationally accepted standards. Specifically, the International Organization for Standardization (ISO) has established two foundational standards ISO 14040 and ISO 14044 that collectively, ensure the consistency, reproducibility, and scientific rigor in LCA studies, across different sectors and regions (ISO14040:2006 2006;

ISO14044:2006 2006). ISO 14040 outlines the fundamental principles and framework of the study, defines key concepts such as goal and scope definition, the Life Cycle Inventory (LCI), the Life Cycle Impact Assessment (LCIA), and their interpretation. ISO 14044 expands on these concepts by further detailing procedural guidelines for executing each phase of the LCA assessment (ISO14040:2006 2006; ISO14044:2006 2006). In practice, conducting an LCA study requires the use of specialized software (e.g., SigmaPro, openLCA, GaBi, and others) with access to well-established LCI databases, containing extensive information on raw materials, their extraction processes, energy consumption, manufacturing processes, transportation, environmental emissions, and end-of-life treatments (Curran 2013; Finnveden et al. 2009; Jacquemin et al. 2012). Among the several available LCI databases, the Ecoinvent (https://ecoinvent.org/), maintained by the Swiss Centre for Life Cycle Inventories, currently contains over 20,000 datasets covering a wide range of sectors, from energy generation, fuel production, chemicals and plastics manufacturing, battery technologies, forestry, pulp and paper industries, and many others.

The LCA methodology is structured into four sequential and interconnected phases:

Phase one, the Goal and Scope Definition, concerns with defining the goal or purpose of the study, its context and boundaries of the system. This primary stage establishes the basis for the entire study. The second stage, the Life Cycle Inventory (LCI), concerns with a systematic assessment of all relevant mass and energy inputs (consumptions), including the raw materials' extraction and outputs (including all types of emissions, wastes, and disposals) within the system boundaries. This step generates a detailed inventory flow throughout the system. The third stage, the Life Cycle Impact Assessment (LCIA), involves translating quantified inventory data into qualitative and quantitative indicators of environmental impacts, such as global warming, ozone depletion, eutrophication, among others, depending on the type of framework being used (Curran 2013; Finnveden et al. 2009; Jacquemin et al. 2012). A widely adopted method for conducting an LCIA is the ReCiPe 2016 framework (Fig. 9.2). ReCiPe was developed by two Dutch institutions, RIVM (Rijksinstituut voor Volksgezondheid en Milieu) and Radboud University to be used within the LCA software. It provides both midpoint indicator (specific environmental issues) and endpoint indicators (which are broader impacts, illustrating the broader consequences of a certain process for instance in human health, ecosystems, or resource availability) (Dekker et al. 2020; Huijbregts et al. 2017). The fourth and final phase concerns with merging the findings from the LCI and LCIA stages into specific actionable insights. This involves an exhaustive evaluation of the gathered data considering the study's original goal, performing sensitivity analyses and drawing strategic decisions to improve the system's environmental sustainability, reduce risks, perform product redesign, among other possibilities (Curran 2013; Finnveden et al. 2009; Jacquemin et al. 2012).

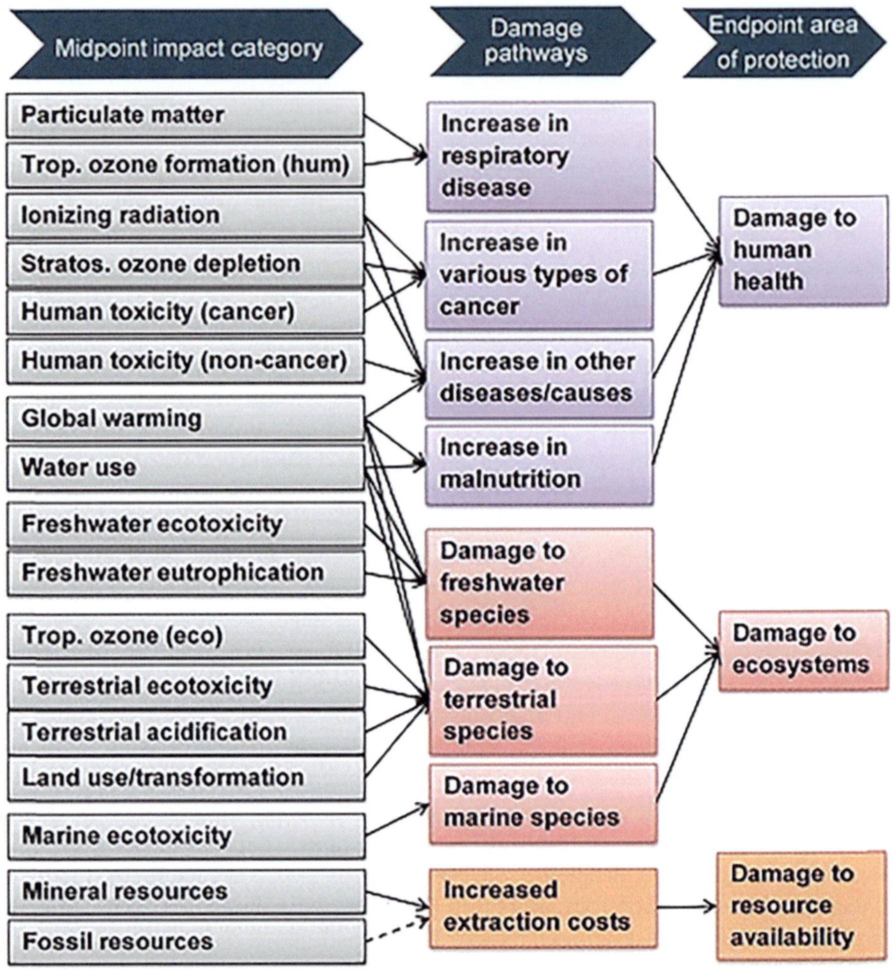

Fig. 9.2 Overview of the impact categories that are covered in the ReCiPe2016 method and their relation to the areas of protection. The dotted line means there is no constant mid-to-endpoint factor for fossil resources. (Figure adapted from: Huijbregts, M.A.J., Steinmann, Z.J.N., Elshout, P.M.F. et al. (2017). ReCiPe2016: a harmonized life cycle impact assessment method at midpoint and endpoint levels. The International Journal of Life Cycle Assessment, 22, 138–147. https://doi.org/10.1007/s11367-016-1246-y Licensed under CC BY 4.0 (Huijbregts et al. 2017))

3.2 Life Cycle Assessment of BC and PMNC

As mentioned above, the production and use of cellulose-based materials, such as PMNC and BC, represent a promising shift towards sustainable alternatives to petroleum-based products. However, the economic viability and environmental benefits of these materials must be carefully evaluated to understand the trade-offs among production costs, material performance, and environmental impacts, particularly given the significant differences in their production processes.

To our knowledge, only four peer-reviewed papers have focused on the LCA of BC. Based on the data from lab-scale results, Aragão et al. (2020) compared the environmental impacts of BC production using standard Hestrin–Schramm (HS) medium and an alternative medium containing corn steep liquor (CSL). In that study, using CSL greatly decreased the environmental impacts, especially those related to freshwater eutrophication, human carcinogenic toxicity, and the potential for global warming (Aragão et al. 2020). Using ReCiPe 2016 methodology, Forte et al. (2021) evaluated the environmental impacts of BC production under a modeled industrial process. Their study demonstrated that the production of raw materials, specifically corn syrup and sodium dihydrogen phosphate, had an important impact in several midpoint categories, such as fossil resource depletion, human toxicity, and climate change. Also, the fermentation process and the downstream processing of the BC were particularly energy-demanding operations (relative to the others), resulting in considerable green gas emissions. In a similar study, using a modeled industrial-scale process, de Araújo e Silva et al. (2020) compared the environmental performance of different BC fermentation methods that were validated at lab-scale. Specifically, they focused on the use of different culture media, hydrolyzed soybean molasses (HSM), diluted soybean molasses, cashew apple juice supplemented with nutrients, and HS culture medium. These authors demonstrated that, as compared to the HS medium, the HSM medium showed the highest environmental performance since a significant reduction of the environmental impacts in several midpoint categories was achieved, for instance in climate change, acidification, and freshwater eutrophication and freshwater ecotoxicity. Aiming at reducing the environmental footprint of the BC production, de Araújo e Silva et al. also showed that replacing sodium hydroxide with potassium hydroxide led to a 50% reduction in impacts related to water scarcity.

A major challenge in the comparison of different LCA studies concerns differences in the defined system boundaries, for instance, Aragão et al. (2020) restricted the system boundary to the BC production process. In contrast, Forte et al. (2021) adopted a broader boundary by including the raw materials extraction, manufacturing, transportation, BC production process, and all emissions (mass and energy) and wastewater treatment. This approach allowed for a more weighted assessment of the environmental impacts, and thus it offered a more integrated view of the BC's sustainability. While de Araújo e Silva et al. (2020) also included upstream processes such as raw material sourcing, input preparation within the system boundary, they excluded other relevant processes such as transport of raw materials, product use, and end-of-life management, making it difficult to establish a comparative analysis.

Another relevant challenge in comparative analysis concerns with the availability and quality of the data. Ideally, an LCA study should rely on primary data, gathered directly from real industrial processes, as it provides the most reliable and relevant information. In certain cases that is not possible, for instance with emerging technologies that are still in early developmental stage, or in certain industrial facilities due to proprietary technology protections. To obviate this, secondary data sources from peer-reviewed publications, industry white papers or publicly available reports and LCA databases, can be used, taking into account, however, that such data often lacks the specificity needed for a precise modeling, thus introducing uncertainties in the analysis (Ciroth 2021). For instance, in the work of Aragão et al. (2020), primary data from lab-scale experiments was used, complemented with information from the literature. While primary, lab-scale data has strong limitations related to projecting the data to industrial-scale operations. In turn, Forte et al. (2021) study relied entirely on secondary data, modeling, and industrial BC facility from peer-reviewed literature and from data from the Ecoinvent database. de Araújo e Silva et al. (2020) also combined primary data from lab-scale fermentations with secondary data from published papers and industry sources, for a more comprehensive but still constrained analysis.

A third critical point of divergence when attempting to perform a comparative LCA analysis is the definition of a functional unit. This is the anchor point of the entire LCA as it defines exactly what is being studied. It represents a quantitative description of the product, not the product itself. Aragão et al. (2020) defined their functional unit as 1 kg of BC pellicles; Forte et al. (2021), 1 kg of dry BC in 138.8 kg of water (since their study considered the production of wet BC) and de Araújo e Silva et al. (2020) adopted two measures of the functional unit given the scale of the analyses: 1 g of dry BC for lab-scale study and 1 metric ton (1000 kg) of dry BC for the industrial-scale model.

A final difficulty in the comparative analysis concerns with the methodologies used in the impact assessment studies. Aragão et al. (2020) ReCiPe Midpoint (I) methodology, containing an earlier version of the midpoint impact indicators, whereas Forte et al. (2021) used the more recent version, ReCiPe 2016 (H). In turn, de Araújo e Silva et al. (2020) integrated two different methodologies: the ILCD Midpoint V1.05 that follows the guidelines set by the European Commission for LCIA, and the AWARE V1.00, a method that was developed to assess impacts related to water scarcity.

Considering the above limitations, in a recent work, Silva et al. (2025) aimed at comparing the LCA of BC and plant celluloses. To circumvent the above-mentioned limitations, Silva et al. standardized the boundaries (Fig. 9.3) among all systems of the study and used a harmonized methodology to compare the modulated BC production with other celluloses (CNC, CNF, and BC). Specifically, for the comparative analysis they used ReCiPe 2016 midpoint (H), CML-IA baseline V3.05, TRACI 2.1 V1.06/US 2008, EF 3.0, IMPACT2002+, Eco-indicator 99, CML, and ILCD 2011 midpoint. As with Forte et al., Silva et al. adopted a cradle-to-gate approach, which includes all processes from raw material extraction to BC production (Fig. 9.3). Their inventory data included water recycling steps. The product was BC pulp with 22% consistency. The functional unit was 1 ton (dry basis) of BC pulp. The study found that, overall, BC pulp had lower environmental impacts than nanocelluloses (CNC and CNF) in categories such as global warming potential, abiotic depletion (fossil fuels), ozone depletion, ozone formation, human toxicity, ecotoxicity (marine and terrestrial ecosystems), acidification, eutrophication, and water depletion (Table 9.5). This suggests that BC pulp is a more sustainable alternative to NC.

According to Forte et al. (2021) and Silva et al. (2025), the preparation of the culture medium (Fig. 9.4) is by far the highest contributor to environmental impacts

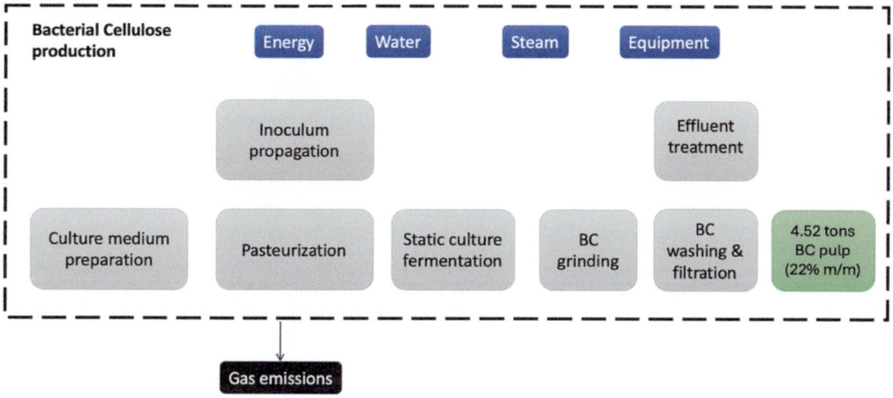

Fig. 9.3 Cradle-to-gate life cycle assessment system boundaries of 1 ton of BC (solids content 22%). (Figure reproduced from: Silva, F.A.G.S., Branco, S., Dourado, F., Neto, B., & Gama, M. (2025). Life cycle assessment of bacterial cellulose and comparison to other cellulosic sources. Journal of Cleaner Production, 493, 144876. https://doi.org/10.1016/j.jclepro.2025.144876 Licensed under CC BY 4.0. Adapted with modifications from the original (Silva et al. 2025))

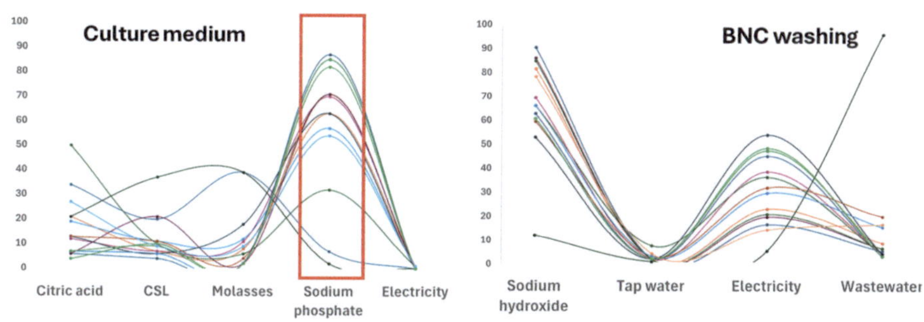

Fig. 9.4 LCA Midpoint impact categories (in %) for the culture medium preparation and BC washing unit operations of the BC-Lyocell process, as estimated using ReCiPe midpoint (H). (Adapted from: Silva, F.A.G.S., Branco, S., Dourado, F., Neto, B., & Gama, M. (2025). Life cycle assessment of bacterial cellulose and comparison to other cellulosic sources. Journal of Cleaner Production, 493, 144876 https://doi.org/10.1016/j.jclepro.2025.144876 Licensed under CC BY 4.0. Adapted with modifications from the original (Silva et al. 2025))

in almost all midpoint categories, followed by the washing process. In a more in-depth analysis of these two unit operations (4), sodium phosphate, even though it is used in a relatively low amount in the culture medium, usually within 0.5–1%, had an overwhelmingly high impact compared to other components of the culture medium. Regarding the washing process, sodium hydroxide, typically used in the range of 1–5%, was the highest contributor. Not only is it used at a higher concentration, but large amounts of water are also used during the washing of BC, which may explain the high impact. In addition, the washing process consumes large amounts of energy.

Table 9.5 Midpoint environmental impacts of the life cycle of Pulp, CNCs, CNFs, and BC using different impacts assessment methods. Comparisons normalized to 1 ton of cellulose (dry mass). (Reproduced from (Silva et al. 2025))

Impact categories	Unit LCIA method used	CNC (de Figueirêdo et al. 2012; Gu et al. 2015; do Nascimento et al. 2016; Rajendran et al. 2023)	CNF (Arvidsson et al. 2015; Gallo Stampino et al. 2021)	BC (Silva et al. 2025)
Global warming potential	E + 03 kg CO_2	122–247	60–930	6.22–6-66
Abiotic depletion	kg Sb eq[2]	0.0349		0.10
	E + 04 MJ eq[2] (fossil fuels)	150		7.1
	E + 03 MJ eq[4] (fossil fuels)	600		7.0
Water depletion	E + 03 m^3 eq[1]	2.3–140	6.5	0.21
	E + 04 m^3 depriv.[5]		180–190	7.0
Land use	E + 03 Pt eq[5]		5200–18,000	8.6
Ozone depletion	E – 03 kg CFC-11 eq[4]	92		0.85
	E – 03 kg CFC-11 eq[5]		120–130	0.82
Ozone formation	kg O_3 eq[4]	14,000		365
	kg NMVOC eq[5]		2000–2300	190
Acidification	kg SO_2 eq[1] (Terrestrial)		450	39.0
	kg SO_2 eq[2]	531		45.6
	kg SO_2 eq[3] (Aquatic)		192	46.9
	kg SO_2 eq[4]	4500		44.4
Eutrophication	kg P eq[1] (Freshwater)		24–56.8	3.1
	kg N eq [1] (Marine)		30.3–65	1.4
	kg PO_4^- eq[2]	65		15.1
	kg PO_4 P-lim[3] (Aquatic)		6	2.7
	Kg N eq[4]	416.7		29.8
Ecotoxicity	E + 05 kg TEG water eq[3] (Marine)	83.9		8.7
	E + 05 kg TEG soil eq[3] (Terrestrial)	19.7		2.3
	E + 03 CTUe eq[4]	810		120
	E + 05 CTUe eq[5] (freshwater)		200–220	2.0
Human toxicity	E + 03 kg1,4-Db eq[1]	35–48		9.1
	E – 03 CTUh eq[4]	79.5		3.5

(continued)

Table 9.5 (continued)

Impact categories	Unit LCIA method used	CNC (de Figueirêdo et al. 2012; Gu et al. 2015; do Nascimento et al. 2016; Rajendran et al. 2023)	CNF (Arvidsson et al. 2015; Gallo Stampino et al. 2021)	BC (Silva et al. 2025)
	E – 03 CTUh eq[5]		9.3–10	0.14
Respiratory effects	kg PM2.5 eq[4]	250		4.7
	Particulate matter disease inc.[5]		0.019–0.026	0.00027

(1) ReCiPe 2016 Midpoint (H); (2) CML; (3) Impact 2002+; (4) TRACI; (5) EF3.0; (6) ILCD 2011
Reproduced from: Silva, F.A.G.S., Branco, S., Dourado, F., Neto, B., & Gama, M. (2025). Life cycle assessment of bacterial cellulose and comparison to other cellulosic sources. Journal of Cleaner Production, 493, 144876. https://doi.org/10.1016/j.jclepro.2025.144876 Licensed under CC BY 4.0. Adapted with modifications from the original (Silva et al. 2025).

From the above, if BC is viewed as an end product, its production can be considered more sustainable than that of plant nanocelluloses. As reported by Forte et al. (2021) cradle-to-gate LCA studies are particularly useful in the early stages of Research and Development (R&D), where the focus is on understanding the environmental impacts of a new technology or production process. These studies are especially valuable when the resulting product may have several downstream applications, some (if not all) of which have not yet been fully developed. Cradle-to-gate LCA results can serve as the foundation for subsequent LCA studies of products that incorporate materials as constituents in downstream applications. In the above comparative analysis, BC and plant cellulose were regarded as products, with their life cycles ending at the factory gate. Within this framework, BC has been shown to have a lower environmental impact than other nanocelluloses. However, considering its industrialization, BC should be regarded as an intermediate product—a raw material to be incorporated into further processing in applications such as cosmetic formulations, food packaging, textiles, or other potential uses. Consequently, the environmental impact of incorporating BC on the production of specific products must be evaluated. One example is its potential use in textiles, specifically for staple fiber production. This topic is more extensively addressed in the work by Silva et al. (2025), where it is demonstrated that, on an LCA perspective, BC cannot be considered more sustainable as compared to other cellulose sources of textiles such as cotton or wood-derived lyocell or viscose.

4 Conclusions

This chapter addressed the urgent need to transition from petroleum-based materials to sustainable alternatives, especially given the escalating issues of plastic pollution and resource depletion worldwide. BC and PMNC have both been pinpointed as highly promising sustainable materials. Yet, a comprehensive techno-economic and environmental sustainability comparison of BC against plant-derived cellulose was missing. This chapter is a thorough evaluation of the economic and environmental viability of BC production in relation to plant-derived cellulose. The results from the techno-economic analysis suggest that BC production can be considered economically viable if established at a projected market price of approximately 41 €/kg. On this basis, the process is shown to yield a positive net present value (NPV) over a 5-year operational period. While the production cost can be relatively high, the physicochemical, mechanical, and functional properties of BC position it well for high-value niche market applications, including those in the medical, cosmetic, and biomedical device sectors. Achieving successful market penetration in higher volume sectors such as food packaging, pulp and paper, textiles, and biodegradable materials will require further process optimization and cost-reduction strategies.

On the environmental analysis, the comparative LCA study showed that when BC is considered as a standalone material, its production has an overall lower environmental impact than that of plant nanocellulose. However, when BC is used as an intermediate product, such as in the production of staple fibers (BC-lyocell), its environmental performance becomes less favorable. The use of sodium phosphate in the culture medium and sodium hydroxide for washing and purification contributes substantially to the high impacts in categories such as toxicity, acidification, and water pollution, making the BC-lyocell process more intensive than cotton, viscose, and lyocell.

Thus, BC holds considerable promise as a sustainable material, particularly for applications that require high-performance biomaterials. However, its widespread commercialization and environmental viability depend heavily on ongoing technological advancements, economic scaling, and eco-design efforts.

References

Alabbosh K, Hazrin-Chong NH, Balawi A (2021) Agricultural wastes as a carbon or nitrogen source for production of bacterial cellulose. A mini review. Pollut Res 40:429–437

Angelis A, Polyakov R, Wouters OJ, Torreele E, McKee M (2023) High drug prices are not justified by industry's spending on research and development. BMJ 380:e071710. https://doi.org/10.1136/bmj-2022-071710

Appels L, Baeyens J, Degrève J, Dewil R (2008) Principles and potential of the anaerobic digestion of waste-activated sludge. Prog Energy Combust Sci 34(6):755–781. https://doi.org/10.1016/j.pecs.2008.06.002

Aragão J, Costa A, Gilson L, Silva G, Silva S, Macêdo J, Galdino C, Milanez F, Sarubbo L (2020) Analysis of the environmental life cycle of bacterial cellulose production. Chem Eng Trans 79:445–450. https://doi.org/10.3303/CET2079075

Arumugam V, Gao Y (2022) Synthesis and applications of organic framework-based cellulosic nanocomposites. In: Barhoum A (ed) Handbook of nanocelluloses: classification, properties, fabrication, and emerging applications. Springer International Publishing, pp 441–472. https://doi.org/10.1007/978-3-030-89621-8_26

Arvidsson R, Nguyen D, Svanström M (2015) Life cycle assessment of cellulose nanofibrils production by mechanical treatment and two different pretreatment processes. Environ Sci Technol 49(11):6881–6890. https://doi.org/10.1021/acs.est.5b00888

Batstone DJ, Virdis B (2014) The role of anaerobic digestion in the emerging energy economy. Curr Opin Biotechnol 27:142–149. https://doi.org/10.1016/j.copbio.2014.01.013

Behera B, Laavanya D, Paramasivan B (2021) Techno-economic feasibility assessment of bacterial cellulose biofilm production during the Kombucha fermentation process. Bioresour Technol 346:126659. https://doi.org/10.1016/j.biortech.2021.126659

Ben-Shahar D, Margalioth Y, Sulganik E (2009) The straight-line depreciation is wanted, dead or alive. J Real Estate Res 31:351–370. https://doi.org/10.1080/10835547.2009.12091257

Bondancia TJ, de Aguiar J, Batista G, Cruz AJG, Marconcini JM, Mattoso LHC, Farinas CS (2020) Production of nanocellulose using citric acid in a biorefinery concept: effect of the hydrolysis reaction time and techno-economic analysis. Ind Eng Chem Res 59(25):11505–11516. https://doi.org/10.1021/acs.iecr.0c01359

Borrelle SB, Ringma J, Law KL, Monnahan CC, Lebreton L, McGivern A, Murphy E, Jambeck J, Leonard GH, Hilleary MA, Eriksen M, Possingham HP, De Frond H, Gerber LR, Polidoro B, Tahir A, Bernard M, Mallos N, Barnes M, Rochman CM (2020) Predicted growth in plastic waste exceeds efforts to mitigate plastic pollution. Science 369(6510):1515–1518. https://doi.org/10.1126/science.aba3656

Cazón P, Vázquez M (2021) Bacterial cellulose as a biodegradable food packaging material: a review. Food Hydrocoll 113:106530. https://doi.org/10.1016/j.foodhyd.2020.106530

Chandana A, Mallick SP, Dikshit PK, Singh BN, Sahi AK (2022) Recent developments in bacterial nanocellulose production and its biomedical applications. J Polym Environ 30(10):4040–4067. https://doi.org/10.1007/s10924-022-02507-0

Chao Y, Sugano Y, Shoda M (2001) Bacterial cellulose production under oxygen-enriched air at different fructose concentrations in a 50-liter, internal-loop airlift reactor. Appl Microbiol Biotechnol 55(6):673–679

Cheng H-P, Wang P-M, Chen J-W, Wu W-T (2002) Cultivation of Acetobacter xylinum for bacterial cellulose production in a modified airlift reactor. Biotechnol Appl Biochem 35(2):125–132. https://doi.org/10.1111/j.1470-8744.2002.tb01180.x

Ciroth A (2021) Data quality in life cycle inventories. In: Ciroth A, Arvidsson R (eds) Life cycle inventory analysis: methods and data. Springer International Publishing, pp 97–122. https://doi.org/10.1007/978-3-030-62270-1_5

Cowie J, Bilek EM, Wegner TH, Shatkin JA (2014) Market projections of cellulose nanomaterial-enabled products—part 2: volume estimates. TAPPI J 13:57

Curran MA (2013) Life cycle assessment: a review of the methodology and its application to sustainability. Curr Opin Chem Eng 2(3):273–277. https://doi.org/10.1016/j.coche.2013.02.002

da Cruz TT, Las-Casas B, Dias IKR, Arantes V (2024) Nanocelluloses as sustainable emerging technologies: state of the art and future challenges based on life cycle assessment. Sustain Mater Technol 41:e01010. https://doi.org/10.1016/j.susmat.2024.e01010

da Silva FAGS, Oliveira JV, Felgueiras C, Dourado F, Gama M, Alves MM (2020) Study and valorisation of wastewaters generated in the production of bacterial nanocellulose. Biodegradation 31(1):47–56. https://doi.org/10.1007/s10532-020-09893-z

Das R, Lindström T, Khan M, Rezaei M, Hsiao BS (2024) Nanocellulose preparation from diverse plant feedstocks, processes, and chemical treatments: a review emphasizing non-woods. Bioresources 19(1):1865–1924. https://doi.org/10.15376/biores.19.1.Das

de Araújo e Silva R, Santa Brígida AI, de Freitas Rosa M, da Silva Neto RM, Spinosa WA, Benício de Sá Filho E, Brito de Figueirêdo MC (2020) An approach for implementing ecodesign at early research stage: a case study of bacterial cellulose production. J Clean Prod 269:122245. https://doi.org/10.1016/j.jclepro.2020.122245

de Figueirêdo MCB, de Freitas Rosa M, Ugaya CML, de Sá Moreira de Souza Filho M, da Silva Braid ACC, de Melo LFL (2012) Life cycle assessment of cellulose nanowhiskers. J Clean Prod 35:130–139. https://doi.org/10.1016/j.jclepro.2012.05.033

Dekker E, Zijp MC, van de Kamp ME, Temme EHM, van Zelm R (2020) A taste of the new ReCiPe for life cycle assessment: consequences of the updated impact assessment method on food product LCAs. Int J Life Cycle Assess 25(12):2315–2324. https://doi.org/10.1007/s11367-019-01653-3

do Nascimento DM, Dias AF, de Araújo Junior CP, de Freitas Rosa M, Morais JPS, de Figueirêdo MCB (2016) A comprehensive approach for obtaining cellulose nanocrystal from coconut fiber. Part II: environmental assessment of technological pathways. Ind Crop Prod 93:58–65

Dourado F, Fontão A, Leal M, Cristina Rodrigues A, Gama M (2016) Process modeling and techno-economic evaluation of an industrial bacterial nanocellulose fermentation process. In: Gama M, Dourado F, Bielecki S (eds) Bacterial nanocellulose: from biotechnology to bio-economy. Elsevier, pp 199–214. https://doi.org/10.1016/B978-0-444-63458-0.00012-3

Dourado F, Fontão AI, Leal M, Rodrigues AC, Gama M (2018) Process modelling and techno-economic evaluation of an industrial air-lift bacterial cellulose fermentation process. In: Lee K-Y (ed) Nanocellulose and sustainability: production, properties, applications, and case studies. CRC Press, Boca Raton, pp 1–16

Erumban AA (2008) Lifetimes of machinery and equipment: evidence from Dutch manufacturing. Rev Income Wealth 54:237–268. https://doi.org/10.1111/j.1475-4991.2008.00272.x

Finnveden G, Hauschild MZ, Ekvall T, Guinée J, Heijungs R, Hellweg S, Koehler A, Pennington D, Suh S (2009) Recent developments in life cycle assessment. J Environ Manag 91(1):1–21. https://doi.org/10.1016/j.jenvman.2009.06.018

Florea M, Hagemann H, Santosa G, Abbott J, Micklem CN, Spencer-Milnes X, de Arroyo Garcia L, Paschou D, Lazenbatt C, Kong D, Chughtai H, Jensen K, Freemont PS, Kitney R, Reeve B, Ellis T (2016) Engineering control of bacterial cellulose production using a genetic toolkit and a new cellulose-producing strain. Proc Natl Acad Sci U S A 113(24):E3431–E3440. https://doi.org/10.1073/pnas.1522985113

Forte A, Dourado F, Mota A, Neto B, Gama M, Ferreira EC (2021) Life cycle assessment of bacterial cellulose production. Int J Life Cycle Assess 26(5):864–878. https://doi.org/10.1007/s11367-021-01904-2

Gallo Stampino P, Riva L, Punta C, Elegir G, Bussini D, Dotelli G (2021) Comparative life cycle assessment of cellulose nanofibres production routes from virgin and recycled raw materials. Molecules 26(9):2558. https://doi.org/10.3390/molecules26092558

Green CD, Grinyer JR, Michaelson R (2002) A possible economic rationale for straight-line depreciation. Abacus 38(1):91–120. https://doi.org/10.1111/1467-6281.00099

Gu H, Reiner R, Bergman R, Rudie A (2015) LCA study for pilot scale production of cellulose nano crystals (CNC) from wood pulp LCA XV paper proceedings—A bright green future

Habibi Y, Lucia LA, Rojas OJ (2010) Cellulose nanocrystals: chemistry, self-assembly, and applications. Chem Rev 110(6):3479–3500. https://doi.org/10.1021/cr900339w

Hu M, Lv X, Wang Y, Zhang Y, Dai H (2024) Guideline for the extraction of nanocellulose from lignocellulosic feedstocks. Food Biomacromolecules 1(1):9–17. https://doi.org/10.1002/fob2.12011

Huijbregts MAJ, Steinmann ZJN, Elshout PMF, Stam G, Verones F, Vieira M, Zijp M, Hollander A, van Zelm R (2017) ReCiPe2016: a harmonised life cycle impact assessment method at midpoint and endpoint level. Int J Life Cycle Assess 22(2):138–147. https://doi.org/10.1007/s11367-016-1246-y

Ioelovich M (2022) Microcellulose vs nanocellulose—a review. World J Adv Eng Technol Sci 5:1–15. https://doi.org/10.30574/wjaets.2022.5.2.0037

ISO14040:2006 (2006) ISO 14040:2006 Environmental management—Life cycle assessment—Principles and framework. International Organization for Standardization (ISO), Geneva, pp 20

ISO14044:2006 (2006) ISO 14044:2006 Environmental management—Life cycle assessment—Requirements and guidelines. International Organization for Standardization (ISO), Geneva, pp 46

Jacquemin L, Pontalier P-Y, Sablayrolles C (2012) Life cycle assessment (LCA) applied to the process industry: a review. Int J Life Cycle Assess 17(8):1028–1041. https://doi.org/10.1007/s11367-012-0432-9

Jozala AF, de Lencastre-Novaes LC, Lopes AM, de Carvalho Santos-Ebinuma V, Mazzola PG, Pessoa-Jr A, Grotto D, Gerenutti M, Chaud MV (2016) Bacterial nanocellulose production and application: a 10-year overview. Appl Microbiol Biotechnol 100(5):2063–2072. https://doi.org/10.1007/s00253-015-7243-4

Kamal T, Islam M, Fatima A, Ullah MW, Manan S (2022) Cost-effective synthesis of bacterial cellulose and its applications in the food and environmental sectors. Gels 8:552. https://doi.org/10.3390/gels8090552

Karan H, Funk C, Grabert M, Oey M, Hankamer B (2019) Green bioplastics as part of a circular bioeconomy. Trends Plant Sci 24(3):237–249. https://doi.org/10.1016/j.tplants.2018.11.010

Kırlı M (2018) Comparison of depreciation methods in "international accounting standard 16 property, plant and equipment" and an application. Econ Appl Inform, pp 91–98

Klemm D, Kramer F, Moritz S, Lindström T, Ankerfors M, Gray D, Dorris A (2011) Nanocelluloses: a new family of nature-based materials. Angew Chem Int Ed 50(24):5438–5466. https://doi.org/10.1002/anie.201001273

Klemm D, Cranston ED, Fischer D, Gama M, Kedzior SA, Kralisch D, Kramer F, Kondo T, Lindström T, Nietzsche S, Petzold-Welcke K, Rauchfuß F (2018) Nanocellulose as a natural source for groundbreaking applications in materials science: today's state. Mater Today 21(7):720–748. https://doi.org/10.1016/j.mattod.2018.02.001

Lau WWY, Shiran Y, Bailey RM, Cook E, Stuchtey MR, Koskella J, Velis CA, Godfrey L, Boucher J, Murphy MB, Thompson RC, Jankowska E, Castillo Castillo A, Pilditch TD, Dixon B, Koerselman L, Kosior E, Favoino E, Gutberlet J et al (2020) Evaluating scenarios toward zero plastic pollution. Science 369(6510):1455–1461. https://doi.org/10.1126/science.aba9475

Li Y, Chen Y, Wu J (2019) Enhancement of methane production in anaerobic digestion process: a review. Appl Energy 240:120–137. https://doi.org/10.1016/j.apenergy.2019.01.243

Lowe D (2023) Don't drug companies spend more on marketing? *Don't drug companies spend more on marketing?* https://www.science.org/content/blog-post/don-t-drug-companies-spend-more-marketing?utm_source=chatgpt.com

MacLeod M, Arp HPH, Tekman MB, Jahnke A (2021) The global threat from plastic pollution. Science 373(6550):61–65. https://doi.org/10.1126/science.abg5433

Mao C, Feng Y, Wang X, Ren G (2015) Review on research achievements of biogas from anaerobic digestion. Renew Sustain Energ Rev 45:540–555. https://doi.org/10.1016/j.rser.2015.02.032

Mata-Alvarez J, Macé S, Llabrés P (2000) Anaerobic digestion of organic solid wastes. An overview of research achievements and perspectives. Bioresour Technol 74(1):3–16. https://doi.org/10.1016/S0960-8524(00)00023-7

Montenegro-Silva P, Ellis T, Dourado F, Gama M, Domingues L (2024) Enhanced bacterial cellulose production in Komagataeibacter sucrofermentans: impact of different PQQ-dependent dehydrogenase knockouts and ethanol supplementation. Biotechnol Biofuels Bioprod 17(1):35. https://doi.org/10.1186/s13068-024-02482-9

OECD (2024) Extended producer responsibility: basic facts and key principles. OECD Environment Policy Papers, Paris

Penloglou G, Basna A, Pavlou A, Kiparissides C (2023) Techno-economic considerations on nanocellulose's future progress: a short review. Processes 11(8):2312. https://doi.org/10.3390/pr11082312

Quijano L, Rodrigues R, Fischer D, Tovar-Castro JD, Payne A, Navone L, Hu Y, Han Y, Pinmanee P, Poon E, Yang J, Barro E (2024) Bacterial cellulose cookbook: a systematic review on sustainable and cost-effective substrates. J Bioresour Bioprod 9:379–409. https://doi.org/10.1016/j.jobab.2024.05.003

Qureshi SS, Nizamuddin S, Xu J, Vancov T, Chen C (2024) Cellulose nanocrystals from agriculture and forestry biomass: synthesis methods, characterization and industrial applications. Environ Sci Pollut Res 31(49):58745–58778. https://doi.org/10.1007/s11356-024-35127-3

R R, Philip E, Thomas D, Madhavan A, Sindhu R, Binod P, Varjani S, Awasthi MK, Pandey A (2021) Bacterial nanocellulose: engineering, production, and applications. Bioengineered 12(2):11463–11483. https://doi.org/10.1080/21655979.2021.2009753

Rajendran N, Runge T, Bergman RD, Nepal P, Houtman C (2023) Techno-economic analysis and life cycle assessment of cellulose nanocrystals production from wood pulp. Bioresour Technol 377:128955. https://doi.org/10.1016/j.biortech.2023.128955

Rodrigues AC, Fontão AI, Coelho A, Leal M, Soares da Silva FAG, Wan Y, Dourado F, Gama M (2019) Response surface statistical optimization of bacterial nanocellulose fermentation in static culture using a low-cost medium. New Biotechnol 49:19–27. https://doi.org/10.1016/j.nbt.2018.12.002

Rodrigues AC, Martins D, Duarte MS, Silva-Carvalho R, Marques S, Cavaleiro AJ, Gama M, Dourado F (2025) Optimizing bacterial nanocellulose production from eucalyptus bark: a circular approach to wastewater management and resource recovery. J Environ Chem Eng 13(2):115442. https://doi.org/10.1016/j.jece.2025.115442

Rosales-Calderon O, Pereira B, Arantes V (2021) Economic assessment of the conversion of bleached eucalyptus Kraft pulp into cellulose nanocrystals in a stand-alone facility via acid and enzymatic hydrolysis. Biofuels Bioprod Biorefin 15(6):1775–1788. https://doi.org/10.1002/bbb.2277

Rosenboom J-G, Langer R, Traverso G (2022) Bioplastics for a circular economy. Nat Rev Mater 7(2):117–137. https://doi.org/10.1038/s41578-021-00407-8

Serra A, González I, Oliver-Ortega H, Tarrès Q, Delgado-Aguilar M, Mutjé P (2017) Reducing the amount of catalyst in TEMPO-oxidized cellulose nanofibers: effect on properties and cost. Polymers (Basel) 9(11):557. https://doi.org/10.3390/polym9110557

Sharma A, Thakur M, Bhattacharya M, Mandal T, Goswami S (2019) Commercial application of cellulose nano-composites—A review. Biotechnol Rep 21:e00316. https://doi.org/10.1016/j.btre.2019.e00316

Silva FAGS, Branco S, Dourado F, Neto B, Gama M (2025) Life cycle assessment of bacterial cellulose and comparison to other cellulosic sources. J Clean Prod 493:144876. https://doi.org/10.1016/j.jclepro.2025.144876

Singh A, Walker K, Ledesma-Amaro R, Ellis T (2020) Engineering bacterial cellulose by synthetic biology. Int J Mol Sci 21:9185. https://doi.org/10.3390/ijms21239185

Taokaew S (2024) Bacterial nanocellulose produced by cost-effective and sustainable methods and its applications: a review. Fermentation 10(6):316. https://www.mdpi.com/2311-5637/10/6/316

The EU Green Deal: 2024 Edition (2024) E. U. I. Florence School of Regulation. https://fsr.eui.eu/publications/?handle=1814/77672

The European Environmental Bureau (2016) Bioplastics in a circular economy: the need to focus on waste reduction and prevention to avoid false solutions. The European Environmental Bureau

Thomas DS (2018) Potential methods and data needs. In: The costs and benefits of advanced maintenance in manufacturing. National Institute of Standards and Technology, pp 17–23

Trache D, Tarchoun AF, Derradji M, Hamidon TS, Masruchin N, Brosse N, Hussin MH (2020) Nanocellulose: from fundamentals to advanced applications. Front Chem 8:392. https://doi.org/10.3389/fchem.2020.00392

United Nations (2018). Single-use plastics: A roadmap for sustainability (rev. ed., pp. vi; 6). U. N. E. P. (UNEP). https://www.unep.org/resources/report/single-use-plastics-roadmap-sustainability

Vanhatalo K, Parviainen K, Dahl O (2014) Techno-economic analysis of simplified microcrystalline cellulose process. Bioresources 9:4741–4755. https://doi.org/10.15376/biores.9.3.4741-4755

Velásquez-Riaño M, Bojacá V (2017) Production of bacterial cellulose from alternative low-cost substrates. Cellulose 24(7):2677–2698. https://doi.org/10.1007/s10570-017-1309-7

Wang J, Tavakoli J, Tang Y (2019) Bacterial cellulose production, properties and applications with different culture methods—A review. Carbohydr Polym 219:63–76. https://doi.org/10.1016/j.carbpol.2019.05.008

Wilkie AC (2005) Anaerobic digestion: biology and benefits. In: Treatment, handling, and community relations dairy manure management, Ithaca

Wu SC, Li MH (2015) Production of bacterial cellulose membranes in a modified airlift bioreactor by Gluconacetobacter xylinus. J Biosci Bioeng 120(4):444–449. https://doi.org/10.1016/j.jbiosc.2015.02.018

Yuzawa M (2017) Move over carbon fiber, here comes cellulose nanofiber. Nikkei Asian Rev. https://asia.nikkei.com/Business/Biotechnology/Move-over-carbon-fiber-here-comes-cellulose-nanofiber. Accessed 6 March

Zhang W, Wang JJ, Gao Y, Zhang LL (2020) Bacterial cellulose synthesized with apple pomace enhanced by ionic liquid pretreatment. Prep Biochem Biotechnol 50(4):330–340. https://doi.org/10.1080/10826068.2019.1692222